Service Parts Management

Nezih Altay · Lewis A. Litteral
Editors

Service Parts Management

Demand Forecasting and Inventory Control

Springer

Editors
Assoc. Prof. Nezih Altay
Department of Management
DePaul University
1 E. Jackson Blvd.
Depaul Center 7000
Chicago, IL 60604
USA
e-mail: naltay@depaul.edu

Assoc. Prof. Lewis A. Litteral
Robins School of Business
University of Richmond
Westhampton Way 28
Richmond, VA 23173
USA
e-mail: llittera@richmond.edu

ISBN 978-0-85729-038-0 e-ISBN 978-0-85729-039-7

DOI 10.1007/978-0-85729-039-7

Springer London Dordrecht Heidelberg New York

British Library Cataloguing in Publication Data
A catalogue record for this book is available from the British Library

Cover design: eStudio Calamar S.L.

Printed on acid-free paper

Springer is part of Springer Science+Business Media (www.springer.com)

Dedications

To my wife and best friend, Ozge, for standing by me without reservation, and my daughter, Ela who I love like the oceans.

N. A.

To the one and only, Anna, my wife and best friend in life; Daniel, Catie, and Drew, the three who make it worthwhile; and Cash, Penny, and Archie who bring joy to all of us.

L. A. L.

Preface

There are 14 distinct contributions to this volume from authors who hail from more than ten countries representing universities from six countries around the world. Although their approaches to the management of spare parts are widely divergent, everyone involved in the project agrees on two things: first, the management of spare parts is important because of its prevailing nature and magnitude and second, the problems associated with the management of spare parts are complex and really very hard. The first point is the motivation for the volume and we think that the second point is moderated somewhat by the talent, experience, and hard work of the authors whose work is presented.

Many industries rely on the management of spare parts including military weapon systems in naval and aircraft applications, commercial aviation, information technology, telecom, automotive, and white goods such as fabrics and large household appliances. Some of these applications involve providing service parts to end users while others involve the maintenance of manufacturing facilities for products like textiles and automobiles.

According to an article in the *McKinsey Quarterly* authored by Thomas Knecht, Ralf Leszinski, and Felix A. Weber, the after-sales business accounts for 10–20% of revenues and a much larger portion of total contribution margin in most industrial companies.Equally important, during a given product's life cycle, after-sales can generate atleast three times the turnover of the original purchase, especially for industrial equipment. A well-run after-sales business can also provide strategic benefits. Customers are usually less concerned about spare part prices than about speed of delivery and availability of service know how, whether on-site or via telephone. The reason is simple: down-time costs typically run at anywhere from 100 to 10,000 times the price of spare parts or service. And that means good performance can boost customer satisfaction, and thus, build repurchase loyalty in the original equipment business.

With specific regard to spare parts management policies used by the armed services of the United States, the General Accounting Office reported in 1997 that the inventory of service parts at non-major locations was valued at over $8.3 billion and that the need for many of the items stored at non-major locations is questionable. Of the $8.3 billion of inventory at the non-major locations, $2.7 billion of it was not needed to meet the services' current operating and war reserve requirements. Maintaining inventory that is not needed is expensive and does not contribute to an effective, efficient, and responsive supply system. Based on GAO's analysis, GAO estimates the services that could save about $382 million annually in inventory holding costs by eliminating inventory at non-major locations that is not needed to meet current operating and war reserve requirements.

Academics and practitioners will find this volume valuable, as a starting point for spare parts research or to augment their current knowledge, for the results presented here as well as the foundation upon which those results are built as indicated in the extensive literature reviews and reference sections of each paper.

There are a number of papers in this volume that provide some level of summary and thus are particularly suited to readers looking for an initial understanding of the field of managing spare parts. One of these is the work by Andrew Huber where he provides a framework for evaluating the application of alternative techniques noting that this is a place where practice often lags theory. Macchi et al. discuss a structured framework of five steps whereby decisions can be made regarding the maintenance of spare parts in the avionic sector. The work of Bucher and Meissner provides a summary of ways that intermittent demand can be categorized, allowing the researcher or practitioner to quickly determine which spare parts management methods to consider given the nature of their particular data. Bartezzaghi and Kalchschmidt present findings on the choice of how the data aggregated can affect the performance of managing inventory. A distinctive contribution to volume is made by Smith and Babai: they present a comprehensive review of bootstrapping methods for spare parts forecasting, a promising area of study that has been developed as a parallel track to parametric methods.

Other papers in this volume present and evaluate various techniques (tabu search, Bayesian analysis, decision trees, and prediction intervals) tested on real and simulated data. Some techniques are theoretical developments while others are heuristics. Even with the computing power available in 2010, some aspects of the spare parts inventory management problem remain so complex that it is impossible to apply traditional techniques to forecasting and inventory management in this context. Various criteria are used to evaluate techniques, and some of which are traditional.

Acknowledgments

This volume represents the work of many individuals. Our greatest debt is to Tricia Fanney of the Robins School of Business at the University of Richmond who has cheerfully and carefully shaped the manuscript into its current form. We are also grateful to each of the authors, many of whom also served as reviewers. Jonathan Whitaker and Steve Thompson of the Robins School served as reviewers and we appreciate their service. Our thanks go to the editorial team at Springer for encouraging and supporting the research in this area. Special thanks are due to Claire Protherough who worked closely with Tricia in bringing this project to publication.

Respectfully submitted:
Nezih Altay and Lewis A. Litteral

Contents

List of Contributors

Nezih Altay, DePaul University, e-mail: naltay@depaul.edu

Natasha Atanackov, Belgrade University, e-mail: Natasha_atanackov@yahoo.co.uk

M. Zied Babai, BEM Bordeaux Management School, e-mail: Mohamed-zied.babai@bem.edu

J. Wesley Barnes, The University of Texas at Austin, e-mail: wbarnes@mail.utexas.edu

Emilio Bartezzaghi, Politecnico di Milano, e-mail: emilio.bartezzaghi@polimi.it

John E. Boylan, Buckinghamshire Chilterns, e-mail: John.Boylan@bucks.ac.uk

David Bucher, Lancaster University, e-mail: d.bucher@lancaster.ac.uk

Sergio Cavalieri, Università degli Studi di Bergamo, e-mail: sergio.cavalieri@unibg.it

Ben Clegg, Aston Business School, e-mail: b.t.clegg@aston.ac.uk

Luca Fumagalli, Politecnico di Milano, e-mail: luca1.fumagalli@polimi.it

Andrew J. Huber, Xerox, e-mail: Andrew.Huber@xerox.com

Matteo Kalchschmidt, Università degli Studi di Bergamo, e-mail: matteo.kalchschmidt@unibg.it

Ibrahim S. Kurtulus, Virginia Commonwealth University, e-mail: ikurtulu@vcu.edu

Erhan Kutanoglu, The University of Texas at Austin, e-mail: erhank@me.utexas.edu

David Lengu, University of Salford, e-mail: d.lengu@salford.ac.uk

Matthew Lindsey, Stephen F. Austin State University, e-mail: lindseymd@sfasu.edu

Lewis A. Litteral, University of Richmond, e-mail: llittera@richmond.edu

Michael MacDonnell, University College Dublin, e-mail: michael.macdonnell @ucd.ie

Marco Macchi, Politecnico di Milano, e-mail: marco.macchi@polimi.it

Joern Meissner, Lancaster University, e-mail: joe@meiss.com

Stefan Minner, University of Vienna, e-mail: stefan.minner@univie.ac.at

David F. Muñoz, Instituto Tecnológico Autónomo de México, e-mail: davidm @itam.mx

Diego F. Muñoz, Stanford University, e-mail: dkedmun@stanford.edu

Robert Pavur, University of North Texas, e-mail: pavur@unt.edu

Roberto Pinto, Università degli Studi di Bergamo, e-mail: roberto.pinto@unibg.it

Marilyn J. Smith, Winthrop University, e-mail: smithm@winthrop.edu

Yi Sui, MicroStrategy, Inc, e-mail: sui1lyi3@yahoo.com

Aris A. Syntetos, University of Salford, e-mail: A.Syntetos@salford.ac.uk

Chapter 1
Intermittent Demand: Estimation and Statistical Properties

Aris A. Syntetos and John E. Boylan

1.1 Introduction

Intermittent demand patterns are very difficult to forecast and they are, most commonly, associated with spare parts' requirements. Croston (1972) proved the inappropriateness of single exponential smoothing (SES) in an intermittent demand context and he proposed a method that relies upon separate forecasts of the inter-demand intervals and demand sizes, when demand occurs. His method for forecasting intermittent demand series is increasing in popularity. The method is incorporated in statistical forecasting software packages (e.g. Forecast Pro), and demand planning modules of component based enterprise and manufacturing solutions (e.g. Industrial and Financial Systems-IFS AB). It is also included in integrated real-time sales and operations planning processes (e.g. SAP Advanced Planning & Optimisation-APO 4.0).

An earlier paper (Syntetos and Boylan 2001) showed that there is scope for improving the accuracy of Croston's method. Since then two bias-corrected Croston procedures have been proposed in the academic literature that aim at advancing the practice of intermittent demand forecasting. These are the Syntetos–Boylan Approximation (SBA, Syntetos and Boylan 2005) and the Syntetos' method (SY, Syntetos 2001; Teunter and Sani 2009).[1]

A. A. Syntetos (✉)
University of Salford, Salford, UK
e-mail: A.Syntetos@salford.ac.uk

J. E. Boylan
Buckinghamshire New University, Buckinghamshire UK
e-mail: John.Boylan@bucks.ac.uk

[1] At this point it is important to note that one more modified Croston procedure has appeared in the literature (Leven and Segerstedt 2004). However, this method was found to be even more biased than the original Croston's method (Boylan and Syntetos 2007; Teunter and Sani 2009) and as such it is not further discussed in this chapter.

N. Altay and L. A. Litteral (eds.), *Service Parts Management*,
DOI: 10.1007/978-0-85729-039-7_1, © Springer-Verlag London Limited 2011

In this paper, these estimators as well as Croston's method and SES are presented and analysed in terms of the following statistical properties: (i) their bias (or the lack of it); and (ii) the variance of the related estimates (i.e. the sampling error of the mean). Detailed derivations are offered along with a thorough discussion of the underlying assumptions and their plausibility. As such, we hope that our contribution may constitute a point of reference for further analytical work in this area as well as facilitate a better understanding of issues related to modelling intermittent demands.

Parametric approaches to intermittent demand forecasting rely upon a lead-time demand distributional assumption and the employment of an appropriate forecasting procedure for estimating the moments of the distribution. However, a number of non-parametric procedures have also been suggested in the literature to forecast intermittent demand requirements (e.g. Willemain et al. 2004; Porras and Dekker 2008). Such approaches typically rely upon bootstrapping procedures that permit a re-construction of the empirical distribution of the data, thus making distributional assumptions redundant. In addition, a number of parametric bootstrapping approaches have been put forward (e.g. Snyder 2002; Teunter and Duncan 2009). These approaches also rely upon bootstrapping but in conjunction with some assumptions about the underlying demand characteristics. Although it has been claimed that all the approaches discussed above may have an advantage over pure parametric methodologies, more empirical research is needed to evaluate the conditions under which one approach outperforms the other. In this chapter, we will be focusing solely on parametric forecasting. In particular, we will be discussing issues related to the estimation procedures that may be used. The issue of statistical distributions for parametric forecasting of intermittent demands is addressed in Chap. 2 of this book. A discussion on non-parametric alternatives may be found in Chap. 6.

The remainder of this chapter is structured around two main sections: in the next section we discuss issues related to the bias of intermittent demand estimates, followed by a discussion on the issue of variance. Some concluding remarks are offered in the last section of the chapter and all the detailed derivations are presented in the Appendices.

1.2 The Bias of Intermittent Demand Estimates

1.2.1 Croston's Critique of Exponential Smoothing

Croston (1972), as corrected by Rao (1973), proved the inappropriateness of exponential smoothing as a forecasting method when dealing with intermittent demands and he expressed in a quantitative form the bias associated with the use of this method when demand appears at random with some time periods showing no demand at all.

He first assumes deterministic demands of magnitude μ occurring every p review intervals. Subsequently the demand Y_t is represented by:

$$Y_t = \begin{cases} \mu, & t = np + 1 \\ 0, & \text{otherwise} \end{cases} \tag{1}$$

where $n = 0{,}1{,}2\ldots$ and $p \geq 1$.

Conventional exponential smoothing updates estimates every inventory review period whether or not demand occurs during this period. If we are forecasting one period ahead, Y'_t, the forecast of demand made in period t, is given by:

$$Y'_t = Y'_{t-1} + \alpha e_t = \alpha Y_t + (1 - \alpha) Y'_{t-1} \tag{2}$$

where α is the smoothing constant value used, $0 \leq \alpha \leq 1$, and e_t is the forecast error in period t.

Under these assumptions if we update our demand estimates only when demand occurs the expected demand estimate per time period is not μ/p, i.e. the population expected value, but rather:

$$\mathrm{E}\left(Y'_t\right) = \frac{\mu}{p} \frac{p\alpha}{1 - (1 - \alpha)^p} = \frac{\mu\alpha}{1 - \beta^p} \tag{3}$$

where $\beta = 1 - \alpha$.

Croston then refers to a stochastic model of arrival and size of demand, assuming that demand sizes, Z_t, are normally distributed, $N(\mu, \sigma^2)$, and that demand is random and has a Bernoulli probability $1/p$ of occurring in every review period (subsequently the inter demand intervals, p_t, follow the geometric distribution with a mean p). Under these conditions the expected demand per unit time period is:

$$\mathrm{E}(Y_t) = \frac{\mu}{p} \tag{4}$$

If we isolate the estimates that are made after a demand occurs, Croston showed that these exponentially smoothed estimates have the biased expected value:

$$\mathrm{E}\left(Y'_t\right) = \mu(\alpha + \beta/p) \tag{5}$$

The error, expressed as a percentage of the average demand, is shown to be $100\alpha(p - 1)$ and reveals an increase in estimation error produced by the Bernoulli arrival of demands as compared with constant inter-arrival intervals.

1.2.2 Croston's Method

Croston, assuming the above stochastic model of arrival and size of demand, introduced a new method for characterising the demand per period by modelling demand from constituent elements. According to his method, separate exponential smoothing estimates of the average size of the demand and the average interval between demand incidences are made after demand occurs. If no demand occurs, the estimates remain the same. If we let:

p'_t = the exponentially smoothed inter-demand interval, updated only if demand occurs in period t so that $E(p'_t) = E(p_t) = p$, and

Z'_t = the exponentially smoothed size of demand, updated only if demand occurs in period t so that $E(Z'_t) = E(Z_t) = z$

then following Croston's estimation procedure, the forecast, Y'_t for the next time period is given by:

$$Y'_t = \frac{Z'_t}{p'_t} \tag{6}$$

and, according to Croston, the expected estimate of demand per period in that case would be:

$$E(Y'_t) = E\left(\frac{Z'_t}{p'_t}\right) = \frac{E(Z'_t)}{E(p'_t)} = \frac{\mu}{p} \tag{7}$$

(i.e. the method is unbiased.)

Now more accurate estimates can be obtained and an advantage of the method is that when demand occurs in every period the method is identical to exponential smoothing. Thus, it can be used not only for the intermittent demand items but for the rest of the Stock Keeping Units (SKUs) as well.

Croston (1972) claimed that the variance of the demand estimates per time period is given by:

$$\mathrm{Var}\left(\frac{Z'_t}{p'_t}\right) = \frac{\alpha}{2-\alpha}\left[\frac{(p-1)}{p^4}\mu^2 + \frac{\sigma^2}{p^2}\right] \tag{8}$$

Rao (1973) pointed out that the right-hand-side of Eq. (8) is only an approximation to the variance. In the following section we show that Croston's equation (8) is not only inexact but also incorrect.

Lead-time replenishment decisions take place only in the time periods following demand occurrence and are based on the equation:

$$R_t = Z'_t + Km_t \tag{9}$$

where

R_t is the replenishment level to which the stock is raised,

m_t is the estimated mean absolute deviation of the demand size forecast errors and

K is a safety factor.

Schultz (1987) proposed a slight modification to Croston's method, suggesting that a different smoothing constant value should be used in order to update the inter-demand interval and the size of demand, when demand occurs. However, this modification to Croston's method has not been widely adopted (an exception may be the study conducted by Syntetos et al. 2009) and it is not discussed further in this chapter.

1.2.3 Assumptions of Croston's Model

Croston advocated separating the demand into two components, the inter-demand time and the size of demand, and analysing each component separately. He assumed a stationary mean model for representing the underlying demand pattern, normal distribution for the size of demand and a Bernoulli demand generation process, resulting in geometrically distributed inter-demand intervals.

Three more assumptions implicitly made by Croston in developing his model are the following: independence between demand sizes and inter-demand intervals, independence of successive demand sizes and independence of successive inter-demand intervals. As far as the last assumption is concerned it is important to note that the geometric distribution is characterised by a 'memory less' process: the probability of a demand occurring is independent of the time since the last demand occurrence, so that this distributional assumption is consistent independent inter-demand intervals.

The normality assumption is the least restrictive one for the analysis conducted by Croston, since the demand sizes may be, theoretically, represented by any probability distribution without affecting the mathematical properties of the demand estimates. The remaining assumptions are retained for the analysis to be conducted in this chapter. These assumptions have been challenged in respect of their realism (see, for example, Willemain et al. 1994) and they have also been challenged in respect of their theoretical consistency with Croston's forecasting method. Snyder (2002) pointed out that Croston's model assumes stationarity of demand intervals and yet a single exponential smoothing (SES) estimator is used, implying a non-stationary demand process. The same comment applies to demand sizes. Snyder commented that this renders the model and method inconsistent and he proposed some alternative models, and suggested a new forecasting approach based on parametric bootstrapping (see also Sect. 1.1). Shenstone and Hyndman (2005) developed this work by examining Snyder's models. In their paper they commented on the wide prediction intervals that arise for non-stationary models and recommended that stationary models should be reconsidered. However, they concluded, "…the possible models underlying Croston's and related methods must be non-stationary and defined on a continuous sample space. For Croston's original method, the sample space for the underlying model included negative values. This is inconsistent with reality that demand is always non-negative" (Shenstone and Hyndman, op. cit., pp. 389–390).

In summary, any potential non-stationary model assumed to be underlying Croston's method must have properties that do not match the demand data being modelled. Obviously, this does not mean that Croston's method and its variants, to be subsequently discussed in this section, are not useful. Such methods do constitute the current state of the art in intermittent demand parametric forecasting. However, an interesting line of further research would be to consider stationary models for intermittent demand forecasting rather than restricting attention to models implying Croston's method. For example, Poisson autoregressive models have been suggested to be potentially useful by Shenstone and Hyndman (2005).

1.2.4 The Size-Interval Method

If there is a random arrival of independent demands, the arrival process can be modelled as a Poisson stream. This idea was explored by Johnston and Boylan (1996). Their analysis was as follows:

If we set,

- W the demand per unit time with mean W_1 and variance W_2
- S the order size with mean S_1 and variance S_2
- I the inter-demand interval with mean I_1 and variance I_2
- N the number of orders per unit time with mean N_1 and variance N_2

then the demand in any period is the sum of the orders in that period and both the individual orders and the number of them in a given period are stochastic variables:

$$W = \sum_{i=1}^{N} S_i \tag{10}$$

Under the assumption that the order arrival process can be modelled as a Poisson stream and combining Clark's calculated mean and variance of the distribution of the summation of a number of stochastic random variables (1957):

$$W_1 = N_1 S_1 \tag{11}$$

$$W_2 = N_1 S_2 + N_2 (S_1)^2 \tag{12}$$

Using Cox's asymptotic equations (1962) for relating the number of orders (N) to the more easily measurable inter-demand interval (I) counting from a random point in time rather than an initial event (i.e. demand occurrence):

$$N_1 = \frac{1}{I_1} \tag{13}$$

$$N_2 \approx \frac{I_2}{(I_1)^3} + \frac{1}{6} + \frac{(I_2)^2}{(I_1)^4} - \frac{I_3}{3(I_1)^3} \tag{14}$$

where I_3 is the third moment about the mean for the inter-order interval.

The authors proposed the following method (Size-Interval) for obtaining accurate (intermittent) demand per period estimates:

$$W_1 = \frac{S_1}{I_1} \tag{15}$$

$$W_2 = \frac{S_2}{I_1} + \frac{(S_1)^2}{I_1} \tag{16}$$

Thus, the forecasts can be generated from estimates of the mean and variance of the order size and the average inter-demand interval.

The SI method was compared with SES on theoretically generated demand data over a wide range of possible conditions. Many different average inter-demand intervals (negative exponential distribution), smoothing constant values, lead times and distributions of the size of demand (negative exponential, Erlang and rectangular), were considered. The comparison exercise was extended to cover not only Poisson but also Erlang demand processes. The results were reported in the form of the ratio of the mean squared error (MSE) of one method to that of another. For the different factor combinations tried in this simulation experiment the SI method was superior to SES for inter-demand intervals greater than 1.25 review periods and in that way the authors showed how intermittent demand needs to be in order to benefit from the SI method (based on Croston's concept) more than SES.

At this stage it is important to note that the estimate of mean demand is identical between Croston's method and the SI method. Thus, later comments on bias of the $\frac{Z_t'}{p_t'}$ (or $\frac{S_1}{I_1}$) estimator hold for both methods.

1.2.5 Expected Estimate of Demand: Croston's Method

We know (assuming that order sizes and intervals are independent) that

$$\mathrm{E}\left(\frac{Z_t'}{p_t'}\right) = \mathrm{E}(Z_t')\mathrm{E}\left(\frac{1}{p_t'}\right) \tag{17}$$

but

$$\mathrm{E}\left(\frac{1}{p_t'}\right) \neq \frac{1}{\mathrm{E}(p_t')} \tag{18}$$

We denote by P_t the inter demand interval that follows the geometric distribution including the first success (i.e. demand occurring period) and by $\frac{1}{p_t}$ the probability of demand occurrence at period t. Now the case of $\alpha = 1$ is analysed since it is mathematically tractable; more realistic α values will be considered in the next sub-section. Assuming that $\alpha = 1$, so that $p_t' = p_t$ we then have:

$$E\left(\frac{1}{p_t}\right) = \sum_{x=1}^{\infty}\frac{1}{x}\frac{1}{p}\left(1-\frac{1}{p}\right)^{x-1}$$

$$= \frac{1}{p}\sum_{x=1}^{\infty}\frac{1}{x}\left(\frac{p-1}{p}\right)^{x-1}$$

[for $p > 1$ (i.e. demand does not occur in every single time period)]

$$= \frac{1}{p}\sum_{x=1}^{\infty}\frac{1}{x}\frac{\left(\frac{p-1}{p}\right)^{x}}{\left(\frac{p-1}{p}\right)^{1}} = \frac{1}{p}\frac{1}{\frac{p-1}{p}}\sum_{x=1}^{\infty}\frac{1}{x}\left(\frac{p-1}{p}\right)^{x}$$

$$= \frac{1}{p-1}\left[\frac{p-1}{p} + \frac{1}{2}\left(\frac{p-1}{p}\right)^{2} + \frac{1}{3}\left(\frac{p-1}{p}\right)^{3} + \cdots\cdots\right]$$

$$= -\frac{1}{p-1}\log\left(\frac{1}{p}\right)$$

Therefore:

$$E\left(\frac{Z'_t}{p'_t}\right) = E(Z'_t)E\left(\frac{1}{p'_t}\right) = \mu\left[-\frac{1}{p-1}\log\left(\frac{1}{p}\right)\right] \tag{19}$$

So if, for example, the average size of demand when it occurs is $\mu = 6$, and the average inter-demand interval is $p = 3$, the average estimated demand per time period using Croston's method (for $\alpha = 1$) is not $\frac{\mu}{p} = \frac{6}{3} = 2$ but it is $6 * 0.549 = 3.295$ (i.e. 64.75% bias implicitly incorporated in Croston's estimate).

The maximum bias over all possible smoothing parameters is given by:

$$\text{Maximum bias} = \mu\left[-\frac{1}{p-1}\log\left(\frac{1}{p}\right)\right] - \frac{\mu}{p} \tag{20}$$

This is attained at $\alpha = 1$. For realistic α values, the magnitude of the error is smaller and it is quantified in the next sub-section.

1.2.6 An Approximation of Croston's Bias

For α values less than 1 the magnitude of the error obviously depends on the smoothing constant value being used. We show, in this sub-section, that the bias associated with Croston's method in practice can be approximated, for all smoothing constant values, by: $\frac{\alpha}{2-\alpha}\mu\frac{(p-1)}{p^2}$

The bias can be conveniently expressed as a percentage of the average demand and it is easily shown to be: $100\frac{\alpha}{2-\alpha}\left(1-\frac{1}{p}\right)$

The above approximation is proven as follows:

We apply Taylor's theorem to a function of two variables, $g(x)$ where:

x is the vector: $x = (x_1, x_2)$ and $g(x) = g(x_1, x_2) = \frac{x_1}{x_2}$ (with $x_1 = Z'_t$ and $x_2 = p'_t$)

$\mathrm{E}(x_1) = \theta_1$, $\mathrm{E}(x_2) = \theta_2$ and

θ is the vector: $\theta = (\theta_1, \theta_2)$ with $g(\theta) = g(\theta_1, \theta_2) = \frac{\theta_1}{\theta_2}$

This is the case for the problem under concern, with $\theta_1 = \mu$ and $\theta_2 = p$.

$$g(x) = g(\theta) + \left[\frac{\partial g}{\partial \theta_1}(x_1 - \theta_1) + \frac{\partial g}{\partial \theta_2}(x_2 - \theta_2)\right] + \frac{1}{2}\left[\frac{\partial^2 g}{\partial \theta_1^2}(x_1 - \theta_1)^2 + 2\frac{\partial^2 g}{\partial \theta_1 \partial \theta_2}(x_1 - \theta_1)(x_2 - \theta_2) + \frac{\partial^2 g}{\partial \theta_2^2}(x_2 - \theta_2)^2\right] + \cdots\cdots. \tag{21}$$

$$\mathrm{E}[g(x)] = \mathrm{E}[g(\theta)] + \mathrm{E}\left[\frac{\partial g}{\partial \theta_1}(x_1 - \theta_1) + \frac{\partial g}{\partial \theta_2}(x_2 - \theta_2)\right] + \frac{1}{2}\mathrm{E}\left[\frac{\partial^2 g}{\partial \theta_1^2}(x_1 - \theta_1)^2 + 2\frac{\partial^2 g}{\partial \theta_1 \partial \theta_2}(x_1 - \theta_1)(x_2 - \theta_2) + \frac{\partial^2 g}{\partial \theta_2^2}(x_2 - \theta_2)^2\right] + \cdots\cdots \tag{22}$$

$$\frac{\partial g}{\partial \theta_1} = \frac{1}{\theta_2} \tag{23}$$

$$\frac{\partial g}{\partial \theta_2} = -\frac{\theta_1}{\theta_2^2} \tag{24}$$

$$\frac{\partial^2 g}{\partial \theta_1^2} = 0 \tag{25}$$

$$\frac{\partial^2 g}{\partial \theta_1 \partial \theta_2} = -\frac{1}{\theta_2^2} \tag{26}$$

$$\frac{\partial^2 g}{\partial \theta_2^2} = -\theta_1\left(-\frac{2}{\theta_2^3}\right) = \frac{2\theta_1}{\theta_2^3} \tag{27}$$

Considering the assumption about independence between demand sizes and inter-demand intervals, Eq. (25) and the fact that the first moment about the mean is always zero, Eq. (22) becomes:

$$\mathrm{E}[g(x)] = \mathrm{E}[g(\theta)] + \frac{1}{2}\frac{\partial^2 g}{\partial \theta_2^2}\mathrm{E}(x_2 - \theta_2)^2 + \cdots\cdots$$

Therefore:

$$\mathrm{E}\left(\frac{x_1}{x_2}\right) = \frac{\theta_1}{\theta_2} + \frac{1}{2}\frac{\partial^2 g}{\partial \theta_2^2} Var(x_2) + \cdots\cdots \tag{28}$$

Assuming that the inter-demand interval series is not auto-correlated and that the inter-demand intervals (p_t) are geometrically distributed with a mean of p and homogeneous variance[2] of $p(p-1)$, it follows that:

$$\mathrm{Var}(x_2) = \mathrm{Var}(p'_t) = \frac{\alpha}{2-\alpha}\mathrm{Var}(p_t) = \frac{\alpha}{2-\alpha}p(p-1)$$

Assuming that demand sizes (z_t) are distributed with a mean, μ, Eq. (28) becomes:

$$\mathrm{E}\left(\frac{x_1}{x_2}\right) \approx \frac{\theta_1}{\theta_2} + \frac{1}{2}\frac{\alpha}{2-\alpha}\frac{2\theta_1}{\theta_2^3}p(p-1) \tag{29}$$

$$\mathrm{E}\left(\frac{z'_t}{p'_t}\right) \approx \frac{\mu}{p} + \frac{\alpha}{2-\alpha}\mu\frac{(p-1)}{p^2} \tag{30}$$

Subsequently, the bias implicitly incorporated in Croston's estimates is approximated by (31):

$$\mathrm{Bias}_{\mathrm{Croston}} \approx \frac{\alpha}{2-\alpha}\mu\frac{(p-1)}{p^2} \tag{31}$$

Syntetos (2001) showed by means of experimentation on a wide range of theoretically generated data that, for $\alpha \leq 0.2$, the difference between the theoretical bias given by (31) and the simulated bias lies within a 99% confidence interval of $\pm 0.2\%$ of the mean simulated demand.

1.2.7 The SY Method

Since Croston's method is biased we consider applying a factor to the estimates produced by his method so that the second order bias term is directly eliminated.

We try to estimate the value of a parameter λ so that:

$$\mathrm{E}(Y'_t) = \mathrm{E}\left(\lambda\frac{z'_t}{p'_t}\right) = \frac{\mu}{p} \tag{32}$$

By applying a factor λ to Croston's updating procedure of sizes and intervals and considering approximation (30) we then have:

---

2 The issue of variance in the geometric distribution is discussed in the next section.

$$\mathrm{E}\left(\lambda \frac{z'_t}{p'_t}\right) \approx \lambda \frac{\mu}{p} + \lambda \frac{\alpha}{2-\alpha} \mu \frac{(p-1)}{p^2}$$

We can then set an approximation to the bias equal to zero in order to specify the value of parameter λ:

$$\text{Bias} \approx (1-\lambda)\frac{\mu}{p} - \lambda \frac{\alpha}{2-\alpha} \mu \frac{(p-1)}{p^2} = 0$$

$$1-\lambda = \frac{\lambda \frac{\alpha}{2-\alpha} \mu \frac{(p-1)}{p^2}}{\frac{\mu}{p}}$$

$$1 = \lambda\left(1 + \frac{\alpha}{2-\alpha} \frac{p-1}{p}\right) \tag{33}$$

$$\lambda = \frac{1}{1 + \frac{\alpha}{2-\alpha} \frac{p-1}{p}} = \frac{1}{\frac{2p-\alpha p+\alpha p-\alpha}{(2-\alpha)p}} = \frac{(2-\alpha)p}{2p-\alpha}$$

$$\lambda = \frac{1-\frac{\alpha}{2}}{1-\frac{\alpha}{2p}}$$

Therefore we propose the following updating procedure for obtaining approximately unbiased intermittent demand estimates:

$$Y'_t = \frac{\left(1-\frac{\alpha}{2}\right) z'_t}{\left(1-\frac{\alpha}{2p'_t}\right) p'_t} = \frac{\left(1-\frac{\alpha}{2}\right) z'_t}{p'_t - \frac{\alpha}{2}} \tag{34}$$

We call this method, for the purpose of our research, the SY method (after Syntetos 2001; for a further discussion see also Teunter and Sani 2009). The expected estimate of mean demand per period for the SY method is given by Eq. (35).

$$\mathrm{E}(Y'_t) = \mathrm{E}\left[\frac{\left(1-\frac{\alpha}{2}\right) z'_t}{p'_t - \frac{\alpha}{2}}\right] \approx \frac{\mu}{p} \tag{35}$$

This approximation is not necessarily accurate when higher order terms are taken into account.

1.2.8 The SBA Method

From Eq. (33) we have:

$$\lambda = \frac{1-\frac{\alpha}{2}}{1-\frac{\alpha}{2p}}$$

But, as $p \to \infty, \lambda \to 1 - \frac{\alpha}{2}$.

Therefore a possible estimation procedure, for intermittent demand data series with a large inter-demand interval, is the following:

$$Y_t' = \left(1 - \frac{\alpha}{2}\right)\frac{z_t'}{p_t'} \tag{36}$$

As in the case of the SY method, the smoothing constant value is considered for generating demand estimates. The above heuristic seems to provide a reasonable approximation of the actual demand per period especially for the cases of very low α values and large p inter-demand intervals. This estimator is known in the literature as the SBA method (after Syntetos–Boylan Approximation, Syntetos and Boylan 2005). The expected estimate of mean demand per period for the SBA method is given by Eq. (37).

$$\mathrm{E}(Y_t') = \mathrm{E}\left(\left(1 - \frac{\alpha}{2}\right)\frac{z_t'}{p_t'}\right) \approx \frac{\mu}{p} - \frac{\alpha}{2}\frac{\mu}{p^2} \tag{37}$$

This approximation is not necessarily accurate when higher order terms are taken into account. For the detailed derivation of (37) see Appendix A. The empirical validity and utility of the SBA have been independently established in work conducted by Eaves and Kingsman (2004) and Gutierrez et al. (2008).

1.2.9 Other Bias Correction Factors

Before we close this section, we need to say that all the work presented above is based upon the assumption of a Bernoulli demand arrival process and SES estimates of sizes and intervals. Boylan and Syntetos (2003) and Shale et al. (2006) presented correction factors to overcome the bias associated with Croston's approach under a Poisson demand arrival process and/or estimation of demand sizes and intervals using a simple moving average (SMA). The correction factors are summarized in the following table (where k is the moving average length and α is the smoothing constant for SES).

At this point it is important to note that SMA and SES are often treated as equivalent when the average age of the data in the estimates is the same (Brown 1963). A relationship links the number of points in an arithmetic average (k) with the smoothing parameter of SES (α) for stationary demand. Hence it may be used to relate the correction factors presented in Table 1.1 for each of the two demand generation processes considered. The linking equation is:

$$k = (2 - \alpha)/\alpha$$

Table 1.1 Bias correction factors

	Demand generation process	
	Bernoulli	Poisson
Estimation		
SES	Equation (34) Syntetos (2001) Equation (36) Syntetos and Boylan (2005)	$1-\frac{\alpha}{2-\alpha}$ Shale et al. (2006)
SMA	$\frac{k}{k+1}$ Boylan and Syntetos (2003)	$\frac{k-1}{k}$ Shale et al. (2006)

1.3 The Variance of Intermittent Demand Estimates

1.3.1 The Variance of Exponentially Smoothed Estimates

According to Croston (1972), and under the stochastic demand model he assumed for his study, the variance of demand per unit time period is:

$$\mathrm{Var}(Y_t) = \frac{p-1}{p^2}\mu^2 + \frac{\sigma^2}{p} \tag{38}$$

and the variance of the exponentially smoothed estimates, updated every period:

$$\mathrm{Var}\left(Y_t'\right) = \frac{\alpha}{2-\alpha}\mathrm{Var}(Y_t) = \frac{\alpha}{2-\alpha}\left[\frac{p-1}{p^2}\mu^2 + \frac{\sigma^2}{p}\right] \tag{39}$$

(where α is the smoothing constant); assuming a stationary mean model and homogeneous variance of demand per unit time period.

If we isolate the estimates that are made just after an issue (which are those that will be used for replenishment purposes by a continuous review stock control system) Croston showed that these estimates have the following variance (as corrected by Rao 1973):

$$\mathrm{Var}\left(Y_t'\right) = \alpha^2\sigma^2 + \frac{\alpha\beta^2}{2-\alpha}\left[\frac{p-1}{p^2}\mu^2 + \frac{\sigma^2}{p}\right] \tag{40}$$

where $\beta = 1 - \alpha$.

1.3.2 The Variance of Croston's Estimates

As discussed in the previous section, Croston suggested estimating the average interval between issues and the average size of an issue when it occurs and to combine those statistics to give an unbiased estimate of the underlying mean demand.

If we let:

p_t = the inter-demand interval that follows the geometric distribution with: $\mathrm{E}(p_t) = p$ and, according to Croston,

$$\mathrm{Var}(p_t) = (p - 1)^2 \tag{41}$$

p'_t = the exponentially smoothed inter-demand interval, updated only after demand occurs

z_t = the demand size, when demand occurs, that follows the normal distribution, $N(\mu, \sigma^2)$, and

Z'_t = the exponentially smoothed size of demand, updated only after demand occurs.

We then have:

$$\mathrm{E}(Z'_t) = \mathrm{E}(z_t) = \mu \tag{42}$$

$$\mathrm{E}(p'_t) = \mathrm{E}(p_t) = p \tag{43}$$

$$\mathrm{Var}(z'_t) = \frac{\alpha}{2-\alpha}\mathrm{Var}(z_t) = \frac{\alpha}{2-\alpha}\sigma^2 \tag{44}$$

$$\mathrm{Var}(p'_t) = \frac{\alpha}{2-\alpha}\mathrm{Var}(p_t) = \frac{\alpha}{2-\alpha}(p-1)^2 \tag{45}$$

The variance of the ratio of two independent random variables x_1, x_2 is given in Stuart and Ord (1994) as follows:

$$\mathrm{Var}\left(\frac{x_1}{x_2}\right) = \left(\frac{\mathrm{E}(x_1)}{\mathrm{E}(x_2)}\right)^2\left[\frac{\mathrm{Var}(x_1)}{(\mathrm{E}(x_1))^2} + \frac{\mathrm{Var}(x_2)}{(\mathrm{E}(x_2))^2}\right] \tag{46}$$

For $x_1 = Z'_t$ and $x_2 = p'_t$, considering Eqs. (42), (43), (44) and (45), the variance of the estimates produced by using Croston's method is calculated by Eq. (47)

$$\mathrm{Var}(Y'_t) = \mathrm{Var}\left(\frac{Z'_t}{p'_t}\right) = \frac{a}{2-a}\left[\frac{(p-1)}{p^4}\mu^2 + \frac{\sigma^2}{p^2}\right] \tag{47}$$

assuming that the same smoothing constant value is used for updating demand sizes and inter-demand intervals and that both demand size and inter-demand interval series are not auto-correlated and have homogeneous variances.

Rao (1973) pointed out that the right-hand side of Eq. (47) is only an approximation to the variance. This follows since Eq. (46) is, in fact, an approximation.

1.3.3 *The Variance of Inter-Demand Intervals*

The number of independent Bernoulli trials (with a specified probability of success) before the first success is represented by the geometric distribution. An alternative form of the geometric distribution involves the number of trials up to and including the first success (demand occurring period). Considering the notation used in this chapter, the variability of the geometrically distributed inter-demand intervals is $p(p-1)$, irrespective of which form of the geometric distribution is utilised. Consequently (41) should be replaced by (48).

$$\mathrm{Var}(p_t) = p(p-1) \tag{48}$$

1.4 The Corrected Variance of Croston's Method Estimates

By taking (48) into consideration, the variance of the demand per period estimates, using Croston's method, would become:

$$\mathrm{Var}\left(\frac{Z'_t}{p'_t}\right) \approx \frac{a}{2-a}\left[\frac{p-1}{p^3}\mu^2 + \frac{\sigma^2}{p^2}\right] \tag{49}$$

indicating that the approximated variance of the estimates produced by Croston's method is in fact greater than that calculated by Croston himself, Eq. (47).[3]

Nevertheless, approximation (49) is still not correct. In fact there is a fundamental problem in directly applying Stuart and Ord's result, given by (46), for the purpose of deriving the variance of the forecasts produced by a biased estimator.

This is proven as follows:

We apply Taylor's theorem to a function of two variables, $g(x)$ where

x is the vector: $x = (x_1, x_2)$ and $g(x) = g(x_1, x_2) = \dfrac{x_1}{x_2}$

with $\mathrm{E}(x_1) = \theta_1$ and $\mathrm{E}(x_2) = \theta_2$.

The vector θ is defined as: $\theta = (\theta_1, \theta_2)$, with $g(\theta) = g(\theta_1, \theta_2) = \dfrac{\theta_1}{\theta_2}$

$$g(x) = g(\theta) + \left[\frac{\partial g}{\partial \theta_1}(x_1 - \theta_1) + \frac{\partial g}{\partial \theta_2}(x_2 - \theta_2)\right] + \cdots\cdots \tag{50}$$

[3] Equation (10) in the original paper.

where $g(\theta) = \frac{\theta_1}{\theta_2}$ is just the first term in the Taylor series and not necessarily the population expected value.

For:

$$\mathrm{E}[g(x)] = g(\theta) + \varepsilon \tag{51}$$

where ε is an error term, which according to Croston, can be neglected, we then have:

$$\begin{aligned}
\mathrm{Var}[g(x)] &= \mathrm{E}\{g(x) - \mathrm{E}[g(x)]\}^2 \\
&= \mathrm{E}[g(x) - g(\theta)]^2 \\
&\approx \mathrm{E}\left[\frac{\partial g}{\partial \theta_1}(x_1 - \theta_1) + \frac{\partial g}{\partial \theta_2}(x_2 - \theta_2)\right]^2 \\
&= \left(\frac{\mathrm{E}(x_1)}{\mathrm{E}(x_2)}\right)^2 \left[\frac{\mathrm{Var}(x_1)}{(\mathrm{E}(x_1))^2} + \frac{\mathrm{Var}(x_2)}{(\mathrm{E}(x_2))^2}\right]
\end{aligned} \tag{52}$$

If we set:
$x_1 = Z'_t$, the estimate of demand size, with $\mathrm{E}(Z'_t) = \mu$
and $x_2 = p'_t$, the estimate of the inter-demand interval, with $\mathrm{E}(p'_t) = p$
so that $g(x) = Y'_t$
it has been proven, in the previous section, that:

$$\mathrm{E}(Y'_t) \neq \frac{\mu}{p} \text{ or } \mathrm{E}[g(x)] \neq g(\theta)$$

Based on that, we argue that the error term in Eq. (51) cannot be neglected and therefore approximation (52) cannot be used to represent the problem in hand.

Our argument is discussed in greater detail in Appendices 2 and 3, where we also derive a correct approximation (to the second order term) of the variance of Croston's estimates. That variance expression is given by (53).

$$\begin{aligned}
\mathrm{Var}\left(\frac{Z'_t}{p'_t}\right) \approx & \frac{\alpha}{2-\alpha}\left[\frac{(p-1)}{p^3}\left(\mu^2 + \frac{\alpha}{2-\alpha}\sigma^2\right) + \frac{\sigma^2}{p^2}\right] \\
& + \frac{\alpha^4}{1-(1-\alpha)^4}\frac{\mu^2}{p^4}\left(1 - \frac{1}{p}\right)\left[9\left(1 - \frac{1}{p}\right)p^2 + 1\right]
\end{aligned} \tag{53}$$

Syntetos (2001) showed, by means of simulation on a wide range of theoretically generated data, that approximation (53) does not increase the accuracy of the calculated variance more than by only considering the first term of this approximation. In fact, for certain cases, approximation (52) was shown to perform worse. Taking that into account the variance of Croston's estimates may be 'safely' approximated by (54).

$$\mathrm{Var}\left(\frac{Z'_t}{p'_t}\right) \approx \frac{\alpha}{2-\alpha}\left[\frac{(p-1)}{p^3}\left(\mu^2 + \frac{\alpha}{2-\alpha}\sigma^2\right) + \frac{\sigma^2}{p^2}\right] \tag{54}$$

1.4.1 The Variance of the SY Method Estimates

The estimation equation for the SY method presented in the previous section is given by:

$$Y'_t = \frac{\left(1-\frac{\alpha}{2}\right)Z'_t}{\left(1-\frac{\alpha}{2p'_t}\right)p'_t} = \frac{\left(1-\frac{\alpha}{2}\right)Z'_t}{p'_t - \frac{\alpha}{2}}$$

and the expected estimate produced by this method was shown to be as follows:

$$\mathrm{E}(Y'_t) = \mathrm{E}\left[\frac{\left(1-\frac{\alpha}{2}\right)Z'_t}{p'_t - \frac{\alpha}{2}}\right] \approx \frac{\mu}{p}$$

In Appendix D we perform a series of calculations in order to find the variance of the estimates of mean demand produced by the SY method. The variance is approximated by Eq. (55).

$$\begin{aligned}\mathrm{Var}\left[\frac{\left(1-\frac{\alpha}{2}\right)Z'_t}{p'_t - \frac{\alpha}{2}}\right] &\approx \frac{\alpha(2-\alpha)}{4}\frac{\left[\left(p-\frac{\alpha}{2}\right)^2\sigma^2 + p(p-1)\mu^2 + \frac{\alpha}{2-\alpha}p(p-1)\sigma^2\right]}{\left(p-\frac{\alpha}{2}\right)^4}\\ &+ \frac{\alpha^4}{1-(1-\alpha)^4}\frac{\left(1-\frac{\alpha}{2}\right)^2\mu^2}{\left(p-\frac{\alpha}{2}\right)^6}p^2\left(1-\frac{1}{p}\right)\left[9\left(1-\frac{1}{p}\right)p^2+1\right]\end{aligned} \tag{55}$$

Similarly to the previous sub-section, Syntetos (2001) showed by means of simulation that consideration of both terms of approximation (55) does not provide, overall, a more reliable estimate of the calculated variance than when only the first term of this approximation is considered. The calculation of the variance of the SY method may be simplified, without sacrificing accuracy, by considering approximation (56).

$$\mathrm{Var}\left[\frac{\left(1-\frac{\alpha}{2}\right)Z'_t}{p'_t - \frac{\alpha}{2}}\right] \approx \frac{\alpha(2-\alpha)}{4}\frac{\left[\left(p-\frac{\alpha}{2}\right)^2\sigma^2 + p(p-1)\mu^2 + \frac{\alpha}{2-\alpha}p(p-1)\sigma^2\right]}{\left(p-\frac{\alpha}{2}\right)^4} \tag{56}$$

1.4.2 The Variance of the SBA Method Estimates

The estimation procedure for the SBA method discussed in the previous section is:

$$Y'_t = \left(1 - \frac{\alpha}{2}\right)\frac{Z'_t}{p'_t}$$

with

$$\mathrm{E}(Y'_t) = \mathrm{E}\left(\left(1 - \frac{\alpha}{2}\right)\frac{z'_t}{p'_t}\right) \approx \frac{\mu}{p} - \frac{\alpha}{2}\frac{\mu}{p^2}$$

The variance of the estimates produced by the SBA is calculated as:

$$\mathrm{Var}(Y'_t) = \mathrm{Var}\left(\left(1 - \frac{\alpha}{2}\right)\frac{z'_t}{p'_t}\right) = \left(1 - \frac{\alpha}{2}\right)^2 \mathrm{Var}\left(\frac{z'_t}{p'_t}\right) \tag{57}$$

Considering approximation (54) we finally have:

$$\mathrm{Var}\left(\left(1 - \frac{\alpha}{2}\right)\frac{Z'_t}{p'_t}\right) \approx \frac{\alpha(2-\alpha)}{4}\left[\frac{(p-1)}{p^3}\left(\mu^2 + \frac{\alpha}{2-\alpha}\sigma^2\right) + \frac{\sigma^2}{p^2}\right] \tag{58}$$

Extensive analysis conducted by Syntetos (2001) justified the choice of (58) instead of (57) for the purpose of approximating the variance of the SBA method.

1.5 Conclusions

Research in the area of forecasting and stock control for intermittent demand items has developed rapidly in recent years with new results implemented into software products because of their practical importance (Fildes et al. 2008). Simple exponential smoothing (SES) is widely used in industry to deal with sales/demand data but is inappropriate in an intermittent demand context. Rather, Croston's method is regarded as the standard estimator for such demand patterns. Recent studies have shown that there is scope for improving Croston's estimates by means of accounting for the bias implicitly incorporated in them. Two such methods have been discussed in this chapter and their statistical properties have been analysed in detail along with those of both Croston's method and SES. We hope that our contribution may constitute a point of reference for further analytical work in this area as well as facilitate a better understanding of issues related to modelling intermittent demands.

Appendix A: The Expectation of the Mean Demand Estimate of SBA

$$\begin{aligned}
E(Y'_t) &= E\left(\left(1-\frac{\alpha}{2}\right)\frac{z'_t}{p'_t}\right)\\
&\approx \left(1-\frac{\alpha}{2}\right)\frac{\mu}{p}+\frac{2-\alpha}{2}\,\frac{\alpha}{2-\alpha}\,\mu\left(\frac{p-1}{p^2}\right)\\
&= \left(1-\frac{\alpha}{2}\right)\frac{\mu}{p}+\frac{2-\alpha}{2}\,\frac{\alpha}{2-\alpha}\,\mu\left(\frac{1}{p}-\frac{1}{p^2}\right)\\
&= \left(1-\frac{\alpha}{2}\right)\frac{\mu}{p}+\frac{\alpha}{2}\frac{\mu}{p}-\frac{\alpha}{2}\frac{\mu}{p^2}\\
&= \frac{\mu}{p}-\frac{\alpha}{2}\frac{\mu}{p}+\frac{\alpha}{2}\frac{\mu}{p}-\frac{\alpha}{2}\frac{\mu}{p^2}\\
&= \frac{\mu}{p}-\frac{\alpha}{2}\frac{\mu}{p^2}
\end{aligned}$$

This proves the result given by Eq. (37).

Appendix B: A Correct Approximation to the Variance of Croston's Estimates

We apply Taylor's theorem to a function of two variables, $g(x)$
where:
x is the vector: $x = (x_1, x_2)$ and

$$g(x) = g(x_1, x_2) = \frac{x_1}{x_2}$$

with $E(x_1) = \theta_1$ and $E(x_2) = \theta_2$.

The vector θ is defined as: $\theta = (\theta_1, \theta_2)$, with $g(\theta) = g(\theta_1, \theta_2) = \frac{\theta_1}{\theta_2}$

$$\begin{aligned}
g(x) = g(\theta) &+ \left[\frac{\partial g}{\partial \theta_1}(x_1-\theta_1)+\frac{\partial g}{\partial \theta_2}(x_2-\theta_2)\right]\\
&+\frac{1}{2}\left[\frac{\partial^2 g}{\partial \theta_1^2}(x_1-\theta_1)^2+2\frac{\partial^2 g}{\partial \theta_1 \partial \theta_2}(x_1-\theta_1)(x_2-\theta_2)+\frac{\partial^2 g}{\partial \theta_2^2}(x_2-\theta_2)^2\right]+\cdots\cdots
\end{aligned} \tag{B.1}$$

where $g(\theta) = \frac{\theta_1}{\theta_2}$ is just the first term in the Taylor series and not the population expected value.

$$\frac{\partial g}{\partial \theta_1} = \frac{1}{\theta_2} \tag{B.2}$$

$$\frac{\partial g}{\partial \theta_2} = -\frac{\theta_1}{\theta_2^2} \tag{B.3}$$

$$\frac{\partial^2 g}{\partial \theta_1^2} = 0 \tag{B.4}$$

$$\frac{\partial^2 g}{\partial \theta_1 \partial \theta_2} = -\frac{1}{\theta_2^2} \tag{B.5}$$

$$\frac{\partial^2 g}{\partial \theta_2^2} = -\theta_1 \left(-\frac{2}{\theta_2^3} \right) = \frac{2\theta_1}{\theta_2^3} \tag{B.6}$$

therefore, considering (B.4), (B.1) becomes:

$$\begin{aligned} g(x) = g(\theta) &+ \left[\frac{\partial g}{\partial \theta_1}(x_1 - \theta_1) + \frac{\partial g}{\partial \theta_1}(x_2 - \theta_2) \right] \\ &+ \left[\frac{\partial^2 g}{\partial \theta_1 \partial \theta_2}(x_1 - \theta_1)(x_2 - \theta_2) + \frac{1}{2}\frac{\partial^2 g}{\partial \theta_2^2}(x_2 - \theta_2)^2 \right] + \cdots\cdots \end{aligned} \tag{B.7}$$

We set:

$x_1 = Z'_t$, the estimate of demand size, with $E(Z'_t) = \mu$
and $x_2 = p'_t$, the estimate of the inter-demand interval, with $E(p'_t) = p$
so that $g(x) = Y'_t$,
It has been proven, in Sect. 1.2, that:

$$E(Y'_t) = E[g(x)] \approx g(\theta) + \frac{1}{2}\frac{\partial^2 g}{\partial \theta_2^2} E(x_2 - \theta_2)^2$$

considering the first three terms in the Taylor series.

Therefore:

$$\begin{aligned} \mathrm{Var}(Y'_t) &= \mathrm{Var}[g(x)] = E[g(x) - E[g(x)]]^2 \\ &\approx E\left\{ \begin{matrix} \frac{\partial g}{\partial \theta_1}(x_1-\theta_1) + \frac{\partial g}{\partial \theta_2}(x_2-\theta_2) + \frac{\partial^2 g}{\partial \theta_1 \partial \theta_2}(x_1-\theta_1)(x_2-\theta_2) + \frac{1}{2}\frac{\partial^2 g}{\partial \theta_2^2}(x_2-\theta_2)^2 \\ -\frac{1}{2}\frac{\partial^2 g}{\partial \theta_2^2}E(x_2-\theta_2)^2 \end{matrix} \right\}^2 \\ &= E\left\{ \left(\frac{\partial g}{\partial \theta_1}\right)^2 (x_1-\theta_1)^2 + \left(\frac{\partial g}{\partial \theta_2}\right)^2 (x_2-\theta_2)^2 + \left(\frac{\partial^2 g}{\partial \theta_1 \partial \theta_2}\right)^2 (x_1-\theta_1)^2 (x_2-\theta_2)^2 \right. \\ &\quad + \frac{1}{4}\left(\frac{\partial^2 g}{\partial \theta_2^2}\right)^2 (x_2-\theta_2)^4 + \frac{1}{4}\left(\frac{\partial^2 g}{\partial \theta_2}\right)^2 \left[E(x_2-\theta_2)^2\right]^2 + 2\frac{\partial g}{\partial \theta_1}(x_1-\theta_1)\frac{\partial g}{\partial \theta_2}(x_2-\theta_2) \end{aligned}$$

$$
\begin{aligned}
&+2\frac{\partial g}{\partial\theta_1}(x_1-\theta_1)\frac{\partial^2 g}{\partial\theta_1\partial\theta_2}(x_1-\theta_1)(x_2-\theta_2)+\frac{\partial g}{\partial\theta_1}(x_1-\theta_1)\frac{\partial^2 g}{\partial\theta_2^2}(x_2-\theta_2)^2\\
&-\frac{\partial g}{\partial\theta_1}(x_1-\theta_1)\frac{\partial^2 g}{\partial\theta_2^2}\mathrm{E}(x_2-\theta_2)^2+2\frac{\partial g}{\partial\theta_2}(x_2-\theta_2)\frac{\partial^2 g}{\partial\theta_1\partial\theta_2}(x_1-\theta_1)(x_2-\theta_2)\\
&+\frac{\partial g}{\partial\theta_2}\frac{\partial^2 g}{\partial\theta_2^2}(x_2-\theta_2)^3-\frac{\partial g}{\partial\theta_2}\frac{\partial^2 g}{\partial\theta_2^2}(x_2-\theta_2)\mathrm{E}(x_2-\theta_2)^2\\
&+\frac{\partial^2 g}{\partial\theta_1\partial\theta_2}(x_1-\theta_1)(x_2-\theta_2)\frac{\partial^2 g}{\partial\theta_2^2}(x_2-\theta_2)^2\\
&-\frac{\partial^2 g}{\partial\theta_1\partial\theta_2}(x_1-\theta_1)(x_2-\theta_2)\frac{\partial^2 g}{\partial\theta_2^2}\mathrm{E}(x_2-\theta_2)^2\\
&\left.-\frac{1}{2}\left(\frac{\partial^2 g}{\partial\theta_2^2}\right)^2(x_2-\theta_2)^2\mathrm{E}(x_2-\theta_2)^2\right\}
\end{aligned}
\tag{B.8}
$$

Assuming that x_1, x_2 are independent:

$$
\begin{aligned}
\mathrm{Var}[g(x)] \approx &\left(\frac{\partial g}{\partial\theta_1}\right)^2\mathrm{E}(x_1-\theta_1)^2+\left(\frac{\partial g}{\partial\theta_2}\right)^2\mathrm{E}(x_2-\theta_2)^2\\
&+\left(\frac{\partial^2 g}{\partial\theta_1\partial\theta_2}\right)^2\mathrm{E}\left((x_1-\theta_1)^2(x_2-\theta_2)^2\right)+\frac{\partial g}{\partial\theta_2}\frac{\partial^2 g}{\partial\theta_2^2}\mathrm{E}(x_2-\theta_2)^3\\
&+\frac{1}{4}\left(\frac{\partial^2 g}{\partial\theta_2^2}\right)^2\mathrm{E}(x_2-\theta_2)^4-\frac{1}{4}\left(\frac{\partial^2 g}{\partial\theta_2^2}\right)^2\left[\mathrm{E}(x_2-\theta_2)^2\right]^2
\end{aligned}
\tag{B.9}
$$

considering Eqs. (B.2), (B.3), (B.5) and (B.6)

$$
\begin{aligned}
\mathrm{Var}[g(x)] \approx &\frac{\mathrm{Var}(x_1)}{\theta_2^2}+\frac{\theta_1^2\mathrm{Var}(x_2)}{\theta_2^4}+\frac{1}{\theta_2^4}\mathrm{Var}(x_1)\mathrm{Var}(x_2)\\
&-\frac{2\theta_1^2\mathrm{E}(x_2-\theta_2)^3}{\theta_2^5}+\frac{\theta_1^2\mathrm{E}(x_2-\theta_2)^4}{\theta_2^6}-\frac{\theta_1^2\left[\mathrm{E}(x_2-\theta_2)^2\right]^2}{\theta_2^6}\\
=&\frac{\mathrm{Var}(x_1)}{(\mathrm{E}(x_2))^2}+\frac{(\mathrm{E}(x_1))^2\mathrm{Var}(x_2)}{(\mathrm{E}(x_2))^4}+\frac{\mathrm{Var}(x_1)\mathrm{Var}(x_2)}{(\mathrm{E}(x_2))^4}-\frac{2(\mathrm{E}(x_1))^2\mathrm{E}(x_2-\theta_2)^3}{(\mathrm{E}(x_2))^5}\\
&+\frac{(\mathrm{E}(x_1))^2\mathrm{E}(x_2-\theta_2)^4}{(\mathrm{E}(x_2))^6}-\frac{(\mathrm{E}(x_1))^2[\mathrm{Var}\,x_2]^2}{(\mathrm{E}(x_2))^6}
\end{aligned}
\tag{B.10}
$$

In Appendix C it is proven that for $n = 2, 3$:

$$
\mathrm{E}\left[x_t' - \mathrm{E}(x)\right]^n = \frac{\alpha^n}{1-(1-\alpha)^n}\mathrm{E}[x_t - \mathrm{E}(x)]^n \tag{B.11}
$$

and also that:

$$\mathrm{E}\left[x'_t - E(x)\right]^4 = \frac{\alpha^4}{1-(1-\alpha)^4}\mathrm{E}[x_t - E(x)]^4 + \frac{\alpha^4}{\left(1-(1-\alpha)^2\right)^2}[\mathrm{Var}(x_t)]^2 \quad (\mathrm{B.12})$$

where:

x_t represents the demand size (z_t) or inter-demand interval (p_t),

x'_t is their exponentially smoothed estimate (z'_t, p'_t) and

E(x) is the population expected value for either series.

Consideration of (B.11) and (B.12) necessitates the adoption of the following assumptions:

no auto-correlation for the demand size and inter-demand interval series

homogeneous moments about the mean for both series

same smoothing constant value is used for both series

Taking also into account that:

$\mathrm{Var}(z_t) = \sigma^2$ and $\mathrm{Var}(p_t) = p(p-1)$

(B.10) becomes:

$$\begin{aligned}\mathrm{Var}\left(\frac{z'_t}{p'_t}\right) \approx\ & \frac{\alpha}{2-\alpha}\frac{\sigma^2}{p^2} + \frac{\alpha}{2-\alpha}\mu^2\frac{p(p-1)}{p^4} + \left(\frac{\alpha}{2-\alpha}\right)^2\frac{\sigma^2 p(p-1)}{p^4} \\ & - \frac{\alpha^3}{1-(1-\alpha)^3}\frac{2\mu^2}{p^5}\mathrm{E}(p_t - p)^3 + \frac{\alpha^4}{1-(1-\alpha)^4}\frac{\mu^2}{p^6}\mathrm{E}(p_t - p)^4 \\ & + \frac{\alpha^4}{\left(1-(1-\alpha)^2\right)^2}\frac{\mu^2}{p^6}p^2(p-1)^2 - \left(\frac{\alpha}{2-\alpha}\right)^2\frac{\mu^2}{p^6}p^2(p-1)^2 \end{aligned} \quad (\mathrm{B.13})$$

The third moment about the mean in the geometric distribution, where: $\frac{1}{p}$ is the probability of success in each trial, is calculated as:

$$\begin{aligned}\mathrm{E}(p_t - p)^3 &= \left(1-\frac{1}{p}\right)\left(1+1-\frac{1}{p}\right)\frac{1}{p^3} = \frac{p-1}{p}\frac{1}{p^3}\left(2-\frac{1}{p}\right) = \frac{p-1}{p}\frac{1}{p^3}\frac{2p-1}{p} \\ &= \frac{(p-1)(2p-1)}{p^5} \end{aligned} \quad (\mathrm{B.14})$$

and the fourth moment:

$$\begin{aligned}\mathrm{E}(p_t - p)^4 &= \frac{9\left(1-\frac{1}{p}\right)^2}{\frac{1}{p^4}} + \frac{1-\frac{1}{p}}{\frac{1}{p^2}} = \left(1-\frac{1}{p}\right)\left[\frac{9\left(1-\frac{1}{p}\right)}{\frac{1}{p^4}} + \frac{1}{\frac{1}{p^2}}\right] \\ &= \left(1-\frac{1}{p}\right)\left[9\left(1-\frac{1}{p}\right)p^4 + p^2\right] \\ &= p^2\left(1-\frac{1}{p}\right)\left[9\left(1-\frac{1}{p}\right)p^2 + 1\right] \end{aligned} \quad (\mathrm{B.15})$$

If we consider (B.14) and (B.15), and also the fact that:

$$\frac{\alpha^4}{\left(1-(1-\alpha)^2\right)^2} = \left(\frac{\alpha}{2-\alpha}\right)^2,$$

(B.13) becomes

$$\begin{aligned}\operatorname{Var}\left(\frac{z_t'}{p_t'}\right) \approx & \frac{\alpha}{2-\alpha}\frac{\sigma^2}{p^2} + \frac{\alpha}{2-\alpha}\mu^2\frac{p(p-1)}{p^4} + \left(\frac{\alpha}{2-\alpha}\right)^2\frac{\sigma^2 p(p-1)}{p^4} \\ & - \frac{\alpha^3}{1-(1-\alpha)^3}\frac{2\mu^2(p-1)(2p-1)}{p^{10}} \\ & + \frac{\alpha^4}{1-(1-\alpha)^4}\frac{\mu^2}{p^4}\left(1-\frac{1}{p}\right)\left[9\left(1-\frac{1}{p}\right)p^2+1\right] \end{aligned} \tag{B.16}$$

Since the fourth part of approximation (B.16) becomes almost zero even for quite low average inter-demand intervals, finally the variance is approximated by (B.17):

$$\begin{aligned}\operatorname{Var}\left(\frac{z_t'}{p_t'}\right) \approx & \frac{a}{2-a}\left[\frac{(p-1)}{p^3}\left(\mu^2+\frac{\alpha}{2-\alpha}\sigma^2\right)+\frac{\sigma^2}{p^2}\right] \\ & + \frac{\alpha^4}{1-(1-\alpha)^4}\frac{\mu^2}{p^4}\left(1-\frac{1}{p}\right)\left[9\left(1-\frac{1}{p}\right)p^2+1\right] \end{aligned} \tag{B.17}$$

This proves the result given by Eq. (53).

Appendix C: The 2nd, 3rd and 4th Moment About the Mean for Exponentially Smoothed Estimates

If we define:

$$x' = \sum_{j=0}^{\infty}\alpha(1-\alpha)^j_{x_{t-j}} \quad \text{(i.e. the EWMA estimate)} \tag{C.1}$$

$$\mathrm{E}(x') = \sum_{j=0}^{\infty}\alpha(1-\alpha)^j\mathrm{E}(x_{t-j})$$

assuming $\mathrm{E}(x_{t-j}) = \mathrm{E}(x)$ for all $j \geq 0$

$$\alpha\mathrm{E}(x)\sum_{j=0}^{\infty}(1-\alpha)^j = \frac{\alpha}{1-(1-\alpha)}\mathrm{E}(x) = \mathrm{E}(x) \tag{C.2}$$

Therefore we can write:

$$x' - \mathrm{E}(x) = \sum_{j=0}^{\infty} \alpha\,(1-\alpha)^j \left[x_{t-j} - \mathrm{E}(x)\right]$$

$$[x' - \mathrm{E}(x)]^n = \left\{\sum_{j=0}^{\infty} \alpha\,(1-\alpha)^j \left[x_{t-j} - \mathrm{E}(x)\right]\right\}^n$$

and

$$\mathrm{E}\,[x' - \mathrm{E}(x)]^n = \mathrm{E}\left\{\sum_{j=0}^{\infty} \alpha\,(1-\alpha)^j \left[x_{t-j} - \mathrm{E}(x)\right]\right\}^n \qquad \text{(C.3)}$$

Assuming series is not auto-correlated, for $n = 2, 3$ we then have:

$$\mathrm{E}\,[x' - \mathrm{E}(x)]^n = \sum_{j=0}^{\infty} \alpha^n\,(1-\alpha)^{nj}\,\mathrm{E}\left[x_{t-j} - \mathrm{E}(x)\right]^n \qquad \text{(C.4)}$$

and assuming $\mathrm{E}[x_{t-j} - \mathrm{E}(x)]^n = \mathrm{E}[x - \mathrm{E}(x)]^n$ for all $j \geq 0$, i.e. homogeneous moments of order n

$$\mathrm{E}\,[x' - \mathrm{E}(x)]^n = \frac{\alpha^n}{1-(1-\alpha)^n}\,\mathrm{E}\,[x - \mathrm{E}(x)]^n \qquad \text{(C.5)}$$

For $n = 2$

$$\mathrm{Var}(x') = \frac{\alpha^2}{1-(1-\alpha)^2}\,\mathrm{Var}(x)$$

and for $n = 3$

$$\mathrm{E}\,[x' - \mathrm{E}(x)]^3 = \frac{\alpha^3}{1-(1-\alpha)^3}\,\mathrm{E}\,[x - \mathrm{E}(x)]^3$$

For $n = 4$, Eq. (C.3) becomes:

$$\mathrm{E}\,[x' - \mathrm{E}(x)]^4 = \mathrm{E}\left\{\sum_{j=0}^{\infty} \alpha\,(1-\alpha)^j \left[x_{t-j} - \mathrm{E}(x)\right]\right\}^4$$

assuming no auto-correlation

$$= \sum_{j=0}^{\infty} \alpha^4\,(1-\alpha)^{4j}\,\mathrm{E}\left[x_{t-j} - \mathrm{E}(x)\right]^4$$

$$+ \sum_{j=0}^{\infty}\sum_{i=0}^{\infty} \alpha^2\,(1-\alpha)^{2j}\,\mathrm{E}\left[x_{t-j} - \mathrm{E}(x)\right]^2 \alpha^2\,(1-\alpha)^{2i}\,\mathrm{E}\left[x_{t-i} - \mathrm{E}(x)\right]^2$$

assuming homogeneous moments about the mean

$$= \alpha^4 \mathrm{E}\left[x_{t-j} - \mathrm{E}(x)\right]^4 \sum_{j=0}^{\infty} (1-\alpha)^{4j} + \alpha^4 \left[\mathrm{Var}(x)\right]^2 \sum_{j=0}^{\infty} \sum_{i=0}^{\infty} (1-\alpha)^{2j} (1-\alpha)^{2i} \tag{C.6}$$

$$\sum_{j=0}^{\infty} (1-\alpha)^{4j} = \frac{1}{1-(1-\alpha)^4} \tag{C.7}$$

$$\begin{aligned} \sum_{j=0}^{\infty} \sum_{i=0}^{\infty} (1-\alpha)^{2j} (1-\alpha)^{2i} = 1 &+ (1-\alpha)^2 + (1-\alpha)^4 + (1-\alpha)^6 + \cdots \\ &+ (1-\alpha)^2 + (1-\alpha)^4 + (1-\alpha)^6 + \cdots \\ &+ (1-\alpha)^4 + (1-\alpha)^6 + \cdots\cdots \end{aligned} \tag{C.8}$$

$$\sum_{j=0}^{\infty} \sum_{i=0}^{\infty} (1-\alpha)^{2j} (1-\alpha)^{2i} = 1 + 2(1-\alpha)^2 + 3(1-\alpha)^4 + 4(1-\alpha)^6 + \cdots\cdots \tag{C.9}$$

If we multiply the first and the second part of Eq. (C.9) with $(1-\alpha)^2$, we then have:

$$(1-\alpha)^2 \sum_{j=0}^{\infty} \sum_{i=0}^{\infty} (1-\alpha)^{2j} (1-\alpha)^{2i} = (1-\alpha)^2 + 2(1-\alpha)^4 + 3(1-\alpha)^6 + \cdots\cdots \tag{C.10}$$

Subtracting (C.9)–(C.10)

$$\begin{aligned} &\left[1-(1-\alpha)^2\right] \sum_{j=0}^{\infty} \sum_{i=0}^{\infty} (1-\alpha)^{2j} (1-\alpha)^{2i} \\ &= 1 + (1-\alpha)^2 + (1-\alpha)^4 + (1-\alpha)^6 + \cdots\cdots = \frac{1}{1-(1-\alpha)^2} \end{aligned}$$

Therefore:

$$\sum_{j=0}^{\infty} \sum_{i=0}^{\infty} (1-\alpha)^{2j} (1-\alpha)^{2i} = \frac{1}{\left(1-(1-\alpha)^2\right)^2} \tag{C.11}$$

Considering Eqs. (C.7) and (C.11), Eq. (C.6) becomes:

$$\mathrm{E}\left[x' - \mathrm{E}(x)\right]^4 = \frac{\alpha^4}{1-(1-\alpha)^4}\mathrm{E}\left[x - \mathrm{E}(x)\right]^4 + \frac{\alpha^4}{\left(1-(1-\alpha)^2\right)^2}\left[\mathrm{Var}(x)\right]^2 \quad \text{(C.12)}$$

This proves the results given by Eqs. (B.11) and (B.12).

Appendix D: The Variance of the SY Method's Estimates

We set, for the problem under concern:

$$x_1 = \left(1 - \frac{\alpha}{2}\right)z'_t$$

with expected value:

$$\mathrm{E}(x_1) = \theta_1 = \mathrm{E}\left[\left(1 - \frac{\alpha}{2}\right)z'_t\right] = \left(1 - \frac{\alpha}{2}\right)\mathrm{E}(z'_t) = \left(1 - \frac{\alpha}{2}\right)\mu \quad \text{(D.1)}$$

and variance:

$$\mathrm{Var}(x_1) = \mathrm{Var}\left[\left(1 - \frac{\alpha}{2}\right)z'_t\right] = \left(1 - \frac{\alpha}{2}\right)^2 \mathrm{Var}(z'_t) = \left(1 - \frac{\alpha}{2}\right)^2 \frac{\alpha}{2-\alpha}\mathrm{Var}(z_t)$$
$$= \left(1 - \frac{\alpha}{2}\right)^2 \frac{\alpha}{2-\alpha}\sigma^2 \quad \text{(D.2)}$$

and

$$x_2 = p'_t - \frac{\alpha}{2}$$

with expected value:

$$\mathrm{E}(x_2) = \theta_2 = \mathrm{E}\left(p'_t - \frac{\alpha}{2}\right) = \mathrm{E}(p'_t) - \frac{\alpha}{2} = p - \frac{\alpha}{2} \quad \text{(D.3)}$$

and variance:

$$\mathrm{Var}(x_2) = \mathrm{Var}\left(p'_t - \frac{\alpha}{2}\right) = \mathrm{Var}(p'_t) = \frac{\alpha}{2-\alpha}\mathrm{Var}(p_t) = \frac{\alpha}{2-\alpha}p(p-1) \quad \text{(D.4)}$$

(for the variance derivations consider also Appendix C).

The third and the fourth moments about the mean for the x_2 variable (consider also Appendices B and C) are calculated as follows:

$$\mathrm{E}\left(x_2 - \theta_2\right)^3 = \mathrm{E}\left(p'_t - \frac{\alpha}{2} - p + \frac{\alpha}{2}\right)^3 = \mathrm{E}\left(p'_t - p\right)^3 = \frac{\alpha^3}{1-(1-\alpha)^3}\mathrm{E}\left(p_t - p\right)^3$$
$$= \frac{\alpha^3}{1-(1-\alpha)^3}\frac{(p-1)(2p-1)}{p^5} \quad \text{(D.5)}$$

$$\begin{aligned}
E\,(x_2-\theta_2)^4 &= E\left(p_t'-\frac{\alpha}{2}-p+\frac{\alpha}{2}\right)^4 = E\,(p_t'-p)^4\\
&= \frac{\alpha^4}{1-(1-\alpha)^4}E\,(p_t-p)^4 + \frac{\alpha^4}{\left(1-(1-\alpha)^2\right)^2}[\mathrm{Var}(p_t)]^2\\
&= \frac{\alpha^4}{1-(1-\alpha)^4}p^2\left(1-\frac{1}{p}\right)\left[9\left(1-\frac{1}{p}\right)p^2+1\right]+\left(\frac{\alpha}{2-\alpha}\right)^2p^2\,(p-1)^2
\end{aligned} \tag{D.6}$$

(assuming that the same smoothing constant value is used for both x_1 and x_2 series and that both series are not auto-correlated and have homogeneous moments about the mean).

Consequently we apply Taylor's theorem to a function of two variables, $g(x)=\frac{x_1}{x_2}$ with:

$$\frac{\partial g}{\partial\theta_1}=\frac{1}{\theta_2} \tag{D.7}$$

$$\frac{\partial g}{\partial\theta_2}=-\frac{\theta_1}{{\theta_2}^2} \tag{D.8}$$

$$\frac{\partial^2 g}{\partial\,{\theta_1}^2}=0 \tag{D.9}$$

$$\frac{\partial^2 g}{\partial\theta_1\partial\theta_2}=-\frac{1}{\theta_2^2} \tag{D.10}$$

$$\frac{\partial^2 g}{\partial\theta_2^2}=-\theta_1\left(-\frac{2}{\theta_2^3}\right)=\frac{2\theta_1}{\theta_2^3} \tag{D.11}$$

If we consider only the first three terms, we have:

$$\begin{aligned}
g(x)=g(\theta)&+\left[\frac{\partial g}{\partial\theta_1}(x_1-\theta_1)+\frac{\partial g}{\partial\theta_2}(x_2-\theta_2)\right]\\
&+\left[\frac{\partial^2 g}{\partial\theta_1\partial\theta_2}(x_1-\theta_1)(x_2-\theta_2)+\frac{1}{2}\frac{\partial^2 g}{\partial\theta_2^2}(x_2-\theta_2)^2\right]+\cdots\cdots
\end{aligned} \tag{D.12}$$

with:

$$E[g(x)]=E\left(\frac{x_1}{x_2}\right)=E\left(\frac{\left(1-\frac{\alpha}{2}\right)z_t'}{p_t'-\frac{\alpha}{2}}\right)\approx\frac{\mu}{p} \tag{D.13}$$

$$g(\theta)=\frac{\theta_1}{\theta_2}=\frac{\left(1-\frac{\alpha}{2}\right)\mu}{p-\frac{\alpha}{2}}\neq\frac{\mu}{p}$$

and (consider also Appendix B)

$$\begin{aligned}\mathrm{Var}[g(x)] &= \mathrm{E}\left[g(x) - \mathrm{E}[g(x)]\right]^2 \\ &\approx \mathrm{E}\left\{\begin{aligned}&\frac{\theta_1}{\theta_2} + \frac{\partial g}{\partial \theta_1}(x_1-\theta_1) + \frac{\partial g}{\partial \theta_2}(x_2-\theta_2) + \frac{\partial^2 g}{\partial \theta_1 \partial \theta_2}(x_1-\theta_1)(x_2-\theta_2) \\ &\qquad + \frac{1}{2}\frac{\partial^2 g}{\partial \theta_2^2}(x_2-\theta_2)^2 - \mathrm{E}\left(\frac{x_1}{x_2}\right)\end{aligned}\right\}^2.\end{aligned} \tag{D.14}$$

In order to simplify somewhat the derivation of the variance of the SY method's estimates we approximate $\mathrm{E}[g(x)]$ by:

$$\mathrm{E}[g(x)] \approx g(\theta) + \frac{1}{2}\frac{\partial^2 g}{\partial \theta_2^2}\mathrm{E}\,(x_2-\theta_2)^2$$

(based on (D.12) and considering Sect. 2.6).

Considering (B.10) and assuming that x_1, x_2 are independent, (D.14) becomes:

$$\begin{aligned}\mathrm{Var}[g(x)] \approx{} & \frac{\mathrm{Var}(x_1))}{(\mathrm{E}(x_2))^2} + \frac{(\mathrm{E}(x_1))^2\,\mathrm{Var}(x_2)}{(\mathrm{E}(x_2))^4} + \frac{\mathrm{Var}(x_1)\mathrm{Var}(x_2)}{(\mathrm{E}(x_2))^4} - \frac{2\,(\mathrm{E}(x_1))^2\,\mathrm{E}\,(x_2-\theta_2)^3}{(\mathrm{E}(x_2))^5} \\ & + \frac{(\mathrm{E}(x_1))^2\,\mathrm{E}\,(x_2-\theta_2)^4}{(\mathrm{E}(x_2))^6} - \frac{(\mathrm{E}(x_1))^2\,[\mathrm{Var}x_2]^2}{(\mathrm{E}(x_2))^6}\end{aligned} \tag{D.15}$$

Finally the variance of the estimates of the SY method is calculated as follows:

$$\begin{aligned}\mathrm{Var}\left[\frac{(1-\frac{\alpha}{2})z'_t}{p'_t-\frac{\alpha}{2}}\right] \approx{} & \left(\frac{\alpha}{2-\alpha}\right)\frac{(1-\frac{\alpha}{2})^2\,\sigma^2}{(p-\frac{\alpha}{2})^2} + \left(\frac{\alpha}{2-\alpha}\right)\frac{(1-\frac{\alpha}{2})^2\,\mu^2 p(p-1)}{(p-\frac{\alpha}{2})^4} \\ & + \frac{(1-\frac{\alpha}{2})^2\left(\frac{\alpha}{2-\alpha}\right)^2\sigma^2 p(p-1)}{(p-\frac{\alpha}{2})^4} - \frac{(1-\frac{\alpha}{2})^2\,\mu^2}{(p-\frac{\alpha}{2})^5}\,\frac{\alpha^3}{1-(1-\alpha)^3} \\ & \times \frac{(p-1)(2p-1)}{p^5} + \frac{\alpha^4}{1-(1-\alpha)^4}\,\frac{(1-\frac{\alpha}{2})^2\,\mu^2}{(p-\frac{\alpha}{2})^6} \\ & \times p^2\left(1-\frac{1}{p}\right)\left[9\left(1-\frac{1}{p}\right)p^2+1\right] + \left(\frac{\alpha}{2-\alpha}\right)^2\frac{(1-\frac{\alpha}{2})^2\,\mu^2}{(p-\frac{\alpha}{2})^6}\,p^2 \\ & \times (p-1)^2 - \left(\frac{\alpha}{2-\alpha}\right)^2\frac{(1-\frac{\alpha}{2})^2\,\mu^2}{(p-\frac{\alpha}{2})^6}\,p^2\,(p-1)^2\end{aligned} \tag{D.16}$$

Since the fourth part of approximation (D.16) becomes almost zero even for quite low average inter-demand intervals, and the last two terms cancel each other, finally the variance is approximated by (D.17):

$$\operatorname{Var}\left[\frac{\left(1-\frac{\alpha}{2}\right)z_t'}{p_t'-\frac{\alpha}{2}}\right] \approx \left(\frac{\alpha}{2-\alpha}\right)\left(1-\frac{\alpha}{2}\right)^2 \frac{\left[\left(p-\frac{\alpha}{2}\right)^2\sigma^2 + p(p-1)\mu^2 + \frac{\alpha}{2-\alpha}p(p-1)\sigma^2\right]}{\left(p-\frac{\alpha}{2}\right)^4} + \frac{\alpha^4}{1-(1-\alpha)^4}\frac{\left(1-\frac{\alpha}{2}\right)^2\mu^2}{\left(p-\frac{\alpha}{2}\right)^6}p^2\left(1-\frac{1}{p}\right)\left[9\left(1-\frac{1}{p}\right)p^2+1\right] \tag{D.17}$$

This proves the result given by Eq. (55).

References

Boylan JE, Syntetos AA (2003) Intermittent demand forecasting: size-interval methods based on average and smoothing. Proceedings of the international conference on quantitative methods in industry and commerce, Athens, Greece

Boylan JE, Syntetos AA (2007) The accuracy of a modified Croston procedure. Int J Prod Econ 107:511–517

Brown RG (1963) Smoothing, forecasting and prediction of discrete time series. Prentice-Hall, Inc., Englewood Cliffs

Clark CE (1957) Mathematical analysis of an inventory case. Oper Res 5:627–643

Cox DR (1962) Renewal theory. Methuen, London

Croston JD (1972) Forecasting and stock control for intermittent demands. Oper Res Q 23:289–304

Eaves AHC, Kingsman BG (2004) Forecasting for the ordering and stock-holding of spare parts. J Oper Res Soc 55:431–437

Fildes R, Nikolopoulos K, Crone S, Syntetos AA (2008) Forecasting and operational research: a review. J Oper Res Soc 59:1150–1172

Gutierrez RS, Solis AO, Mukhopadhyay S (2008) Lumpy demand forecasting using neural networks. Int J Prod Econ 111:409–420

Johnston FR, Boylan JE (1996) Forecasting for items with intermittent demand. J Oper Res Soc 47:113–121

Levén E, Segerstedt A (2004) Inventory control with a modified Croston procedure and Erlang distribution. Int J Prod Econ 90:361–367

Porras EM, Dekker R (2008) An inventory control system for spare parts at a refinery: an empirical comparison of different reorder point methods. Eur J Oper Res 184:101–132

Rao AV (1973) A comment on: forecasting and stock control for intermittent demands. Oper Res Q 24:639–640

Schultz CR (1987) Forecasting and inventory control for sporadic demand under periodic review. J Oper Res Soc 38:453–458

Shale EA, Boylan JE, Johnston FR (2006) Forecasting for intermittent demand: the estimation of an unbiased average. J Oper Res Soc 57:588–592

Shenstone L, Hyndman RJ (2005) Stochastic models underlying Croston's method for intermittent demand forecasting. J Forecast 24:389–402

Snyder R (2002) Forecasting sales of slow and fast moving inventories. Eur J Oper Res 140:684–699

Stuart A, Ord JK (1994) Kendall's advanced theory of statistics (vol 1, Distribution theory, 6th edn). Edward Arnold, London

Syntetos AA (2001) Forecasting of intermittent demand. Unpublished PhD thesis, Buckinghamshire Chilterns University College, Brunel University, UK

Syntetos AA, Boylan JE (2001) On the bias of intermittent demand estimates. Int J Prod Econ 71:457–466

Syntetos AA, Boylan JE (2005) The accuracy of intermittent demand estimates. Int J Forecast 21:303–314

Syntetos AA, Babai MZ, Dallery Y, Teunter R (2009) Periodic control of intermittent demand items: theory and empirical analysis. J Oper Res Soc 60:611–618

Teunter R, Duncan L (2009) Forecasting intermittent demand: a comparative study. J Oper Res Soc 60:321–329

Teunter R, Sani B (2009) On the bias of Croston's forecasting method. Eur J Oper Res 194:177–183

Willemain TR, Smart CN, Shockor JH, DeSautels PA (1994) Forecasting intermittent demand in manufacturing: a comparative evaluation of Croston's method. Int J Forecast 10:529–538

Willemain TR, Smart CN, Schwarz HF (2004) A new approach to forecasting intermittent demand for service parts inventories. Int J Forecast 20:375–387

Chapter 2
Distributional Assumptions for Parametric Forecasting of Intermittent Demand

Aris A. Syntetos, M. Zied Babai, David Lengu and Nezih Altay

2.1 Introduction

Parametric approaches to stock control rely upon a lead-time demand distributional assumption and the employment of an appropriate forecasting procedure for estimating the moments of such a distribution. For the case of fast demand items the Normality assumption is typically sufficient. However, Stock Keeping Units (SKUs) often exhibit intermittent or irregular demand patterns that may not be represented by the normal distribution. This is perhaps not true when lead times are very long, in which case the Normality assumption may be plausible due to the Central Limit Theorem. This issue is further discussed later in this chapter.

Intermittent demand appears at random, with some time periods having no demand at all. Moreover, demand, when it occurs, is not necessarily for a single unit or a constant demand size. In the academic literature, intermittent demand is often referred to as lumpy, sporadic or erratic demand. A conceptual framework that serves the purpose of distinguishing between such non-normal demand patterns has been discussed by Boylan et al. (2007). A demand classification framework has also been presented by Lengu and Syntetos (2009) and this is

A. A. Syntetos (✉) · D. Lengu
University of Salford, Salford, UK
e-mail: A.Syntetos@salford.ac.uk

M. Z. Babai
BEM Bordeaux Management School, Bordeaux, France
e-mail: Mohamed-zied.babai@bem.edu

D. Lengu
e-mail: d.lengu@edu.salford.ac.uk

N. Altay
DePaul University, Chicago, IL, USA
e-mail: naltay@depaul.edu

N. Altay and L. A. Litteral (eds.), *Service Parts Management*,
DOI: 10.1007/978-0-85729-039-7_2,

further discussed in Sect. 2.5 of the chapter. Intermittent demand items may be engineering spares (e.g. Mitchell 1962; Hollier 1980; Strijbosch et al. 2000), spare parts kept at the wholesaling/retailing level (e.g. Sani 1995), or any SKU within the range of products offered by all organisations at any level of the supply chain (e.g. Croston 1972; Willemain et al. 1994). Such items may collectively account for up to 60% of the total stock value (Johnston et al. 2003) and are particularly prevalent in the aerospace, automotive and IT sectors. They are often the items at greatest risk of obsolescence.

Research in the area of forecasting and stock control for intermittent demand items has developed rapidly in recent years with new results implemented into software products because of their practical importance (Fildes et al. 2008). Key issues remaining in this area relate to (i) the further development of robust operational definitions of intermittent demand for forecasting and stock control purposes and (ii) a better modelling of the underlying demand characteristics for the purpose of proposing more powerful estimators useful in stock control. Both issues link directly to the hypothesised distribution used for representing the relevant demand patterns. Surprisingly though, not much has been contributed in this area in the academic literature.

Classification for forecasting and stock control entails decisions with respect to an appropriate estimation procedure, an appropriate stock control policy and an appropriate demand distributional assumption. The subtle linkages between operationalized SKU classification procedures and distributional assumptions have not been adequately explored. In addition, the compound nature of intermittent demand necessitates, conceptually at least, the employment of compound distributions, such as the negative binomial distribution (NBD). Although this area has attracted some academic attention (please refer also to the second section of this chapter) there is still more empirical evidence needed on the goodness-of-fit of these distributions to real data.

The objective of this work is three-fold: first, we conduct an empirical investigation that enables the analysis of the goodness-of-fit of various continuous and discrete, compound and non-compound, two-parameter statistical distributions used in the literature in the context of intermittent demand; second, we critically link the results to theoretical expectations and the issue of classification for forecasting and stock control; third, we provide an agenda for further research in this area. We use three empirical datasets for the purposes of our analysis that collectively constitute the individual demand histories of approximately 13,000 SKUs. Two datasets come from the military sector (Royal Air Force, RAF UK and US Defense Logistics Agency, DLA) and one from the Electronics industry. In all cases the SKUs are spare/service parts.

At this point it is important to note that some non-parametric procedures have also been suggested in the literature to forecast intermittent demand requirements (e.g. Willemain et al. 2004; Porras and Dekker 2008). Such approaches typically rely upon bootstrapping procedures that permit a re-construction of the empirical distribution of the data, thus making distributional assumptions redundant. Although it has been claimed that such approaches have an advantage over

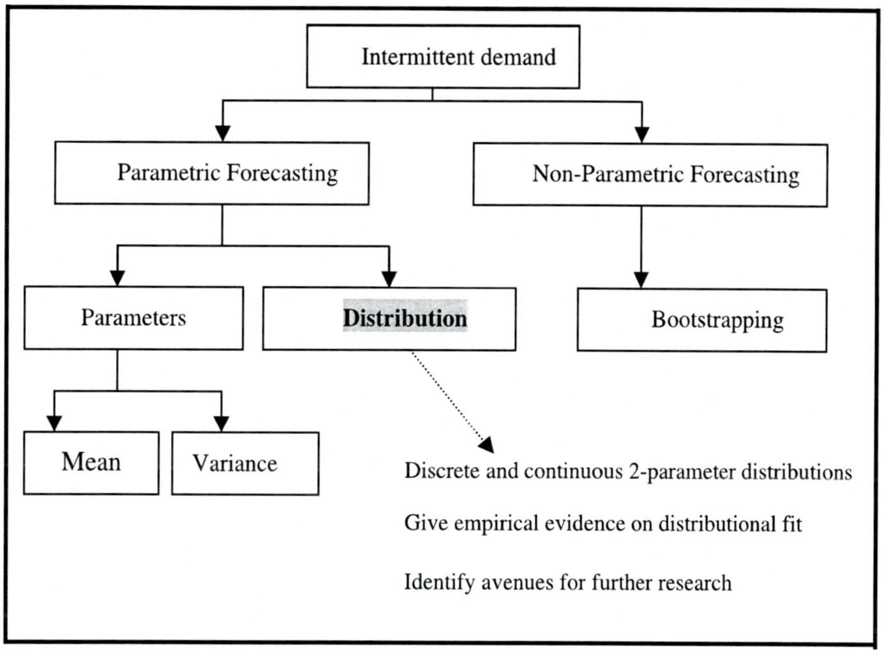

Fig. 2.1 Focus of the research

parametric methodologies, more empirical research is needed to evaluate the conditions under which one approach outperforms the other. In this chapter, we will be focusing solely on parametric forecasting. The focus of our research is presented in Fig. 2.1.

The remainder of this chapter is organized as follows. In Sect. 2.2, a brief research background dealing with forecasting and stock control issues in the context of intermittent demand is presented along with a review on the demand distributions discussed in the literature and/or used by practitioners. In Sect. 2.3, we present the datasets used for the purpose of this empirical investigation, the statistical goodness-of-fit tests that we have performed and the empirical results. A critical discussion of the empirical findings follows in Sects. 2.4 and 2.5. Finally, the conclusions of our research along with some natural extensions for further work in this area are given in Sect. 2.6.

2.2 Research Background

In this section, a brief review of the literature on issues related to parametric forecasting of intermittent demand is presented. First we address the issue of estimating the mean and variance of intermittent demands, followed by a discussion of various suggestions that have been made in the literature with regards to the hypothesized distribution of such demands.

2.2.1 Parametric Forecasting

Practical parametric approaches to inventory management rely upon estimates of some essential demand distribution parameters. The decision parameters of the inventory systems (such as the re-order point or the order-up-to-level) are then based on these estimates.

Different inventory systems require different variables to be forecasted. Some of the most cited, for example (R, s, S) policies (Naddor 1975; Ehrhardt and Mosier 1984), require only estimates of the mean and variance of demand. (In such systems, the inventory position is reviewed every R periods and if the stock level drops to the re-order point s enough is ordered to bring the inventory position up to the order-up-to-level S.)

In other cases, and depending on the objectives or constraints imposed on the system, such estimates are also necessary, although they do not constitute the 'key' quantities to be determined. We may consider, for example, an (R, S) or an (s, Q) policy operating under a fill-rate constraint—known as P_2. (In the former case, the inventory position is reviewed periodically, every R periods, and enough is ordered to bring it up to S. In the latter case, there is a continuous review of the inventory position and as soon as that drops to, or below, s an order is placed for a fixed quantity Q.) In those cases we wish to ensure that $x\%$ of demand is satisfied directly off-the-shelf and estimates are required for the probabilities of any demands exceeding S or s (for the (R, S) an (s, Q) policy, respectively). Such probabilities are typically estimated indirectly, based on the mean demand and variance forecast in conjunction with a hypothesized demand distribution. Nevertheless, and as discussed in the previous section, a reconstruction of the empirical distribution through a bootstrapping (non-parametric) procedure would render such forecasts redundant; this issue is further discussed in this Handbook in Chap.6. Similar comments apply when these systems operate under a different service driven constraint: there is no more than $x\%$ chance of a stock-out during the replenishment cycle (this service measure is known as P_1). Consequently, we need to estimate the $(100 - x)$th percentile of the demand distribution.

In summary, parametric approaches to forecasting involve estimates of the mean and variance of demand. In addition, a demand distribution needs also to be hypothesized, in the majority of stock control applications, for the purpose of estimating the quantities of interest. Issues related to the hypothesized demand distribution are addressed in the following sub-section. The estimation of the mean and variance of demand is addressed in Chap.1 of this Handbook.

2.2.2 The Demand Distribution

Intermittent demand patterns are characterized by infrequent demands, often of variable size, occurring at irregular intervals. Consequently, it is preferable to model demand from constituent elements, i.e. the demand size and inter-demand interval. Therefore, compound theoretical distributions (that explicitly take into account the size-interval combination) are typically used in such contexts of application. We first discuss some issues related to modelling demand arrivals and hence inter-demand intervals. We then extend our discussion to compound demand distributions.

If time is treated as a discrete (whole number) variable, demand may be generated based on a Bernoulli process, resulting in a geometric distribution of the inter-demand intervals. When time is treated as a continuous variable, the Poisson demand generation process results in negative exponentially distributed inter-arrival intervals.

There is sound theory in support of both geometric and exponential distribution for representing the time interval between successive demands. There is also empirical evidence in support of both distributions (e.g. Dunsmuir and Snyder 1989; Kwan 1991; Willemain et al. 1994; Janssen 1998; Eaves 2002). With Poisson arrivals of demands and an arbitrary distribution of demand sizes, the resulting distribution of total demand over a fixed lead time is compound Poisson. Inter-demand intervals following the geometric distribution in conjunction with an arbitrary distribution for the sizes, results in a compound binomial distribution.

Regarding the compound Poisson distributions, the stuttering Poisson, which is a combination of a Poisson distribution for demand occurrence and a geometric distribution for demand size, has received the attention of many researchers (for example: Gallagher 1969; Ward 1978; Watson 1987). Another possibility is the combination of a Poisson distribution for demand occurrence and a normal distribution for demand sizes (Vereecke and Verstraeten 1994), although the latter assumption has little empirical support. Particularly for lumpy demands, the demand size distribution is heavily skewed to the right, rendering the normality assumption far from appropriate. Quenouille (1949) showed that a Poisson-Logarithmic process yields a negative binomial distribution (NBD). When event arrivals are assumed to be Poisson distributed and the order size is not fixed but follows a logarithmic distribution, total demand is then negative binomially distributed over time.

Another possible distribution for representing demand is the gamma distribution. The gamma distribution is the continuous analogue of the NBD and "although not having a priori support [in terms of an explicit underlying mechanism such as that characterizing compound distributions], the gamma is related to a distribution which has its own theoretical justification" (Boylan 1997, p. 168). The gamma covers a wide range of distribution shapes, it is defined for non-negative values only and it is generally mathematically tractable in its inventory control applications (Burgin and Wild 1967; Burgin 1975; Johnston 1980). Nevertheless if it is assumed that demand is discrete, then the gamma can be only an approximation to the distribution of demand. At this point it is important to note that the use of both NBD and gamma distributions requires estimation of the mean and variance of demand only. In addition, there is empirical evidence in support of both distributions (especially the former) and therefore they are recommended for practical applications.

If demand occurs as a Bernoulli process and orders follow the Logarithmic-Poisson distribution (which is not the same as the Poisson-Logarithmic process that yields NBD demand) then the resulting distribution of total demand per period is the log-zero-Poisson (Kwan 1991). The log-zero-Poisson is a three parameter distribution and requires a rather complicated estimation method. Moreover, it was found by Kwan (1991) to be empirically outperformed by the NBD. Hence, the log-zero Poisson cannot be recommended for practical applications. One other compound binomial distribution appeared in the literature is that involving normally distributed demand sizes (Croston 1972, 1974). However, and as discussed above, a normality assumption is unrealistic and therefore the distribution is not recommended for practical applications.

Despite the inappropriateness of the normal distribution for representing demand sizes it may in fact constitute a reasonable assumption for lead time demand itself, when lead times are long (see also Syntetos and Boylan 2008). This is because long lead times permit central limit theorem effects for the sum of demands over the corresponding period, thus making the normality assumption more plausible. In addition, the assumption of normality may also be likely to be good when the coefficient of variation (CV) of the distribution of demand per period is small. Finally, algorithms based on normality are simple to implement making the normal distribution a very commonly assumed one among practitioners.

For very slow moving items, such as those commonly encountered in a military context for example, the Poisson distribution is known to offer a very good fit and much of the stock control theory in this area has been developed upon the explicit assumption that demand per period is Poisson distributed (see, for example, Silver et al. 1998). In this case demand is assumed to arrive as a Poisson process couple with unit-sized transactions. In an early work, Friend (1960) also discussed the use of a Poisson distribution for demand occurrence, combined with demands of constant size. Vereecke and Verstraeten (1994) presented an algorithm developed for the implementation of a computerised stock control system for spare parts in a chemical plant. The demand was assumed to occur as a Poisson process with a package of several pieces being requested at each demand occurrence. The resulting distribution of demand per period was called a 'Package Poisson' distribution. The same distribution has appeared in the literature under the name 'hypothetical SKU' (h-SKU) Poisson distribution (Williams 1984), where demand is treated as if it occurs as a multiple of some constant, or 'clumped Poisson' distribution, for multiple item orders for the same SKU of a fixed 'clump size' (Ritchie and Kingsman 1985). The 'Package Poisson' distribution requires, as the Poisson distribution itself, an estimate of the mean demand only.

The short review of the literature presented above indicates that it is worthwhile testing the empirical goodness-of-fit of the following distributions: (i) Poisson; (ii) NBD; (iii) stuttering Poisson; (iv) Gamma; and (v) Normal. In the next section we conduct such tests and we comment on the plausibility of the relevant assumptions for applications in an intermittent demand context.

2.3 Empirical Investigation

In this section, we first describe the datasets used for the purposes of this empirical investigation, followed by a discussion of the statistical goodness-of-fit tests conducted and the empirical results.

2.3.1 Empirical Data

The empirical databases available for the purposes of our research come from the US Defense Logistics Agency (DLA), Royal Air Force (RAF) and Electronics Industry and they consist of the individual monthly demand histories of 4,588, 5,000 and 3,055 SKUs, respectively. Some information regarding these datasets is presented in Table 2.1, followed by detailed descriptive statistics on the demand data series characteristics for each of the datasets presented in Tables 2.2, 2.3, and 2.4. At this point it should be noted that the time series considered have not been tested for stationarity.

2.3.1.1 Statistical Goodness-of-Fit Tests

Two tests have been mainly used and discussed in the literature for checking statistically significant fit, namely: the Chi-Square test and the Kolmogorov–Smirnov (K–S) test (see, for example, Harnett and Soni 1991). These tests measure

Table 2.1 Empirical datasets

#	Country	Industry	No of SKUs	Time bucket	History length	Lead-time info	Cost info
1	USA	Military/DLA	4,588	Month	60	No	No
2	UK	Military/RAF	5,000	Month	84	Yes	Yes
3	Europe	IT	3,055	Month	48	Constant = 3	Yes

Table 2.2 Dataset #1—US Defense Logistics Agency

4,588 SKUs	Demand intervals		Demand sizes		Demand per period	
	Mean	SD	Mean	SD	Mean	SD
Min	1.000	0.000	1.000	0.000	0.083	0.279
25%	1.967	1.665	2.894	2.314	0.650	1.672
Median	3.278	3.236	5.375	5.142	1.750	3.749
75%	5.600	6.049	11.940	12.435	4.550	9.403
Max	12	24.597	1326.875	1472.749	783.917	1219.012

Table 2.3 Dataset #2—Royal Air Force

5,000 SKUs	Demand intervals		Demand sizes		Demand per period	
	Mean	SD	Mean	SD	Mean	SD
Min	3.824	0.000	1.000	0.000	0.036	0.187
25%	7.273	5.431	1.556	0.815	0.155	0.538
Median	9.000	6.930	3.833	3.062	0.369	1.452
75%	11.571	8.630	11.333	9.315	1.155	4.434
Max	24.000	16.460	668.000	874.420	65.083	275.706

Part of Dataset #2 has been used in the following study: Syntetos et al. (2009)

Table 2.4 Dataset #3—electronics

3,055 SKUs	Demand intervals		Demand sizes		Demand per period	
	Mean	SD	Mean	SD	Mean	SD
Min	1.000	0.000	1.000	0.000	0.042	0.245
25%	1.500	1.011	3.462	3.011	0.896	2.215
Median	2.556	2.285	5.900	6.220	2.104	4.501
75%	4.700	4.389	12.122	13.863	6.010	10.480
Max	24.000	32.527	5366.188	9149.349	5366.188	3858.409

Dataset #3 has been used in the following study: Babai et al. (2009)

the degree of fit between observed and expected frequencies. Problems often arise with the standard Chi-Square test through the requirement that data needs to be grouped together in categories to ensure that each category has an expected frequency of at least a minimum of a certain number of observations. Some modifications of this test have also been considered in the literature. A modified Chi-Square test has been developed for the purpose of testing the goodness-of-fit for intermittent demands (Eaves 2002). This test differs in that boundaries are specified by forming a certain number of categories with similar expected frequencies throughout, rather than combining groups just at the margins. However, the implementation of this test requires the specification of the number of categories to be used. We encountered a difficulty in using the standard or modified Chi-Square test in our research, namely that of deciding how to specify the categories' intervals or the number of categories. On the other hand, the K–S test does not require grouping of the data in any way, so no information is lost; this eliminates the troublesome problem of categories' intervals specification.

In an inventory context one could argue that measures based on the entire distribution can be misleading (Boylan and Syntetos 2006). A good overall goodness-of-fit statistic may relate to the chances of low demand values, which can mask poor forecasts of the chances of high-demand values. However, for inventory calculations, attention should be restricted to the upper end of the distribution (say the 90th or 95th percentiles). The development of modified goodness-of-fit tests for application in inventory control, and even more specifically in an intermittent demand context, is a very important area but not one considered as part of this research. Consequently, we have selected the K–S test for the purpose of assessing goodness-of-fit.

The K–S test assumes that the data is continuous and the standard critical values are exact only if this assumption holds. Several researchers (e.g. Noether 1963, 1967; Walsh 1963; Slakter 1965) have found that the standard K–S test is conservative when applied to data that is discrete. The standard exact critical values provided for the continuous data are larger than the true exact critical values for discrete data. Consequently, the test is less powerful if the data is discrete as in the case of this research; it could result in accepting the null hypothesis at a given significance level while the correct decision would have been to reject the null hypothesis. Conover (1972) proposed a method for determining the exact critical levels for discrete data.

As discussed in the previous section, we are considering five distributions the fit of which is tested on the demand data related to 12,643 SKUs. The distribution of the demand per period has been considered rather than the distribution of the lead-time demand; this is due to the lack of information on the actual lead times associated with the dataset 1. (Although this may be very restrictive regarding the performance of the normal distribution, this would still be expected to perform well on the time series that are associated with a small coefficient of variation of demand per period.)

Critical values have been computed based on K–S statistical tables for 1 and 5% significance levels. We consider that:

- There is a 'Strong Fit' if the *P*-value is less than both critical values;
- There is 'Good Fit' if the *P*-value is less than the critical value for 1% but larger than the one for 5%;
- There is 'No Fit' if the *P*-value is larger than both critical values.

2.3.1.2 Empirical Results

In Table 2.5 we present the percentage of SKUs that satisfy the various degrees of goodness-of-fit taken into account in our research, for each of the datasets and statistical distributions considered.

As shown in Table 2.5, the discrete distributions, i.e. Poisson, NBD and stuttering Poisson provide, overall, a better fit than the continuous ones, i.e. Normal and Gamma. More precisely, and with regards to 'Strong Fit, the stuttering Poisson distribution performs best in all three datasets considered in our research. This is followed by the NBD and then by the Poisson distribution. On the other hand, the normal distribution is judged to be far from appropriate for intermittent demand items; this is partly due to the experimental structure employed for the purposes of our investigation that relied upon the distribution of demand per time period rather than the distribution of the lead time demand.

Contrary to our expectations, the gamma distribution has also been found to perform poorly. This may be explained in terms of the inconsistency between the distribution under concern, which is continuous in nature, and the discreteness of the

Table 2.5 Goodness-of-fit results

Dataset #	No of SKUs	Distribution	Percentage of SKUs (%)		
			Strong fit	Good fit	No fit
1	4,588	Poisson	39.45	5.51	55.04
		NBD	71.19	3.86	24.95
		Stuttering Poisson	84.18	3.64	12.18
		Normal	11.84	14.25	73.91
		Gamma	13.84	3.88	82.28
2	5,000	Poisson	59.84	2.94	37.22
		NBD	82.48	2.7	14.82
		Stuttering Poisson	98.64	0.48	0.88
		Normal	12.2	18.12	69.68
		Gamma	19.2	12.32	68.48
3	3,055	Poisson	32.64	7.4	59.96
		NBD	73.94	5.31	20.75
		Stuttering Poisson	79.05	4.49	16.46
		Normal	9.92	14.34	75.74
		Gamma	11.69	3.83	84.48

(demand) data employed in our goodness-of-fit tests. We return to this issue in the last section of the chapter where the next steps of our research are discussed in detail.

2.4 Linking the Goodness-of-Fit to Demand Characteristics

Johnston and Boylan (1996) offered for the first time an operationalised definition of intermittent demand for forecasting purposes (demand patterns associated with an average inter-demand interval (p) greater than 1.25 forecast revision periods). The contribution of their work lies on the identification of the average inter-demand interval as a demand classification parameter rather than the specification of an exact cut-off value. Syntetos et al. (2005) took this work forward by developing a demand classification scheme that it relies upon both p and the squared coefficient of variation of demand sizes (CV^2), i.e. the contribution of their work lies in the identification of an additional categorisation parameter for demand forecasting purposes. Nevertheless, inventory control issues and demand distributional assumptions were not addressed. Boylan et al. (2007) assessed the stock control implications of the work discussed above by means of experimentation on an inventory system developed by a UK-based software manufacturer. The researchers demonstrated, empirically, the insensitivity of the p cut-off value, for demand classification purposes, in the approximate range 1.18–1.86 periods.

In this section, we attempt to explore the potential linkages between demand distributional assumptions and the classification scheme developed by Syntetos et al. (2005). In the following figures we present for dataset #1 and each of the distributions considered, the SKUs associated with a 'Strong Fit' as a function of the inter-demand intervals (p) and the squared demand coefficient of variation (CV^2). The relevant results for the other two datasets are presented in the Appendix.

As shown in the figures presented below/in the Appendix and theoretically expected, both the stuttering Poisson and the Negative Binomial distribution perform comparatively better for all the datasets considered. This is true both for the SKUs with high inter-demand intervals (e.g. SKUs with p being up to 12 in dataset #1 or SKUs with a p value up to 24 in datasets #2 and #3) and low demand intervals (e.g. SKUs with p values starting from 1 in datasets #1 and #3). Moreover, it should be noted that there is a strong fit of NBD and stuttering Poisson to all the SKUS that are also associated with a strong fit of the Poisson distribution, which is expected since both distributions under concern are compound Poisson ones. The SKUs where there is commonly a strong fit of those three distributions are the ones characterized by relatively low CV^2 values (Figs. 2.2, 2.3, and 2.4).

Furthermore, the normal distribution performs well for the SKUs with relatively low inter-demand intervals (e.g. SKUs with p values close to 1 in datasets #1 and #3 and $p = 3.82$ in the dataset #2). However, there are also a few SKUs with high inter-demand intervals (p going up to 12 in dataset #1, 24 in dataset #2 and 15 in dataset #3) for which the normal distribution provides a strong fit. Those latter SKUs have a minimum CV^2 (i.e. $CV^2 = 0$) which can be explained by the fact that their demand is very low (in most of the cases, the demand is equal to zero and one) and can fit to the normal distribution with low mean (i.e. equivalently high values of p) and variance. As shown in Figs. 2.5 and 2.6, in addition to the SKUs where there is a fit to the normal distribution (those with low values of p), the

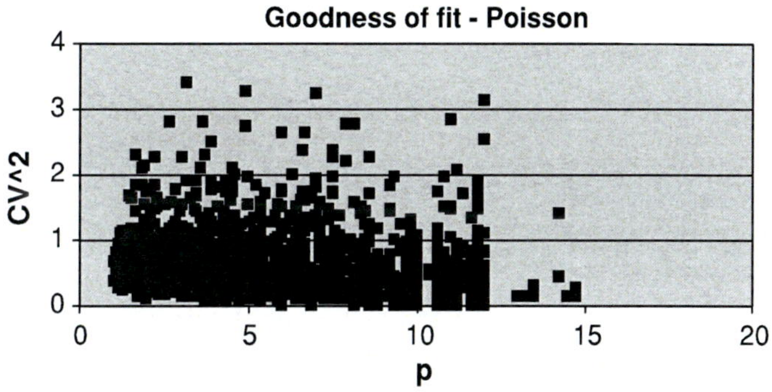

Fig. 2.2 Dataset #1—goodness-of-fit results for the Poisson distribution

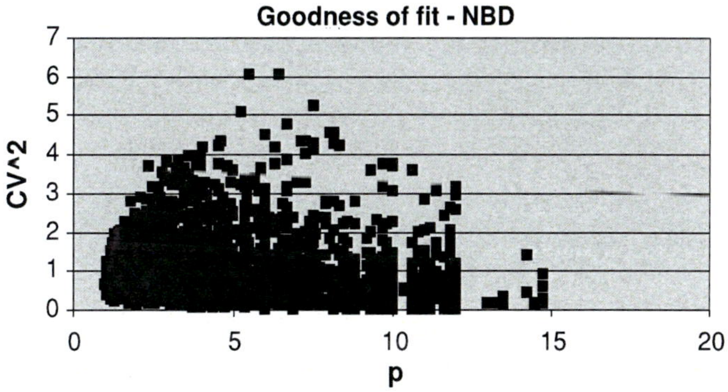

Fig. 2.3 Dataset #1—goodness-of-fit results for the NBD

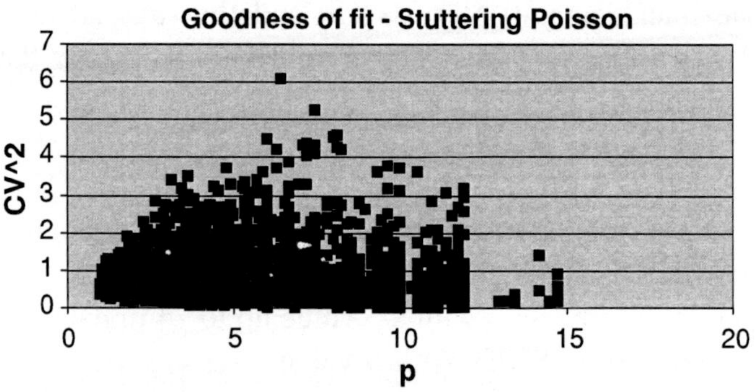

Fig. 2.4 Dataset #1—goodness-of-fit results for the stuttering Poisson

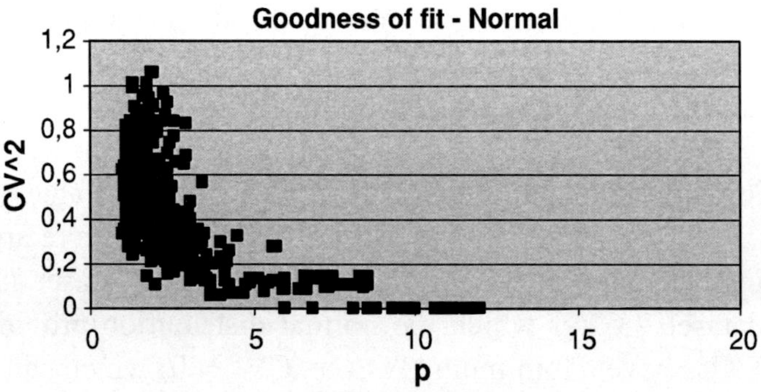

Fig. 2.5 Dataset #1—goodness-of-fit results for the normal distribution

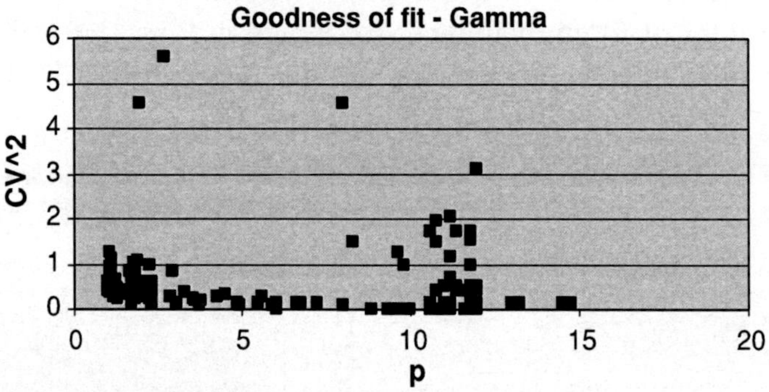

Fig. 2.6 Dataset #1—goodness-of-fit results for the gamma distribution

gamma distribution provides also a strong fit to the SKUs with very high values of p (i.e. SKUs with an inter-demand interval going up to 12 periods in dataset #1 and 24 periods in datasets #2 and #3) and high CV^2 values (i.e. SKUs with CV^2 up to 6 in dataset #1, $CV^2 = 10$ in the dataset #2 and $CV^2 = 8$ in the dataset #3). This is also expected since the gamma distribution is known to be very flexible in terms of its mean and variance, so it can take high values for its p and CV^2 and can be reduced to the normal distribution for certain parameters of the mean and the variance.

Based on the goodness-of-fit results presented in this section, we have attempted to derive inductively an empirical rule that suggests which distribution should be used under particular values of the inter-demand interval and squared coefficient of variation of the demand sizes. That is to say, we have explored the possibility of extending the classification scheme discussed by Syntetos et al.

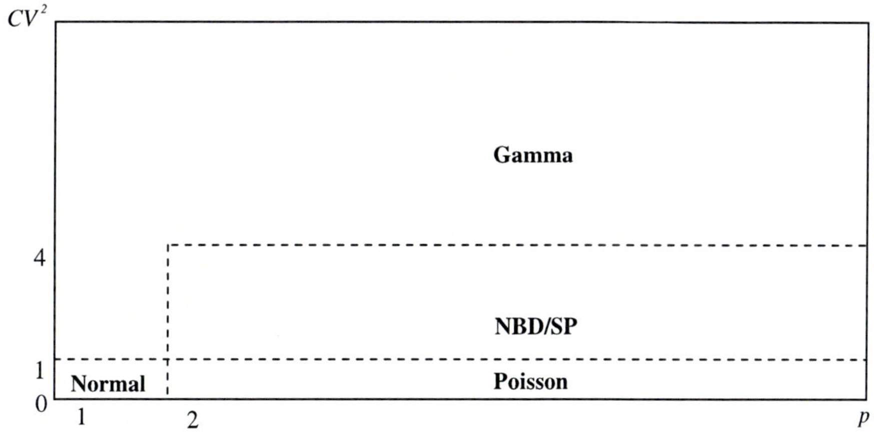

Fig. 2.7 Demand distributional assumptions: an inductive classification rule

(2005) to demand distributional assumptions. An inductive *Rule* has been developed (please refer to Fig. 2.7) based on the reported empirical performance of the distributions considered in our research in relation to specific values of p and CV^2. The *Rule* suggests appropriate regions for the selection of these distributions, i.e. Normal is used for SKUs with 'low' p and CV^2, Poisson is used for SKUs with low CV^2, Gamma is used for SKUs with 'extreme' values of p and CV^2, NBD and stuttering Poisson (SP) are used for the other ranges. The stock control implications of using such a rule were evaluated through the use of the Syntetos–Boylan Approximation (Syntetos and Boylan, 2005) for forecasting purposes and the standard order-up-to-level stock control policy for a specified target cycle service level. Inventory volumes and achieved service levels were compared against those obtained from the same inventory management system that relies though upon a single demand distributional assumption, i.e. NBD. However, the results indicated no superior empirical performance of the 'new' approach. This may be explained in terms of the construction of the goodness-of-fit testing that considers the entire demand distribution whereas stock control performance is explicitly dependant upon the fit on the right-hand tail of the distributions. This is an important issue in Inventory Management and one that has not received adequate attention in the academic literature. We return to this issue in the last section of this chapter.

2.5 Theoretical Expectations

Lengu and Syntetos (2009) proposed a demand classification scheme based on the underlying demand characteristics of the SKUs (please refer to Fig. 2.8). SKUs are first categorised as *non-qualifying* if the variance of the demand per period is less than the mean or *qualifying* if the variance is at least equal to the mean. Compound Poisson distributions can be used to model the demand series of qualifying SKUs but they are regarded as not appropriate for modelling the demand of non-qualifying SKUs. Let us assume that demand is generated from a compound Poisson

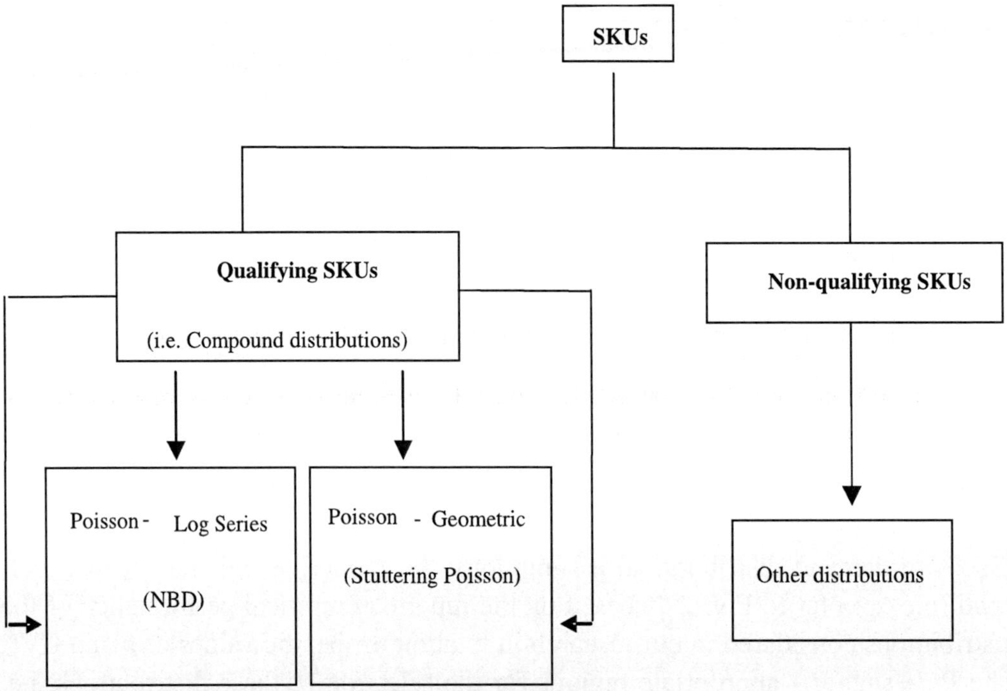

Fig. 2.8 Compound Poisson modelling of intermittent series

model (i.e. demand 'arrives' according to a Poisson process and, when demand occurs, it follows a specified distribution). If we let μ denote the mean demand per unit time period and σ^2 denote the variance of demand per unit period of time, then

$$\mu = \lambda \mu_z \tag{1}$$

$$\sigma^2 = \lambda\left(\mu_z^2 + \sigma_z^2\right) \tag{2}$$

where λ is the rate of demand occurrence, and μ_z and σ_z^2 the mean and variance, respectively, of the transaction size when demand occurs. Note that

$$\frac{\sigma^2}{\mu} = \frac{\lambda(\mu_z^2 + \sigma_z^2)}{\lambda \mu_z} \geq 1 \tag{3}$$

since $\mu_z^2 \geq \mu_z$ (the transaction size is at least of 1 unit) and σ_z^2 is non-negative. The compound Poisson demand model is therefore not appropriate for SKUs associated with $\sigma^2/\mu < 1$ (non-qualifying). Note that the actual rate of demand occurrence λ does not affect the classification of SKUs as to whether they are qualifying or not.

2.5.1 Poisson-Geometric Compound Distribution (stuttering Poisson)

The Geometric distribution Ge(π_G) is a discrete monotonically decreasing distribution with $0 \leq \mathrm{CV}^2 \leq 1$ and mode $\tilde{m} = 1$. It can model transaction sizes that are usually equal to one but can also take higher values. The Poisson-Geometric compound distribution also accommodates the case of clumped demand since the Poisson distribution is a special case of the Poisson-Geometric distribution. Specifically, if the parameter of the Geometric distribution Ge(π_G) is 1, then the transaction size can only take one value (transaction size 1). With the transaction size being clumped, the demand model is now reduced to a standard Poisson distribution. In the empirical goodness-of-fit tests, the Poisson-Geometric distribution provided the most frequent fit of all the distributions considered (see Table 2.5).

2.5.2 Poisson-Logarithmic Series Compound Distribution (NBD)

The Logarithmic series distribution Log(π_L) is a discrete monotonically decreasing distribution with an unbounded CV^2 and $\tilde{m} = 1$. Just like the Geometric distribution, the Logarithmic distribution can model transaction sizes that are constant or monotonically decreasing. However, unlike the Geometric distribution the parameter CV^2 does not have an upper bound. The Poisson-Logarithmic series compound distribution is therefore more flexible and can accommodate SKUs with exceptionally large transaction sizes. In the empirical goodness-of-fit tests, the Poisson-Logarithmic series distribution provided the second most frequent fit after the stuttering Poisson distribution.

The work discussed in this section has been developed under the assumption that demand arrivals follow a Poisson process. Similar results would be obtained if demand was assumed to occur according to a Bernoulli process since when the probability of more than one occurrence per period is negligible the Poisson and Bernoulli distributions are nearly identical. In such cases, the Poisson distribution, $P_0(\lambda)$, is approximately equal to the Bernoulli distribution with:

$$P(0) = \exp(-\lambda) \quad \text{and} \quad P(1) = 1 - \exp(-\lambda).$$

2.5.3 Non-Qualifying SKUs

While qualifying SKUs can be reasonably modelled using compound distributions, modelling non-qualifying SKUs is more challenging. Adan et al. (1995) proposed using a Binomial distribution-based model for what is termed as non-qualifying SKUs for the purposes of our research. Note that for the binomial distribution $\mathrm{Bi}(n,p)$, $\sigma^2/\mu = npq/np = q < 1$; the binomial distribution can therefore accommodate non-qualifying SKUs. We are not aware of any empirical studies conducted to determine whether the model proposed by Adan et al. may provide adequate fit for real-life demand series. Moreover, it is not possible from that model to distinguish between the demand occurrence process and the transaction size distribution. Such a model could however be useful for modeling slow-moving non-qualifying SKUs and we will consider it in the next steps of our research.

2.5.4 *Other Considerations*

The normal distribution and the gamma distribution seem to be the least promising of all the distributions considered in the empirical part of this chapter. For either distribution, the variance can be less than, equal to or larger than the mean. The two distributions can therefore be used to model both qualifying and non-qualifying SKUs. Furthermore, the normal distribution and the gamma distributions have been studied extensively and tables of the critical values for both distributions are widely available. However, in the empirical study, the two distributions provided the least frequent fit and there is no clear pattern associated with the SKUs for which the distributions provided a good fit. The normal distribution and the gamma distribution might be convenient to use but that should be contrasted to their rather poor empirical performance.

As we have mentioned in Sect. 2.2, that the K–S test assumes that the data is continuous and the test is less powerful if the data is discrete as in the case of this research. The standard exact critical values provided for the continuous data are larger than the true exact critical values for discrete data. Conover (1972) and Pettitt and Stephens (1977) proposed a method for determining the exact critical levels for the K–S test for discrete data. Choulakian et al. (1994) proposed a method of calculating the critical values of the Cramér–von Mises test and the Anderson–Darling test for discrete data. These tests have one significant drawback because of their sensitivity: their critical values are very much dependent upon the model being tested. Different tables of the critical values are therefore required for each demand model being tested. Steele and Chaselling (2006) have compared the power of these different goodness-of-fit tests for discrete data but their study was not extensive enough to indicate which test is the most powerful for our purposes.

2.6 Conclusions and Further Research

Parametric approaches to forecasting rely upon an explicit demand distributional assumption. Although the normal distribution is typically adequate for 'fast' demand items this is not true when demand is intermittent. Some research has been conducted with regards to the hypothesised distributions needed for representing such patterns and a number of distributions have been put forward as potential candidates on the basis of: (i) theoretical arguments, (ii) intuitive appeal; (iii) empirical support. A review of the literature though reveals that: (i) more empirical studies are required in order to develop our understanding on the adequacy of these distributions under differing underlying intermittent demand structures; (ii) there is some scope for linking demand distributional assumptions to classification for forecasting and stock control purposes. Both these issues are explored as part of the research work presented in this chapter. The empirical databases available for the purposes of our investigation come from the US DLA, RAF and Electronics Industry and they consist of the individual monthly demand histories of 4,588, 5,000 and 3,055 SKUs, respectively.

The empirical goodness-of-fit of five distributions (of demand per period) has been assessed by means of employing the Kolmogorov–Smirnov (K–S) test. These distributions are: Poisson, Negative Binomial Distribution (NBD), stuttering Poisson, Normal and Gamma. The results indicate that both the NBD and stuttering Poisson provide the most frequent fit. Both these distributions are compound in nature, meaning that they account explicitly for a demand arrival process (Poisson) and a different distribution for the transaction sizes (Log series and Geometric for the NBD and stuttering Poisson, respectively). Despite previous claims, the gamma distribution does not perform very well and the same is true for the normal distribution. This may be attributed to the continuous nature of these distributions (since their fit is tested on discrete observations) but also to the fact that we model demand per unit time period as opposed to lead time demand. Upon reflection, this is viewed as a limitation of our work since lead time demand could have been considered for two of the three datasets available to us (in those cases the actual lead time was available). If that was the case, both the Normal and gamma distribution would be associated potentially with a better performance. The Poisson distribution provides a 'reasonable' fit and this is theoretically expected for slow moving items.

Some recent work on the issue of demand classification (Syntetos et al. 2005) has focused on both the demand arrival pattern and distribution of the demand sizes. In this chapter, we have attempted empirically to link the goodness-of-fit of the above discussed distributions to the classification scheme proposed by Syntetos et al. (2005). Although some of the results were matched indeed by relevant theoretical expectations this was not the case when the inventory implications of the proposed scheme were considered. Goodness-of-fit tests focus on the entire demand distribution whereas stock control performance is explicitly dependant upon the fit on the right-hand tail of a distribution. This is an important issue in Inventory Management and one that has not received adequate attention in the academic literature. The empirical results discussed above have also been contrasted to some theoretical expectations offered by a conceptual demand classification framework presented by Lengu and Syntetos (2009). The framework links demand classification to some underlying characteristics of intermittent demand patterns and although it seems capable of explaining a number of empirical results it may not be utilized in an operationalised fashion yet.

The work presented in this chapter has revealed a number of interesting themes for further research. Distributional assumptions play a critical role in practical inventory management applications and further work on the following issues should prove to be valuable both from a theoretical and practitioner perspective:

- The development of modified goodness-of-fit tests for application in inventory control, and even more specifically in an intermittent demand context, is a very important area. In particular, putting more emphasis on the right-hand tail of the distribution seems appropriate for stock control applications.
- Quantifying the effect that the inconsistency between the discrete nature of demand data and the continuous nature of certain distributions may have on goodness-of-fit statistics constitutes an interesting research question.
- The inconsistency between the discrete nature of demand observations and the implicit assumption of continuous data employed by various goodness-of-fit tests should be further explored.
- Replication of the analysis conducted in this chapter in larger demand datasets coupled with the assessment of the goodness-of-fit of various distributions to the lead time demand as opposed to demand per period should help advance knowledge in this area.

Acknowledgements The research described in this chapter has been partly supported by the Engineering and Physical Sciences Research Council (EPSRC, UK) grants no. EP/D062942/1 and EP/G006075/1. More information on the former project may be obtained at http://www.business.salford.ac.uk/research/ommss/projects/Forecasting/. In addition, we acknowledge the financial support received from the Royal Society, UK: 2007/Round 1 Inter. Incoming Short Visits—North America.

Appendix

Goodness-of-Fit Results

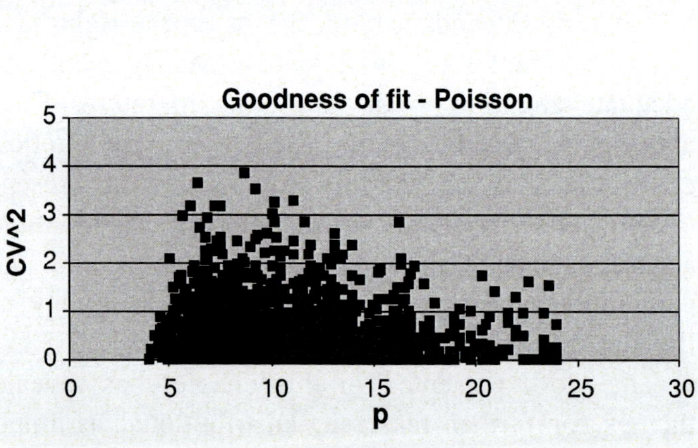

Fig. A1 Dataset #2—goodness-of-fit results for Poisson distribution

Fig. A2 Dataset #2—goodness-of-fit results for the NBD

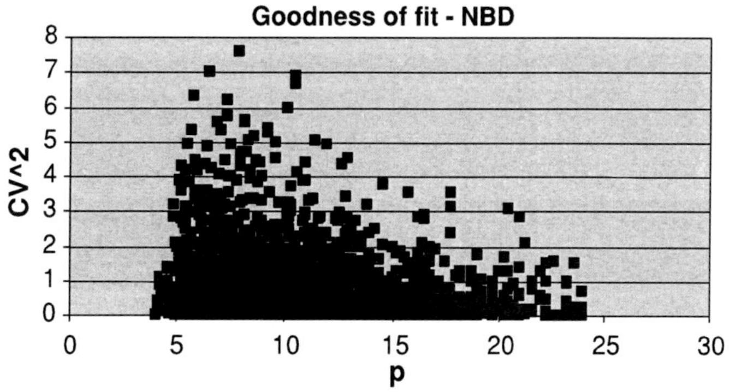

Fig. A3 Dataset #2—goodness-of-fit results for the stuttering Poisson

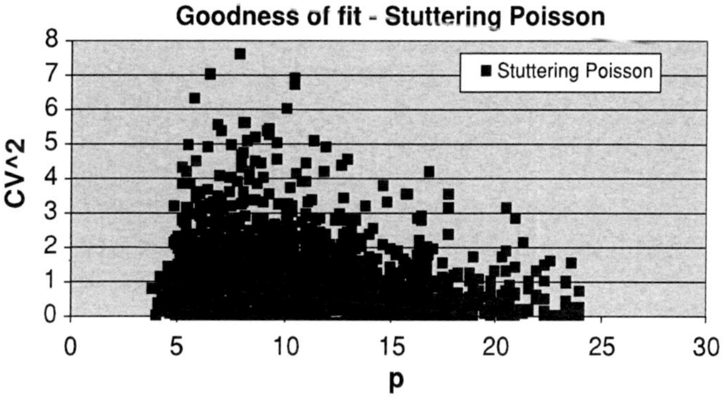

Fig. A4 Dataset #2—goodness-of-fit results for the normal distribution

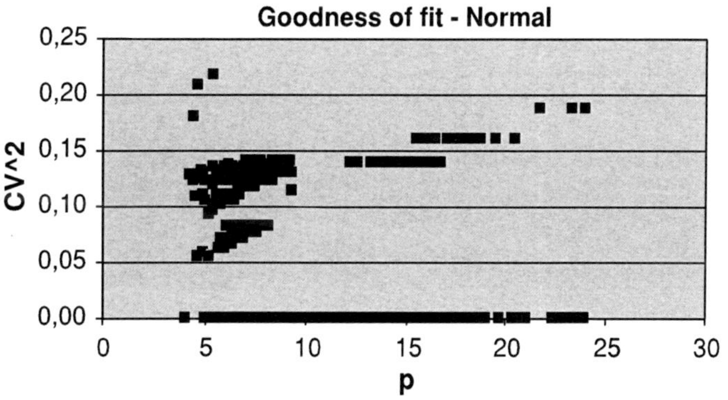

Fig. A5 Dataset #2—goodness-of-fit results for gamma distribution

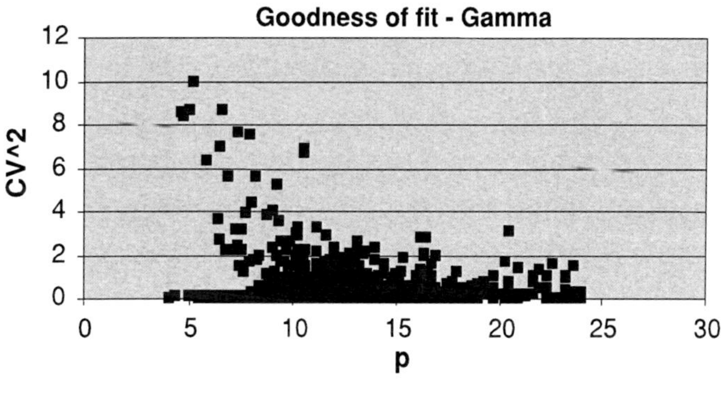

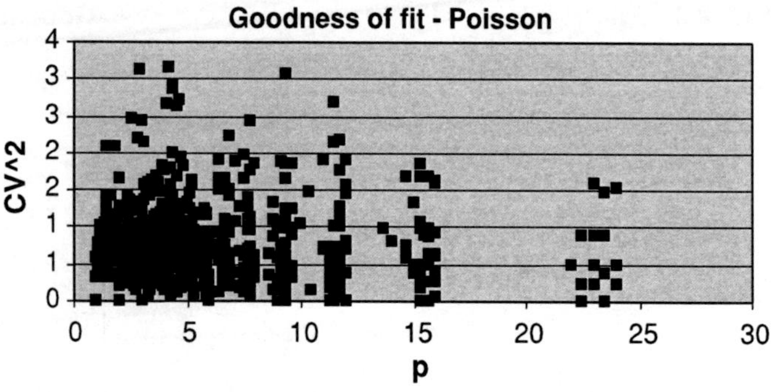

Fig. A6 Dataset #3—goodness-of-fit results for the Poisson distribution

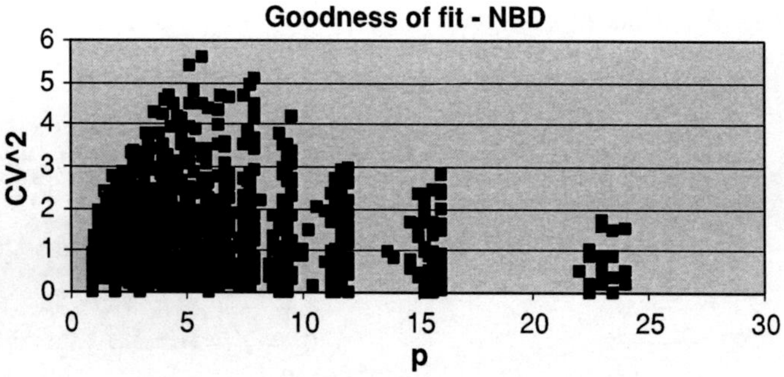

Fig. A7 Dataset #3—goodness-of-fit results for the NBD

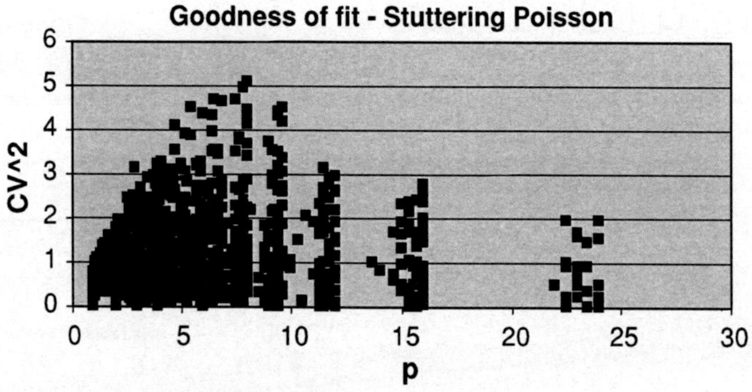

Fig. A8 Dataset #3—goodness-of-fit results for the stuttering Poisson

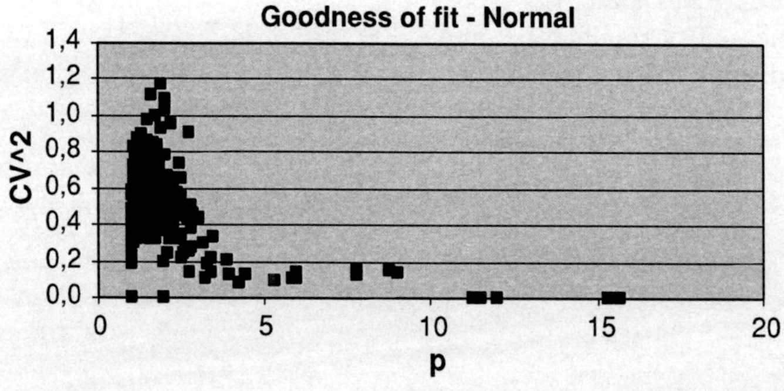

Fig. A9 Dataset #3—goodness-of-fit results for the normal distribution

Fig. A10 Dataset #3—goodness-of-fit results for gamma distribution

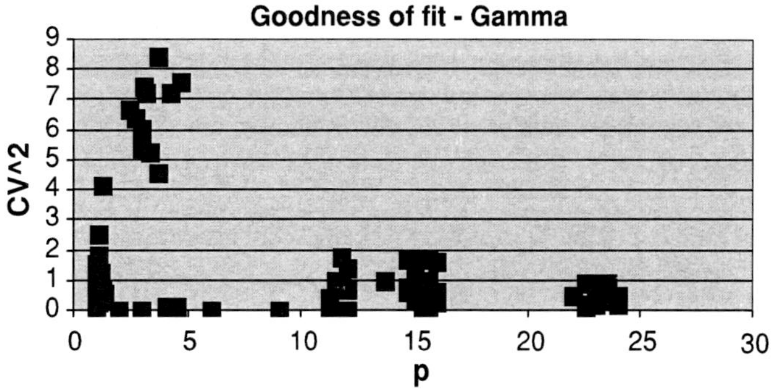

References

Adan I, van Eeenige M, Resing J (1995) Fitting discrete distributions on the first two moments. Probab Eng Inf Sci 9:623–632

Babai MZ, Syntetos AA, Teunter R (2009) On the empirical performance of (T, s, S) heuristics. Eur J Oper Res (in press)

Boylan JE (1997) The centralisation of inventory and the modelling of demand. Unpublished PhD thesis, University of Warwick, UK

Boylan JE, Syntetos AA (2006) Accuracy and accuracy-implication metrics for intermittent demand. FORESIGHT: Int J Appl Forecast 4:39–42

Boylan JE, Syntetos AA, Karakostas GC (2007) Classification for forecasting and stock control: a case study. J Oper Res Soc 59:473–481

Burgin TA (1975) The gamma distribution and inventory control. Oper Res Q 26:507–525

Burgin TA, Wild AR (1967) Stock control experience and usable theory. Oper Res Q 18:35–52

Choulakian V, Lockhart RA, Stephens MA (1994) Cramér–von Mises statistics for discrete distributions. Can J Stat 22:125–137

Conover WJ (1972) A Kolmogorov goodness-of-fit test for discontinuous distributions. J Am Stat Assoc 67:591–596

Croston JD (1972) Forecasting and stock control for intermittent demands. Oper Res Q 23:289–304

Croston JD (1974) Stock levels for slow-moving items. Oper Res Q 25:123–130

Dunsmuir WTM, Snyder RD (1989) Control of inventories with intermittent demand. Eur J Oper Res 40:16–21

Eaves A (2002) The forecasting for the ordering and stock holding of consumable spare parts. Unpublished PhD thesis, Lancaster University, UK

Ehrhardt R, Mosier C (1984) A revision of the power approximation for computing (s, S) inventory policies. Manag Sci 30:618–622

Fildes R, Nikolopoulos K, Crone S, Syntetos AA (2008) Forecasting and operational research: a review. J Oper Res Soc 59:1150–1172

Friend JK (1960) Stock control with random opportunities for replenishment. Oper Res Q 11:130–136

Gallagher DJ (1969) Two periodic review inventory models with backorders and stuttering Poisson demands. AIIE Trans 1:164–171

Harnett DL, Soni AK (1991) Statistical methods for business and economics, 4th edn. Addison Wesley, New York

Hollier RH (1980) The distribution of spare parts. Int J Prod Res 18:665–675

Janssen FBSLP (1998) Inventory management systems; control and information issues. Published PhD thesis, Centre for Economic Research, Tilburg University, The Netherlands

Johnston FR (1980) An interactive stock control system with a strategic management role. J Oper Res Soc 31:1069–1084

Johnston FR, Boylan JE, Shale EA (2003) An examination of the size of orders from customers, their characterization and the implications for inventory control of slow moving items. J Oper Res Soc 54:833–837

Kwan HW (1991) On the demand distributions of slow moving items. Unpublished PhD thesis, Lancaster University, UK

Lengu D, Syntetos AA (2009) Intermittent demand: classification and distributional assumptions. Working Paper (WP) 333/09, Management and Management Sciences Research Institute (MaMS RI), University of Salford, UK

Mitchell GH (1962) Problems of controlling slow-moving engineering spares. Oper Res Q 13:23–39

Naddor E (1975) Optimal and heuristic decisions in single and multi-item inventory systems. Manag Sci 21:1234–1249

Noether GE (1963) Note on the Kolmogorov statistic in the discrete case. Metrika 7:115–116

Noether GE (1967) Elements of nonparametric statistics. Wiley, New York

Pettitt AN, Stephens MA (1977) The Kolmogorov–Smirnov goodness-of-fit statistic with discrete and grouped data. Technometrics 19:205–210

Porras EM, Dekker R (2008) An inventory control system for spare parts at a refinery: an empirical comparison of different reorder point methods. Eur J Oper Res 184:101–132

Quenouille MH (1949) A relation between the logarithmic, Poisson and negative binomial series. Biometrics 5:162–164

Ritchie E, Kingsman BG (1985) Setting stock levels for wholesaling: performance measures and conflict of objectives between supplier and stockist. Eur J Oper Res 20:17–24

Sani B (1995) Periodic inventory control systems and demand forecasting methods for low demand items. Unpublished PhD thesis, Lancaster University, UK

Silver EA, Pyke DF, Peterson R (1998) Inventory management and production planning and scheduling, 3rd edn. Wiley, New York

Slakter MJ (1965) A comparison of the Pearson chi-square and Kolmogorov goodness-of-fit tests with respect to validity. J Am Stat Assoc 60:854–858

Stephens MA (1974) EDF statistics for goodness of fit and some comparisons. J Am Stat Assoc 69:730–737

Stephens MA (1976) Asymptotic results for goodness-of-fit statistics with unknown parameters. Ann Stat 4:357–369

Stephens MA (1977) Goodness of fit for the extreme value distribution. Biometrika 64:583–588

Strijbosch LWG, Heuts RMJ, van der Schoot EHM (2000) A combined forecast-inventory control procedure for spare parts. J Oper Res Soc 51:1184–1192

Syntetos AA, Boylan JE (2005) The accuracy of intermittent demand estimates. Int J Forecast 21:303–314

Syntetos AA, Boylan JE (2008) Smoothing and adjustments of demand forecasts for inventory control. IMA J Manag Math 19:175–192

Syntetos AA, Boylan JE, Croston JD (2005) On the categorisation of demand patterns. J Oper Res Soc 56:495–503

Syntetos AA, Babai MZ, Dallery Y, Teunter R (2009) Periodic control of intermittent demand items: theory and empirical analysis. J Oper Res Soc 60:611–618

Vereecke A, Verstraeten P (1994) An inventory management model for an inventory consisting of lumpy items, slow movers and fast movers. Int J Prod Econ 35:379–389

Walsh JE (1963) Bounded probability properties of Kolmogorov–Smirnov and similar statistics for discrete data. Ann Inst Stat Math 15:153–158

Ward JB (1978) Determining re-order points when demand is lumpy. Manag Sci 24:623–632

Watson RB (1987) The effects of demand-forecast fluctuations on customer service and inventory cost when demand is lumpy. J Oper Res Soc 38:75–82

Willemain TR, Smart CN, Shockor JH, DeSautels PA (1994) Forecasting intermittent demand in manufacturing: a comparative evaluation of Croston's method. Int J Forecast 10:529–538

Willemain TR, Smart CN, Schwarz HF (2004) A new approach to forecasting intermittent demand for service parts inventories. Int J Forecast 20:375–387

Williams TM (1984) Stock control with sporadic and slow-moving demand. J Oper Res Soc 35:939–948

Chapter 3
Decision Trees for Forecasting Trended Demand

Natasha N. Atanackov and John E. Boylan

3.1 Introduction

Service parts, by their very nature, are often subject to trends in demand. The trends may be short-term, as a consequence of changing market conditions, or long-term, related to the life-cycle of the part. Fortuin (1980) suggested that there are three phases of a service part's life history: initial, normal and final. The initial phase is often a time of growth in demand, as more original equipment begins to fail. The normal phase may be more stable, but still subject to shorter-term trends, perhaps reflecting wider market trends. The final phase is generally one of long-term decline in demand, as the original equipment is replaced by newer models and the service parts are required less frequently.

The nature of the trend in demand is not always clear, particularly during the normal phase of demand, when noise in demand may mask trends. Therefore, forecasting may be enhanced by improving the selection process between trended and non-trended series, and between different types of trended series.

In many organisations, service parts are voluminous, amounting to thousands or tens of thousands of items. This has led to the implementation of automatic methods in forecasting and inventory management software. One approach to the selection of methods is to use a ‚pick best' method, whereby a competition of methods is conducted for each series. However, this approach is not suitable for forecasting trended demand when the length of demand history is short, making it difficult to discern growth or decline. Another approach is to simply allow users to choose a forecasting technique from a menu. This presupposes that users are aware

N. N. Atanackov (✉)
Belgrade University, Belgrade, Serbia
e-mail: Natasha_atanackov@yahoo.co.uk

J. E. Boylan
Buckinghamshire New University, High Wycombe, UK
e-mail: John.Boylan@bucks.ac.uk

N. Altay and L. A. Litteral (eds.), *Service Parts Management*,
DOI: 10.1007/978-0-85729-039-7_3, © Springer-Verlag London Limited 2011

of the strengths and weaknesses of the methods on the menu, and know how to conduct appropriate accuracy comparisons. This is not always a reasonable assumption, particularly for those new to demand forecasting, or insufficiently well-trained.

An alternative approach is to develop decision rules for method selection based on the analysis of the models that underpin forecasting methods. Such selection methods, known as protocols, can be implemented quite straightforwardly in forecasting software. Analysing the problem from this perspective may bring new insights and offer the opportunity of more accurate forecasts.

The aim of this chapter is to assess protocols for method selection, based on rules derived from the time series features. The pool of methods consists of well known non-seasonal exponential smoothing methods, namely Single Exponential Smoothing (SES), Holt's linear method, and damped Holt's method. The damped Holt's method, introduced by Gardner and McKenzie (1985), includes a damping parameter in Holt's method to give more control over trend extrapolation. It is included in many service parts applications and has been found to perform well in forecasting competitions on real data. In this chapter, a distinction will be maintained between a forecasting method, which is a computational procedure to calculate a forecast, and a forecasting model, which is a stochastic demand process. A model-based method takes the stochastic process as given, and is based on an appropriate estimation procedure for the parameters specified in the model. A schematic representation is shown in Fig. 3.1.

The models considered in this chapter are restricted to non-intermittent demand, for which the standard smoothing methods are appropriate. Intermittent demand occurs when there is no demand at all in some time periods and requires different smoothing approaches, such as Croston's method. The selection of forecasting methods for intermittent items is an important problem and has been a subject of growing interest in the academic literature. Boylan and Syntetos (2008) review recent developments and offer a framework for classification of service parts, including consideration of intermittence and erraticness of demand. The approaches suggested for the selection of methods for intermittent demand are quite different to those that have been developed for non-intermittent demand. Both categories of demand are vital for service parts management but, in this chapter, we shall concentrate on non-intermittent demand.

The reminder of this chapter will develop an approach to select from the models that underpin SES, Holt's and damped Holt's methods. Then, the approach will be

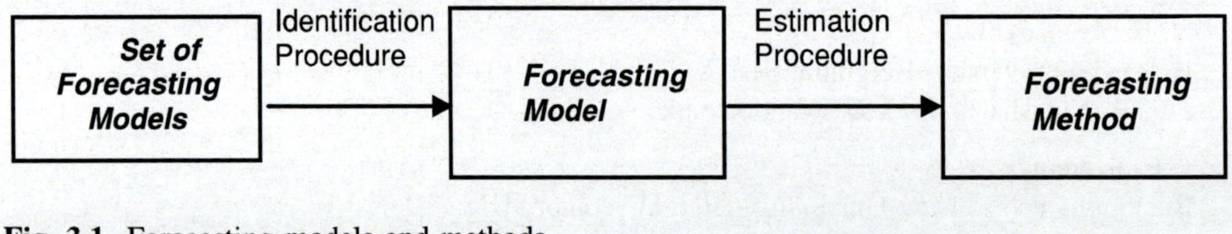

Fig. 3.1 Forecasting models and methods

examined and tested on real-world data. Three datasets will be employed. The first two are ‚standard data sets', having been used in previous forecasting competitions, while the third has not been analysed previously. The third dataset consists of weekly demand histories; forecasts of weekly demand are required for inventory management purposes.

3.2 Trend Forecasting Methods

Exponential smoothing (ES) is an excellent choice for automatic forecasting systems involving many thousands of time series. Such systems are commonly used for the inventory management of service parts. The implementation of three of the most commonly applied exponential smoothing methods is discussed in this section, including optimisation of smoothing parameters.

3.2.1 Single Exponential Smoothing Method (SES)

The single exponential smoothing method can be written as follows:

$$\hat{l}_t = \alpha d_t + (1 - \alpha)\hat{l}_{t-1} \tag{2.1}$$

$$\hat{d}_{t+n} = \hat{l}_t \tag{2.2}$$

where d_t represents the observation for the current time period t; $\hat{l}_t$ represents an estimate of the underlying mean level of the series; $\hat{d}_{t+n}$ denotes a forecast made at time t for n-periods ahead; and α represents a smoothing parameter. Equation (2.1) shows how the smoothed level is updated when a new observation becomes available. Equation (2.2) represents the n-step ahead forecast using observations up to and including time t.

The choice of values of smoothing parameters in ES methods can be made in a number of ways. Firstly, the parameter values can be set arbitrarily to values suggested in the academic literature. Secondly, the estimation period of the time series (discussed in Sect. 3.6.2) can be used to establish the parameters by selecting values that minimise Mean Squared Error (or other error measure). The values will be kept constant regardless of the new observations that become available. Finally, the smoothing parameters can be optimised at each period, before the forecasts are generated, meaning that all the available time series information is used. In this study, the second approach is used, allowing for optimisation of parameters, but without the computational burden of re-optimisation every period. For SES, the smoothing constant is optimised from the range 0.05 to 0.95, with a step value of 0.05.

3.2.2 Holt's Linear Method

When trend is present in the time series, the SES method needs to be generalised. Holt's linear trend method (Holt 1957, 2004a, b) produces estimates of the local level and of the local growth rate. The method may be written as follows:

$$\hat{l}_t = \alpha d_t + (1-\alpha)\left(\hat{l}_{t-1} + \hat{b}_{t-1}\right) \tag{2.3}$$

$$\hat{b}_t = \beta\left(\hat{l}_t - \hat{l}_{t-1}\right) + (1-\beta)\hat{b}_{t-1} \tag{2.4}$$

$$\hat{d}_{t+n} = \hat{l}_t + n\hat{b}_t \tag{2.5}$$

where $\hat{b}_t$ represents an estimate of the underlying trend of the series; α and β represent smoothing parameters; and the remaining notation is unchanged. Equation (2.3) shows how the smoothed level of the series is updated when a new observation becomes available, while (2.4) serves the same purpose for the trend estimate. Equation (2.5) gives a straight line trend-projection for the n-step ahead forecast.

As for single exponential smoothing, the parameters can be chosen arbitrarily, or they can be estimated from the time series. In this chapter, the two parameters are optimised once, with both parameters being required to lie in the interval 0.05–0.95 (with step value of 0.05).

3.2.3 Damped Holt's Method

It is not clear when the idea of damped trend was introduced for the first time in the academic literature. Roberts (1982) introduced a predictor for sales forecasting with an incremental growth estimate and a damping parameter, and proved its optimality for an ARIMA (1, 1, 2) process. Gardner and McKenzie (1985) discussed Holt's linear method, and suggested a generalised version of the method by adding an autoregressive-damping parameter, ϕ, to give more control over trend extrapolation. The damped Holt's method may be written as follows:

$$\hat{l}_t = \alpha d_t + (1-\alpha)\left(\hat{l}_{t-1} + \phi\hat{b}_{t-1}\right) \tag{2.6}$$

$$\hat{b}_t = \beta\left(\hat{l}_t - \hat{l}_{t-1}\right) + (1-\beta)\phi\hat{b}_{t-1} \tag{2.7}$$

$$\hat{d}_{t+n} = \hat{l}_t + \sum_{i=1}^{n}\phi^i\hat{b}_t \tag{2.8}$$

In the above equations, ϕ is a damping parameter between zero and one, and the remaining notation is unchanged. Depending on the value of the damping parameter ϕ, Gardner and McKenzie (1985) distinguished between four types of trend projected by the damped Holt's method:(i) If $\phi = 0$, there is no trend in the time series and the method is equivalent to Single Exponential Smoothing; (ii) If $0 < \phi < 1$, the trend is damped and the forecast approaches an asymptote given by the horizontal line $\hat{l}_t + \hat{b}_t \frac{\phi}{1-\phi}$; iii) If $\phi = 1$, the method is equivalent to the standard Holt's method and the trend is linear; (iv) if $\phi > 1$, the trend is exponential. The last option is considered to be dangerous in an automatic forecasting system (Gardner and McKenzie 1985) and it will not be discussed further in this chapter. Additionally, the ϕ parameter can have a negative value. In that case, the process would have an oscillatory character, because the sign of the damping parameter would be different in every subsequent time period. This situation is rarely seen in practice and it will not be elaborated further.

The three parameters required for the Damped Holt's method are optimised in the same way as the other two smoothing methods. The range for the two smoothing parameters is the same as for Holt's method. The damping parameter is optimised from the range 0.1 to 0.9 (step 0.1). The influence of the parameter values will be discussed in Sects. 3.6 and 3.7.

3.2.4 Other Methods

Pegels (1969) included multiplicative trend forecasting within his classification of exponential smoothing methods. This method was extended by Taylor (2003), who introduced a version of exponential smoothing with a damped multiplicative trend. Multiplicative trend methods are not commonly available in forecasting packages for service parts. Therefore, these methods will not be analysed further in this chapter, although their investigation would be a worthwhile topic for further research.

Another method that could be considered for analysis is the Theta Method (Assimakopoulos and Nikolopoulos 2000). This method, as applied in the M3 competition (Makridakis and Hibon 2000), has been shown to be a member of the exponential smoothing family of methods (Hyndman and Billah 2003). Like the damped multiplicative trend method, the Theta method has yet to be implemented in commercial software that may be applied to service parts, although academic software has been developed (Makridakis et al. 2008).

In summary, the work presented in this chapter does not aim to provide selection protocols between all exponential smoothing methods (see Hyndman et al. 2008, for a comprehensive taxonomy of methods). Rather, an approach is developed for three of the most commonly employed methods. A natural extension to this work would be to include exponential smoothing methods for seasonal data, discussed further in the final section of this chapter.

3.3 Approaches to Method Selection

Extrapolating trends is risky because if the trend forecast is in the wrong direction, the resulting forecast will be less accurate than the random walk. This may lead to significant over-stocking or under-stocking of service parts, particularly if lead-times are long. Therefore, it is desirable to examine methods which can discern when trend is present and when it is absent.

Numerous approaches to model and method selection have been discussed over the last 40 years. Box and Jenkins (1970) proposed the analysis of the auto-correlation function (ACF) and the partial auto-correlation function (PACF), after appropriate differencing of series, to select between ARIMA models. Originally, this procedure depended on careful analysis of the ACF and PACF of each individual series, and was not suited to automatic forecasting systems. Since then, automatic selection of models has been incorporated in forecasting packages such as Autobox.

A less sophisticated approach is prediction validation (see, for example, Makridakis et al. 1998). A subset of the dataset is withheld and various methods are compared on this ‚out-of-sample' subset, using an error criterion such as Mean Square Error or Mean Absolute Percentage Error. Billah et al. (2005) analysed simulated and empirical data, and showed that prediction validation can be improved upon by approaches based on ‚encompassing' and information criteria.

An ‚encompassing approach' relies on the most general method being applied to all series, where all other methods under consideration are special cases of the general method. Billah et al. (2005) compared SES, Holt's method and Holt–Winters' method, using Holt's as the encompassing approach for non-seasonal data and Holt–Winters' for seasonal data.

Information criteria are often recommended for the selection of an appropriate forecasting method. These criteria penalise the likelihood by a function of the number of parameters in the model. They have stimulated much academic interest and are used in service parts computer applications such as Forecast Pro. The following information criteria are well-established in the literature: Akaike's Information Criterion (Akaike 1974), the bias-corrected AIC (Hurvich and Tsai 1989), the Bayesian Information Criterion (Schwarz 1978), and the Hannan–Quinn Information Criterion (Hannan and Quinn 1979). More recently, linear empirical information criteria have been proposed by Billah et al. (2005). Gardner (2006) comments that studies comparing the performance of different information criteria should also include a comparison with universal application of damped Holt's method, as this offers a benchmark which is difficult to beat. This offers a more general encompassing benchmark approach than Holt's method (for non-seasonal data) and is adopted in this study.

An alternative approach to model selection is to employ expert systems. Such systems are based on sets of rules recommended by experts in the field. Collopy and Armstrong (1992) gave rules to select between a random walk, time-series regression, Brown's double exponential smoothing method and Holt's linear method.

Vokurka et al. (1996) gave a fully automatic expert system to distinguish between SES, the damped trend method, classical decomposition and a combination of all methods. Gardner (1999) tested the expert systems of Collopy and Armstrong (1992) and Vokurka et al. (1996) and found them to be less accurate than universal application of the damped trend method. A subsequent study by Adya et al. (2001) enhanced the rules of Collopy and Armstrong (1992) and found the new rules to perform better than universal damped trend for annual data and about the same for seasonally adjusted quarterly and monthly data. In the light of these studies, and the comments by Gardner (2006) on the analysis of methods based on information criteria, discussed previously, universal application of damped Holt's method will be used as a benchmark for comparison with the performance of the protocols and decision trees tested in this study.

Another approach to method selection is based on time-series characteristics. Shah (1997) developed a rule for selecting the best forecasting method for each individual series, using discriminant analysis. He demonstrated that a choice of forecasting method using summary statistics for an individual series was more accurate than using any single method for all series considered. Meade (2000) designed an experiment for testing the properties of 25 summary statistics. Nine forecasting methods were used, divided into three groups, and tested on two data sets, the M1-competition data and Fildes' telecommunications data. The author concluded that the summary statistics can be used to select a good forecasting method or set of methods, but not necessarily the best.

Gardner and McKenzie (1988) proposed an approach based on comparison of the variances of the original series, the once-differenced and the twice-differenced series. Tashman and Kruk (1996) analysed three protocols for method selection on real time series: Gardner and McKenzie's (1988) variance procedure, the set of rules from Rule-Based Forecasting (Collopy and Armstrong 1992), and a method-switching procedure developed by Goodrich (1990). Tashman and Kruk found that the protocols are effective in selecting the appropriate applications of strong trend methods but do not effectively distinguish applications of non-trended and weak-trended methods. This issue will be addressed in this chapter. A modification to the Gardner and McKenzie protocol will be suggested, and tested on simulated and real data.

3.4 Trend Forecasting Models

Mathematical models underpinning exponential smoothing methods will be reviewed in this section. State-space models are investigated, rather than ARIMA models, which were the original motivation of the work by Gardner and McKenzie (1988). This offers a new perspective for the comparison of protocols based on differencing of series.

Multiple source of error (MSOE) models are used in this chapter, rather than single source of error (SSOE). The essential difference between the two forms of state-space models is that the error terms are assumed to be independent in MSOE models and perfectly correlated in SSOE models. Detailed discussions on these model forms are provided by Harvey (2006), who advocates MSOE models, and by Hyndman et al. (2008) who advocate SSOE models. A detailed comparison of the effect of model forms on method selection methods is beyond the scope of this chapter, but would form an interesting follow-up study.

3.4.1 Steady State Model (SSM)

Under the Steady State Model (SSM) assumption, the mean level of the time series fluctuates stochastically over time. The SSM model is given in the following form:

$$d_t = l_t + \varepsilon_t \tag{4.1}$$

$$l_t = l_{t-1} + \gamma_t \tag{4.2}$$

where d_t represents the observation for the current time period t; l_t represents the unobserved series mean level at time t; ε_t and γ_t are uncorrelated normally distributed random variables each with zero mean $E(\varepsilon_t) = E(\gamma_t) = 0$ and constant variance $V(\varepsilon) =$ const. and $V(\gamma) =$ const.; ε_t and γ_t are also serially uncorrelated. The above model is often referred to as the local level model (eg Commandeur and Koopman 2007). It corresponds to an ARIMA (0, 1, 1) process.

3.4.2 Linear Growth Model (LGM)

Theil and Wage (1964) analysed generating processes for trended time series. Harrison (1967) extended their work and formulated a probabilistic Linear Growth Model (LGM) of the following form:

$$d_t = l_t + \varepsilon_t \tag{4.3}$$

$$l_t = l_{t-1} + b_t + \gamma_t \tag{4.4}$$

$$b_t = b_{t-1} + \delta_t \tag{4.5}$$

where b_t denotes the underlying trend; ε_t, γ_t, and δ_t are uncorrelated normally distributed random variables each with zero mean $E(\varepsilon_t) = E(\gamma_t) = E(\delta_t) = 0$ and constant variance $V(\varepsilon_t) =$ const., $V(\gamma_t) =$ const. and $V(\delta_t) =$ const; ε_t, γ_t, and δ_t are also serially uncorrelated. The above model is often referred to as the local linear trend model (e.g. Commandeur and Koopman 2007) and it corresponds to an ARIMA (0, 2, 2) process.

3.4.3 Damped Trend Model (DTM)

In the first Damped Trend Model (DTM 1), it is assumed that the dampening starts in the second time period after the end of the historical data:

$$d_t = l_t + \varepsilon_t \tag{4.6}$$

$$l_t = l_{t-1} + b_t + \gamma_t \tag{4.7}$$

$$b_t = \phi b_{t-1} + \delta_t \tag{4.8}$$

where d_t represents the observation for the current time period t; l_t represents the unobserved series level at time t; b_t denotes the underlying trend; ε_t, γ_t, and δ_t are uncorrelated normally distributed random variables each with zero mean $E(\varepsilon_t) = E(\gamma_t) = E(\delta_t) = 0$ and constant variance $V(\varepsilon_t) = \text{const.}$, $V(\gamma_t) = \text{const.}$ and $V(\delta_t) = \text{const}$; ε_t, γ_t, and δ_t are also serially uncorrelated; ϕ represents a damping parameter.

In the second Damped Trend Model (DTM 2), it is assumed that the dampening starts in the first time period after the end of the historical data:

$$d_t = l_t + \varepsilon_t \tag{4.9}$$

$$l_t = l_{t-1} + \phi b_t + \gamma_t \tag{4.10}$$

$$b_t = \phi b_{t-1} + \delta_t \tag{4.11}$$

where d_t, l_t, b_t, ε_t, γ_t, δ_t and ϕ are as defined in the previous sub-section. In this model, the damping parameter is applied one period earlier than in DTM 1.

Atanackov (2004) analysed the Serial Variation Curves, discussed in the next section, for both model forms. She showed that the Serial Variation Curve for DTM 1, for the special case of $\phi = 0$, is not consistent with the Serial Variation Curve for the Steady State Model (SSM). On the other hand, the Serial Variation Curve for DTM 2 is consistent with SSM for $\phi = 0$ and is consistent with the Linear Growth Model for $\phi = 1$. Therefore, for the remainder of this chapter, DTM 2 will be adopted.

3.5 Method Selection Protocols and Decision Trees

Protocols and decision trees for method selection purposes will be analysed in this section. The aim is to propose a more coherent approach that exhibits good performance in practical applications. The research process will be based on the selection of mathematical models, discussed in Sect. 3.4, that underpin the methods discussed in Sect. 3.2, namely SES, linear Holt's and damped Holt's methods.

Harrison's (1967) Serial Variation Curves will be examined and extended, to form a protocol for method selection. Also, the existing Gardner and McKenzie's (1988) procedure, based on the analysis of the variances of time series, will be examined and a modification of the procedure will be suggested.

Taking into account the performance of these protocols, two types of decision trees will be designed to select between non-trended and trended time series and, subsequently, between different types of underlying trend (damped or undamped). Decision trees are completely automatic and human judgement is not necessary. As such, the trees can be utilised for short-term forecasting of service parts, particularly for inventory control and production planning for forecasting thousands of items on a regular basis.

3.5.1 The SVC Protocol: Harrison's Serial Variation Curves

Harrison (1967) suggested testing the lagged second differences of the data in order to assess the suitability of a data generating process. The idea was originally applied to the Linear Growth Model (LGM) only, and has not been extended since. In this chapter, the SVC procedure will be applied for the first time to the Damped Trend Model (DTM). It will be shown that the SVC approach applied to the LGM and DTM models can form an independent diagnostic, from now on called the SVC protocol, to select between the two trended models.

Defining $D1(t) = d_t - d_{t-1}$ as the first difference of the observations at time t, and defining $\Delta_n(t) = D1(t) - D1(t-n)$ as the second difference of the observations lagged by n periods, and taking the squared expectations of the above differences, Harrison (1967) showed that, for $n \geq 2$:

$$E[\Delta_n(t)]^2 = 4V(\varepsilon) + 2V(\gamma) + nV(\delta) \tag{5.1}$$

Thus, the variances of the lagged second differences are a linear function of the lag, with gradient $V(\delta)$. This relationship is called a Serial Variation Curve (SVC).

Using the same notation as above, the SVC of the DTM 2 has been derived (Atanackov 2004) for different lag lengths ($n \geq 2$):

$$E[\Delta_n(t)]^2 = 4V(\varepsilon) + 2V(\gamma) + \frac{2\phi^2(1-\phi^n)}{1-\phi^2}V(\delta) \tag{5.2}$$

Equation (5.1) is a special case of (5.2) when $\phi = 1$ (by application of L'Hôpital's rule to (5.2)). Equation (5.2) shows that the Serial Variation Curve of the lagged second differences of the data for the DTM 2 ($0 < \phi < 1$) is expected to be a curve, as illustrated in Fig. 3.2.

Because the Serial Variation Curve (SVC) of the LGM model is linear, and the SVC of the DTM model is non-linear, a protocol can be designed to distinguish between series that follow a weak trended model (DTM) and series that follow a strong trended model (LGM). The new protocol, called the SVC protocol, is based

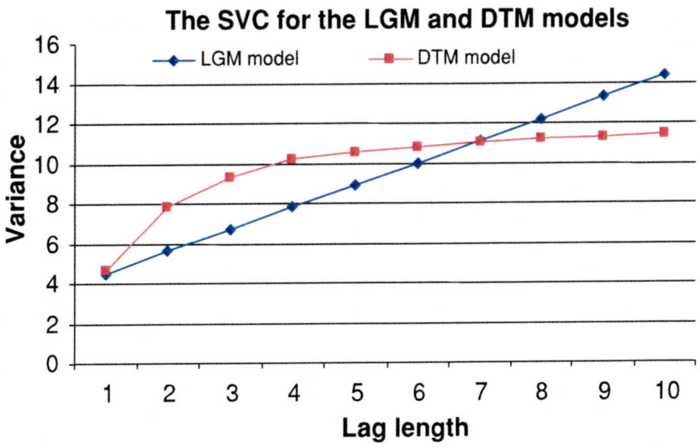

Fig. 3.2 The SVC of the two trended models: LGM and DTM ($\phi = 0.8$)

on fitting a straight line, using least squares regression, through the SVC points (that is through the variances of the lagged second differences) and on testing the significance of the autocorrelation (AC) of the residuals.

From Fig. 3.2 it follows that there are two cases: (i) for the LGM model, there is no significant AC of the residuals; (ii) for the DTM model, there is significant positive AC of the residuals. The one-sided Durbin–Watson statistic (Durbin and Watson 1951) is appropriate for testing the autocorrelation of the residuals.

In order to use the SVC protocol for real life applications it is necessary to establish its operational rules. The lag length to be used for the SVC fitting is not known in advance. Therefore, it will be tested as part of the simulation analysis. Six different lag lengths of 15, 20, 25, 30, 35, and 40 are examined under two significance levels: 1 and 5%. Their performance is compared in terms of the percentage of correctly identified time series models and the best operational rules for the SVC protocol are established. The findings are summarised in Sect. 3.6.5.

3.5.2 *Gardner and McKenzie's Variance Procedure*

The original Gardner–McKenzie (GM) protocol is based on a comparison of the variances of the original data, first-differenced data and second-differenced data. Gardner and McKenzie (1988) recommend:

- If the variance of the original time series $V(0)$ is minimum, a constant mean model is diagnosed, and the SES method should be used;
- If the variance of the first difference of the series $V(1)$ is minimum, a weak trend model is diagnosed and damped Holt's method is recommended;
- If the variance of the second difference of the time series $V(2)$ is minimum, a strong trend model is diagnosed and Holt's linear method should be applied.

The diagnostics and the models for which the recommended methods are optimal are summarised in Table 3.1.

Table 3.1 Gardner and McKenzie's diagnostics and recommended methods

Model	ARIMA (0, 1, 1)	ARIMA (1, 1, 2)	ARIMA (0, 2, 2)
Diagnostic	$V(0)$ minimum	$V(1)$ minimum	$V(2)$ minimum
Diagnosis	No trend	Weak trend	Strong trend
Recommended method	SES	Damped Holt's	Holt's linear

The original GM protocol has its roots in the Box–Jenkins methodology of transforming a non-stationary series into a stationary series by differencing the data. The middle coefficient in the ARIMA processes (see Table 3.1) refers to the level of differencing required in order to achieve stationarity in a given time series.

Since SES and damped Holt's are both optimal for ARIMA series with first-differencing, both forecasting methods will be suitable when the variance $V(1)$ is minimum. However, Gardner and McKenzie's protocol selects SES for stationary series, the case in which no differencing is necessary, and differencing the series results in an increase in the series variance. Tashman and Kruk (1996) criticise this selection, arguing that a stationary ARMA procedure would outperform SES on such stationary series. They further note that the GM protocol would select the damped trend method for time series that follow an ARIMA (0, 1, 1) model, and therefore exhibit non-trended behaviour. These series, however, would be ideally forecasted by the SES forecasting method. Of course, the damped Holt's method incorporates SES as a special case, when the ϕ parameter takes a zero value. However, there is no assurance that a forecasting algorithm will optimise the damping parameter ϕ at the zero value.

3.5.3 *V*(0)*V*(2) Protocol

Following the above discussion, a modification of Gardner and McKenzie's variance procedure is suggested in this section. Its purpose is to distinguish between non-trended series (assumed to follow a Steady State Model) and trended series (assumed to follow a Linear Growth Model). The new protocol does not attempt to diagnose a Damped Trend Model. However, the protocol is included in Decision Trees, discussed in the next two sub-sections, which will further distinguish between damped and undamped trend.

Tashman and Kruk (1996) found that the GM protocol could distinguish effectively between weak and strong trend (accomplished by a $V(1)$ vs. $V(2)$ comparison) but was less effective at distinguishing between no trend and weak trend. It is proposed that a distinction should be made between non-trended and trended series by comparing the variance of the original series, $V(0)$, with the variance of the second differences of the series, $V(2)$. If $V(0)$ is lower, the series is categorised as ‚non-trended' (Steady State Model), whereas if $V(2)$ is lower, it is categorised as ‚trended' (Linear Growth Model). This rule is called the $V(0)V(2)$ protocol.

If $V(2)$ is minimum, the categorisation of the series as following the Linear Growth Model is consistent, from an ARIMA perspective, with the requirement of twice-differencing to achieve stationarity. If $V(0)$ is minimum, the categorisation of the series as following the Steady State Model is not consistent with the requirement of differencing once to achieve stationarity. Of course, the use of $V(1)$ is required for consistency with this requirement but, as shown by Tashman and Kruk (1996) and as discussed later, $V(1)$ should be reserved for the effective identification of damped trend series.

The state-space models introduced in Sect. 3.4 offer a different lens through which these rules may be viewed. By analysing variance expressions, the conditions under which the correct models will be identified by the $V(0)V(2)$ protocol may be identified. For the SSM model, comparison of $V(0)$ and $V(2)$ gives the following inequality (Atanackov 2004):

$$V(0) < V(2) \quad \text{if and only if} \quad n < \frac{V(\varepsilon)}{V(\gamma)} + 2 \tag{5.3}$$

where n is the length of the time series. Further analysis of the above expression shows that the $V(0)V(2)$ protocol can detect the SSM time series for low values of the smoothing parameter α and for short time series. This result is confirmed by simulation experiments.

For the LGM model, comparison of the variances $V(0)$ and $V(2)$ gives the following inequality (Atanackov 2004):

$$V(0) > V(2) \quad \text{if and only if} \quad (n-2)\frac{V(\gamma)}{V(\delta)} + \frac{2n^3 + 3n^2 - 23n + 18}{6} > \frac{V(\varepsilon)}{V(\delta)} \tag{5.4}$$

where the notation is unchanged. Further analysis of the above expression shows that the $V(0)V(2)$ protocol can detect the LGM time series in a high percentage of cases for the higher values of the α parameter, and for time series containing more than 20 observations. For low α values and for short time series the $V(0)V(2)$ protocol detects non-trended series.

The performance of the $V(0)V(2)$ protocol will be analysed further using simulation experiments on theoretically generated data, and tested on real data sets.

3.5.4 Decision Tree A

Two new decision trees are presented in this section. The first, Decision Tree A, incorporates an extension of Harrison's Serial Variation Curves. The second, Decision Tree B, is based solely on the idea of comparing the variances of the original series and the once- and twice-differenced series.

Decision Tree A is based on protocols to distinguish between the non-trended and trended time series and, subsequently, between damped (weak) and linear (strong) trend. The $V(0)V(2)$ protocol is applied as a trend detector and then the SVC protocol is employed to select the type of the trend. Decision Tree A, shown below, selects between non-trended and trended models first, and then between damped trend (DTM) and linear trend (LGM), as shown below:

3.5.5 Decision Tree B

Decision Tree B is based on the analysis of the three variances, $V(0)$—variance of the time series, $V(1)$—variance of the differenced series, and $V(2)$—variance of the twice differenced series. The comparison is based on the same variables as suggested by Gardner and McKenzie (1988), but the rules are different. Firstly, as in Tree A, if $V(0) < V(2)$ then the time series is considered to be non-trended, indicating a steady state model (SSM) and hence, SES as the optimal forecasting method.

On the other hand, if the series exhibits variance $V(2)$ less than $V(0)$ it is considered as evidence of trend. The type of the trend still needs to be detected. Roberts (1982) proved that the ARIMA (1, 1, 2) represents a damped trend model while the ARIMA (0, 2, 2) is equivalent to the linear growth model. Therefore, the comparison between the variances $V(1)$ and $V(2)$ will be an indicator of the appropriate trend model. Since the middle number in the ARIMA triplet indicates the level of differencing, it follows that minimum $V(1)$ reveals a damped trend model and hence, damped Holt's method for generating forecasts while the minimum $V(2)$ shows strong trend in time series, and hence Holt's linear method for generating forecasts.

3.6 Simulation Experiment

The simulation experiment serves two purposes. The first purpose is to identify suitable parameters for the SVC protocol and, subsequently, for the Decision Tree A, namely the lag length and the relevant significance level, that cannot be established otherwise. The second purpose is to assess the classification and forecasting performance of the protocols under controlled conditions, where the underlying models are known. The protocols and decision trees to be tested are as follows: (i) The SVC protocol—extended Harrison's Serial Variation Curves; (ii) Original GM protocol—Gardner and McKenzie's (1988) variance procedure; (iii) $V(0)V(2)$ protocol; (iv) Decision Tree A; and (v) Decision Tree B.

The three mathematical models outlined in Sect. 3.4 (SSM, LGM and DTM2) will be employed in the first part of the experiment to simulate time series exhibiting known characteristics. Choosing different combinations of smoothing parameters, and making connections between the method and model parameters, it will be possible to generate time series data that follow a given model for which the optimal forecasting method is known in advance. Eight different lengths of time series, namely 10, 20, 30, 40, 50, 75, 100, and 150 observations per series, will be generated for each of the three models. This covers all the series lengths in the empirical study, to be presented in the following section. For every length, 10,000 replications will be made.

The protocols are tested for classification performance. Protocols will be applied to the relevant data sets (for example, the SVC is relevant for the LGM and DTM models) and the percentage of series for which the model is correctly identified will be recorded. The aim of this classification is to establish the operational rules necessary for the protocols' practical application, and to support the analysis of the forecasting performance.

Finally, the simulation is designed to test the forecasting performance of the protocols. Forecasting performance will be evaluated over a six-period horizon. Since the methods are MSE-optimal for the corresponding models, the Mean Square Error (MSE) is used as an accuracy measure, and the Geometric Root Mean Square Error (GRMSE) will be used in the simulation experiment to summarise the accuracy results. In addition, the Mean Absolute Error (MAE) will be employed in the simulation exercise because it is intended to use this error measure in the empirical analysis. This error measure is less sensitive to outlying or extreme observations. The error measures are discussed in more detail in Sect. 3.6.4.

3.6.1 Generating the Models

The objective here is to establish the relationships between the method and the model parameters for the three methods, to generate data sets that follow specific models and exhibit specific forecasting methods as optimal. In order to cover many different combinations of the smoothing parameter values, while keeping the simulation experiment manageable, five different α values will be used for generating the SSM model, twelve pairs of α and β parameters will be chosen for generating the time series that follow the LGM model, and eighteen triplets of smoothing parameters and the damping parameter ϕ will be used for the DTM2 model generation. Altogether, thirty five data sets, each containing 80,000 time series will be generated for experimental purposes.

In order to simulate the time series that follow the SSM model and genuinely have the SES as the optimal forecasting method, the connection between the method and the model parameters needs to be established. In this particular case, there is only one method parameter, the smoothing parameter α, and two model parameters, the variances of the error terms ε_t and γ_t, i.e. $V(\gamma)$ and $V(\varepsilon)$. Following Harrison (1967), the connection between the optimal smoothing parameter for the SES method and the variance parameters for the SSM model can be expressed as follows:

$$V(\varepsilon) = \frac{1-\alpha}{\alpha^2} V(\gamma) \tag{6.1}$$

The above formula will be used in the experiment to generate time series data that follow the SSM model. Five different smoothing parameter values (0.1, 0.3, 0.5, 0.7, and 0.9) will be used, to cover the range between 0 and 1. Finally, in order to generate the SSM process, γ_t and ε_t will be chosen from two sets of independently normally distributed random numbers, with zero means and constant variances $V(\gamma)$ and $V(\varepsilon)$, respectively. The initial level will be set at 100. The run-in procedure (also known as ,warming up') will be used, to allow the time series to become stable.

To simulate the LGM process, four parameters will be taken into account: two smoothing parameters α and β for Holt's linear exponential smoothing method, and two variance ratios characterising the LGM model. The relationship between the model and optimal method parameters, obtained by Harrison (1967) and further discussed by Harvey (1984), can be expressed as follows:

$$V(\gamma) = \frac{\alpha^2 + \alpha^2\beta - 2\alpha\beta}{\alpha^2\beta^2} V(\delta) \tag{6.2}$$

$$V(\varepsilon) = \frac{1-\alpha}{\alpha^2\beta^2} V(\delta) \tag{6.3}$$

Twelve pairs of the smoothing parameter values, α and β, are selected, to assess the influence of the parameters on the forecasting performance of Holt's linear method for the series following the LGM model, while covering a wide region of possible combinations. The random components ε_t, γ_t, and δ_t will be chosen from three different sets of independently normally distributed random numbers, with zero means and constant variances $V(\varepsilon)$, $V(\gamma)$ and $V(\delta)$, respectively. The initial level will be set at 100 and the run-in period will be applied.

The relationships between method and model parameters for damped Holt's method, given the DTM 2 model, can be expressed as follows (Atanackov 2004):

$$\frac{V(\gamma)}{V(\delta)} = \frac{(\alpha^2 + \phi\alpha^2\beta - \alpha\beta - \phi\alpha\beta)\phi^2}{\phi^2\alpha^2\beta^2 + \alpha\beta(1-\phi) - \phi^2\alpha\beta(1-\phi) + \phi\alpha^2\beta(1-\phi^2)} \tag{6.4}$$

$$\frac{V(\varepsilon)}{V(\delta)} = \frac{(1-\alpha)\phi^2}{\phi^2\alpha^2\beta^2 + \alpha\beta(1-\phi) - \phi^2\alpha\beta(1-\phi) + \phi\alpha^2\beta(1-\phi^2)} \tag{6.5}$$

The value of the damping parameter ϕ has been assumed to be in the interval (0, 1). The smoothing parameter values, α and β, will be chosen in such a way as to test the influence of the parameters on the forecasting performance of damped Holt's method for time series that follow the DTM model, while encompassing a wide region of possible combinations. The random components ε_t, γ_t, and δ_t will be chosen from three different sets of independently normally distributed random numbers, with zero means and constant variances $V(\varepsilon)$, $V(\gamma)$ and $V(\delta)$, respectively. The initial level will be set at 100 and the run-in period will be applied (Figs. 3.3, 3.4).

3.6.2 In-sample and Out-of-Sample Characteristics

In order to conduct the empirical evaluation of the forecasting performances of the selection protocols, the time series will be divided in two parts (see Fig. 3.5). The first part, called the estimation period, is used to calibrate the forecasting method to the historical data, while the second part, called the out-of-sample period, is used to test the accuracy of the selected forecasting method and the forecasting performances of the employed protocol.

The estimation period, often called the in-sample period, consists of two parts: the initialisation period and the calibration period. The observations from the initialisation period are used to initialise the smoothing method. The calibration period, on the other hand, is used for the estimation of the optimal method parameters. The process is carried out by generating the forecasts for each combination of the smoothing parameters and, subsequently, the Mean Squared Error (MSE) of the forecasts is computed and recorded. The procedure is repeated for

Fig. 3.3 Decision tree A

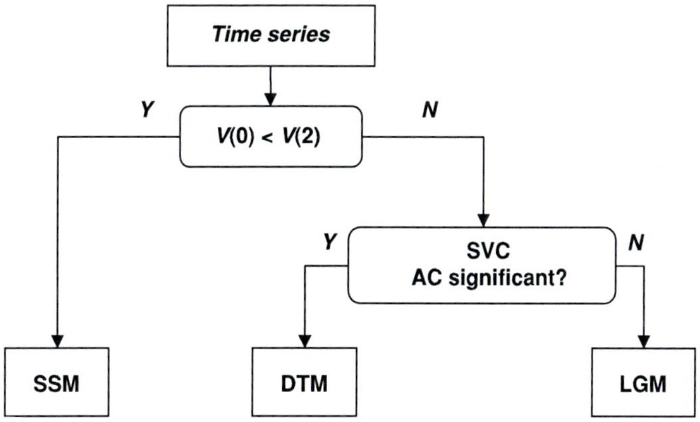

Fig. 3.4 Decision tree B

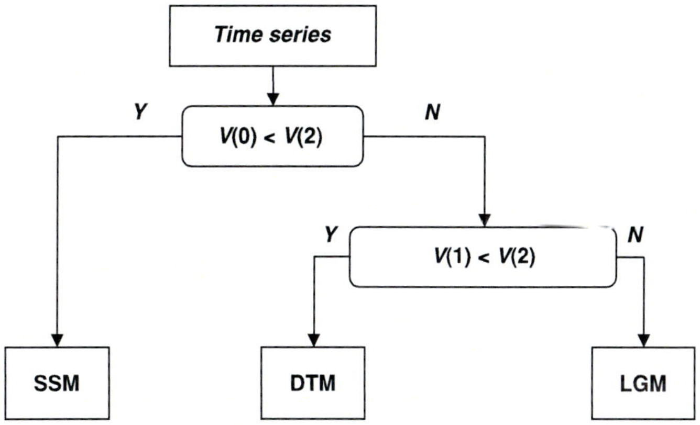

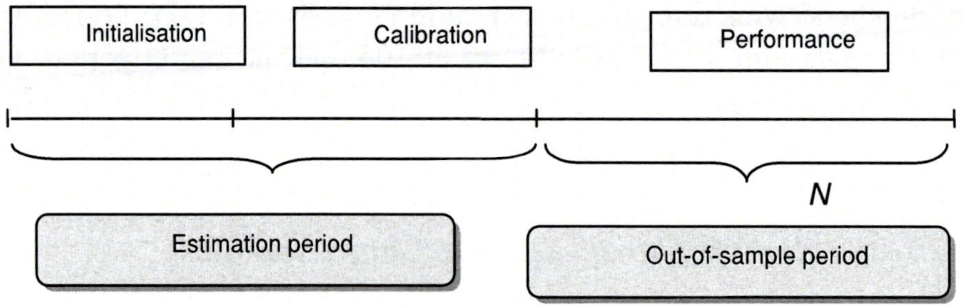

Fig. 3.5 Series splitting design

every combination of smoothing parameters relevant for the method under concern and, finally, the errors are compared. The parameters that produce the minimum error will be chosen as the optimal ones.

As argued by many researchers (e.g. Fildes 1992; Tashman 2000; Armstrong 2001) forecasting methods should be assessed for accuracy using the out-of-sample period rather than goodness of fit to the past data. Therefore, in this study the estimation period will not be used for comparison of the forecasting performances of relevant protocols. During this research, the series will be split as shown in Table 3.2.

3.6.3 Fixed Versus Rolling Forecasting Origin

After selecting a method and optimising its parameters, the actual observations from the out-of-sample period will not be taken into account when selecting and optimising a method. The last observation in the estimation period (Fig. 3.5) is called the forecasting origin. There are two ways of performing an out-of-sample test regarding the forecasting origins: (i) Single (or fixed-origin), and (ii) Multiple (or rolling-origin).

Given the length of the performance period (Fig. 3.5) the single forecasting origin will produce only one forecast for each of the relevant forecasting horizons

Table 3.2 The time series splitting rules

Time series length	Initialisation	Calibration	Performance
14–15	2	6	6–7
16–18	3	6	7–9
19–21	4	6	9–11
22–24	5	7	10–12
25–27	6	12	7–9
28–30	8	12	8–10
31–33	8	13	10–12
34–36	8	16	10–12
37–39	8	19	10–12
>40	8	$n - 8 - 12$	12

At the bottom of the table, n represents the series length

$((P + 1), (P = 2), \ldots, (P = N))$. Therefore, the forecast error value and, consequently, the performance of a given forecasting method, or the protocol in this particular case, will heavily depend on the choice of the starting point. If the selected point is an irregular observation (i.e. outlier) then the errors for every subsequent forecasting horizon might not reflect reality and completely distort the picture of the pattern in the time series, and finally offer a misleading evaluation of the method performance. Therefore, throughout the simulation experiment either with the theoretically generated data or the real data, multiple time origins will be utilised.[1]

3.6.4 Accuracy Measures

In order to report on the accuracy for a given method, the errors are firstly averaged over the length of the performance period and, subsequently, averaged across all series for a given length. Two error measures, the Geometric Root Mean Square Error (GRMSE) and the Mean Absolute Error (MAE), will be employed to express the accuracy of the protocols and universal application of forecasting methods using theoretically generated time series data.

The Geometric Root Mean Squared Error (GRMSE) was first used in forecasting competitions by Newbold and Granger (1974) and is defined as follows:

$$\mathrm{GRMSE} = \sqrt[2n]{\prod_{t=1}^{n} (Y_t - F_t)^2}$$

where Y_t represents the observation at period t, and F_t represents the forecast made for the period t, n is the number of data points in the performance period.

The Mean Square Error (MSE) is scale dependent and sensitive to outliers (Chatfield 1988, 1992). This difficulty is mitigated by the Geometric Root Mean Square Error (GRMSE) and eliminated when used as a relative measure (i.e. calculating the GRMSE of one method to that of another) under an assumption of multiplicative random error terms (Fildes 1992). However, relative measures will not be used in simulation, since the purpose of the experiment is to test the protocols using theoretically generated time series rather than real life series. This error measure is well-behaved and has a straightforward interpretation, (Fildes 1992; Sanders 1997)

Another error measure, considered to be suitable for the simulation experiment, will be the Mean Absolute Error (MAE). An accuracy measure must be robust from one data set to another, and not be unduly influenced by outliers. From a practical point of view, it must make sense, be easily understood, and convey as much information about accuracy as possible. The MAE satisfies all the above conditions. Therefore, the Mean Absolute Error will be employed in order to allow the results achieved in the simulation experiment, concerning the comparison of

[1] A consequence of this choice is that a direct comparison with M3-competition results will not be possible.

the protocols, to be tested on real life time series. The MAE will provide a ground for comparison between the performance of the protocols in the theoretically made environment and real forecasting applications. In the empirical analysis, the MAE will be complemented by the Mean Absolute Percentage Error (MAPE).

3.6.5 The SVC Operational Rules

The first stage in the simulation experiment is to examine the protocol classification performance and to establish the SVC operational rules. Durbin and Watson (1951) statistics are used to test the significance of the autocorrelation of the residuals. Six different lengths 15, 20, 25, 30, 35, and 40 were tested, under 5 and 1% significance levels.[2] The best simulation results for both LGM and DTM models are presented in Figs. 3.6 and 3.7.

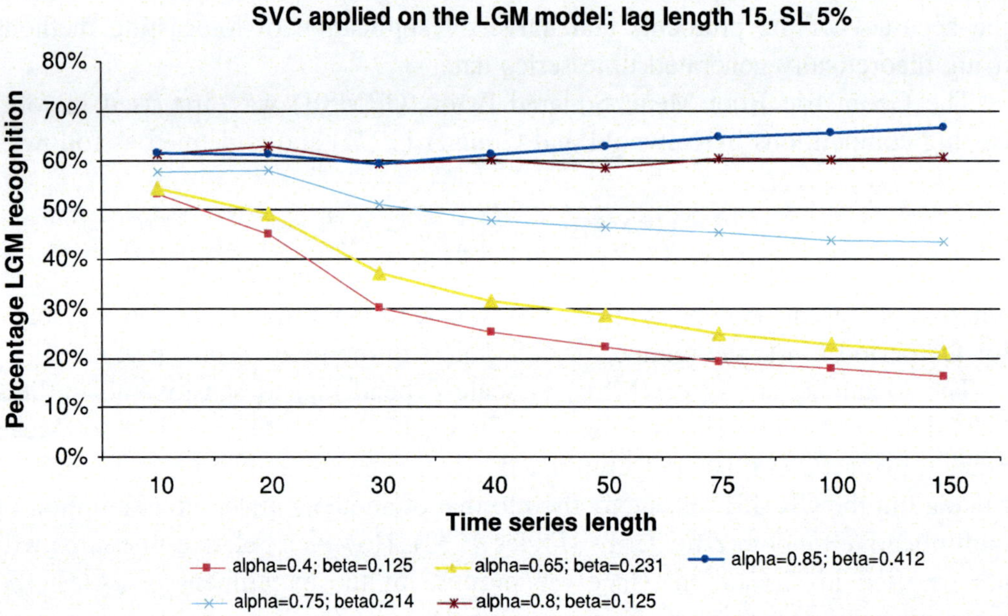

Fig. 3.6 Percentage of the LGM series recognised by SVC protocol for different smoothing parameters, as a function of time series length (lag length 15, 5% significance level)

[2] A drawback of the DW statistic has been taken into account. The Durbin-Watson statistic has a gap between the significant positive autocorrelation, representing the DTM model, and not significant autocorrelation, representing the LGM model. Therefore, if the result belongs to that gap it follows that the DW test is inconclusive. Therefore, an operational rule had to be adopted in order to overcome the problem. There were two possibilities: either to allocate the inconclusive time series to the LGM model or to the DTM model. Having analysed the above issue in both cases during the simulation experiment, it was concluded that the penalty in terms of forecast accuracy is lower, if the inconclusive time series are allocated to the DTM model. Since the LGM model is a special case of the DTM model (for $\phi = 1$), it follows that the LGM could be detected, but not vice versa.

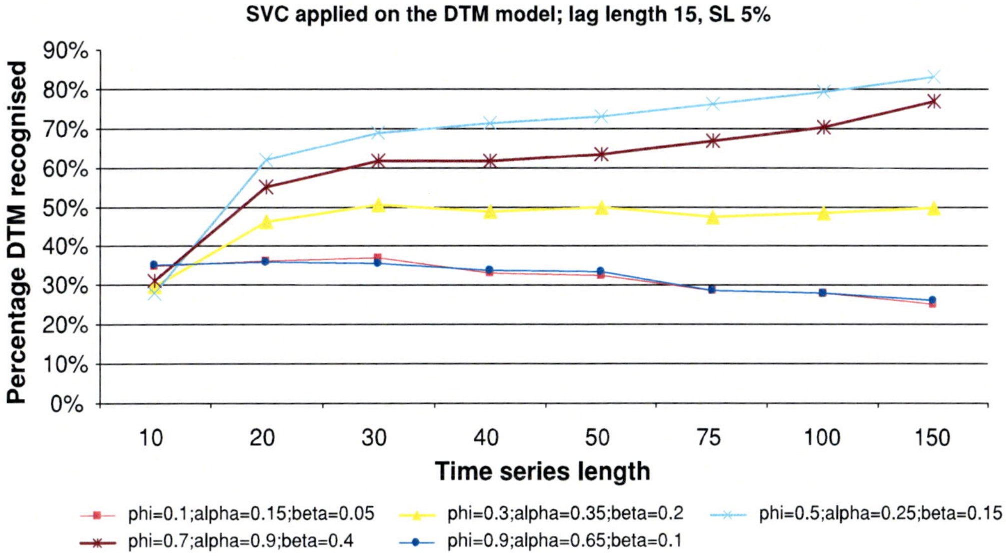

Fig. 3.7 Percentage of the DTM series recognised by SVC protocol for different smoothing parameters, as a function of time series length (lag length 15, 5% significance level)

Five sets of smoothing parameters for the LGM model were tested and the lag length 15 was the best performing one. A significance level of 5% was selected. Although a significance level of 1% for testing the difference between the curve (DTM) and a straight line (LGM) performed well on the LGM time series, recognising a high percentage of series, it was not chosen because it did not perform well on the DTM time series.

From Fig. 3.6, it follows that the SVC protocol detects the LGM model in a higher percentage of cases for higher values of the smoothing parameters, regardless of the length of the time series. According to Fildes et al. (1998) this result is not surprising. On the other hand, for low values of the smoothing parameters the SVC detects the LGM model only for the short time series. For longer series, the LGM model is missed and the DTM model is detected in a high percentage of cases.

From Fig. 3.7, it follows that the SVC protocol with lag length 15 and significance level 5% performs better, in terms of percentage recognition of the DTM time series, for the middle values of the damping parameter ϕ (between 0.3 and 0.7 inclusive) and the series containing more than 30 observations. When using the same combinations of the smoothing parameters and the damping parameter ϕ but using 1% significance level instead of 5%, the DTM time series percentage recognition by the SVC protocol drops heavily for every combination tested in the experiment. Therefore, the 5% significance level will be adopted as a rule for the SVC practical applications.

Based on the above analysis, the operational rules for the SVC protocol derived from the theoretically generated time series can be summarised as follows: (i) The lag length for the SVC fitting should be 15; (ii) The significance level to be employed should be 5%; (iii) The number of observations in the time series should be 30 or more.

The Decision Tree A, designed to use the SVC protocol, will use these operational rules. The original Gardner and McKenzie variance procedure and its modified version, called the $V(0)V(2)$ protocol, as well as the Decision Tree B do not require operational rules for their practical application.

3.6.6 Results of the Simulation Experiment

Having generated 35 data sets and established the operational rules for the relevant protocols, the next stage involves testing the protocols' classification performance. Each protocol will be applied on the time series following the models relevant for the protocol of interest. The percentage recognition of the specific model, and the forecast accuracy of the recommended method will be recorded for comparison purposes.

Single exponential smoothing (SES) needs one parameter to be estimated from the time series in order to generate forecasts. The smoothing parameter α will be selected based on the outcome of the comparison of the MSE errors for the estimation period. The method will be initialised using an average of up to 8 observations from the beginning of data history, and values of the smoothing parameter will be chosen from a range of 0.05 up to 0.95 with a step of 0.05. In this study, Holt's linear and damped Holt's methods will be initialised using simple linear regression, using up to the first eight observations. In testing the performance of the universal application of forecasting methods, the values of the smoothing parameters α and β will be chosen from a range of 0.05 up to 0.95 with a step of 0.05.

Forecasts from 1-, up to 6-periods ahead will be computed and the performance of the forecasting methods and, consequently, selection protocols will be compared for every separate length of the horizons. Finally, throughout this study, minimum n-step ahead MSE will be used during the calibration period in order to select the smoothing parameters for the length of forecasting horizon equals n.

3.6.6.1 Steady State Model (SSM)

This section reports on the protocols' performance when time series follow the Steady State Model (SSM). Five different values of smoothing parameter α were used in order to generate five sets of the SSM time series. The relevant protocols to be applied for the series following the SSM model are the original GM protocol and the $V(0)V(2)$ protocol. Both trees perform the same as the $V(0)V(2)$ protocol, while the SVC protocol is not applicable since it deals with trended series only.

Table 3.3 shows that both protocols perform less well for higher smoothing parameter values. Such series exhibit behaviour that may appear to be trended in the short-term. The declining performance of both protocols as a function of series

Table 3.3 Percentage of SSM series recognised by the protocols

Protocols	Time series length							
	10	20	30	40	50	75	100	150
SSM model, α = 0.1								
$V(0)V(2)$, both trees	91.4	82.4	71.7	61.8	52.3	33.4	21.5	8.4
Original GM	65.8	37.3	20.3	11.4	6.4	1.3	0.3	0
SSM model, α = 0.3								
$V(0)V(2)$, both trees	72.0	47.0	29.4	18.5	11.1	3.4	0.9	0.1
Original GM	39.0	12.1	3.3	0.9	0.2	0.0	0.0	0.0
SSM model, α = 0.5								
$V(0)V(2)$, both trees	67.7	42.2	24.6	14.5	8.0	2.3	0.4	0.0
Original GM	35.9	9.2	2.3	0.7	0.2	0.0	0.0	0.0
SSM model, α = 0.7								
$V(0)V(2)$, both trees	66.7	40.5	23.8	12.6	7.7	1.8	0.4	0.0
Original GM	35.3	8.7	2.2	0.5	0.1	0.0	0.0	0.0
SSM model, α = 0.9								
$V(0)V(2)$, both trees	65.5	39.3	23.0	13.0	7.1	1.6	0.3	0.0
Original GM	34.6	8.8	2.2	0.6	0.1	0.0	0.0	0.0

length is due to the greater tendency for long-term drifts in the mean level to emerge, which are misdiagnosed as trend.

The original GM protocol exhibits poor classification characteristics when detecting series with no trend (Table 3.3) as it chooses the DTM in a high percentage of cases for longer time series (Table 3.4).

Both protocols and the two decision trees were tested for forecasting accuracy (see Table 3.5). Protocols' performance are compared with universal application of the SES method, as it is the best predictor for the series following the SSM model ($\alpha = 0.1$).

Even though the percentage recognition of the adequate mathematical model is identical, Tree B produces more accurate forecasts than Tree A, for all forecasting horizons tested in the experiment and for lengths of data history of 20 observations or more. The main difference is in the models that are identified. According to

Table 3.4 Percentage of SSM series misidentified by the protocols (SSM model, $\alpha = 0.1$)

Protocols	Time series length							
	10	20	30	40	50	75	100	150
Percentage DTM identified								
Tree A	2.7	6.7	13.7	20.6	28.3	41.6	51.6	62.6
Tree B	8.0	17.6	28.3	38.2	47.7	66.7	78.5	91.6
Original GM	33.5	62.7	79.7	88.6	93.6	98.7	99.7	100.0
Percentage LGM identified								
Tree A	5.9	10.9	14.6	17.6	19.4	25.0	26.9	29.0
Tree B	0.7	0.0	0.0	0.0	0.0	0.0	0.0	0.0
Original GM	0.7	0.0	0.0	0.0	0.0	0.0	0.0	0.0

Table 3.5 Mean Absolute Errors for the universal application of the SES method, $V(0)V(2)$ protocol, Decision Trees and the original GM protocol (SSM model, $\alpha = 0.1$)

Protocols	Time series length							
	10	20	30	40	50	75	100	150
MAE 1-step ahead								
SES	13.74	14.18	13.74	13.21	13.02	12.79	12.76	12.76
$V(0)V(2)$	14.16	14.56	14.13	13.53	13.32	13.09	13.02	13.05
Tree A	13.97	14.49	14.05	13.43	13.22	12.96	12.89	12.88
Tree B	14.01	14.36	13.92	13.34	13.14	12.88	12.81	12.79
Original GM	14.33	14.77	14.33	13.57	13.26	12.91	12.81	12.80
MAE 3-step ahead								
SES	17.38	18.37	18.42	18.03	17.80	17.49	17.27	17.21
$V(0)V(2)$	18.55	19.47	19.58	19.20	19.00	18.61	18.44	18.37
Tree A	17.68	19.18	19.28	18.82	18.58	18.09	17.84	17.69
Tree B	17.94	18.74	18.77	18.39	18.15	17.75	17.48	17.35
Original GM	18.44	19.38	19.24	18.69	18.34	17.82	17.49	17.35
MAE 6-step ahead								
SES	–	25.53	22.92	23.12	23.02	22.73	22.51	22.38
$V(0)V(2)$	–	28.46	25.19	25.57	25.64	25.42	25.41	25.34
Tree A	–	27.50	24.58	24.72	24.57	24.07	23.77	23.50
Tree B	–	25.70	23.45	23.73	23.62	23.20	22.89	22.62
Original GM	–	25.41	24.29	24.03	23.75	23.22	22.91	22.62

Table 3.4, Tree A selects the LGM model more often, while Tree B selects the DTM model more often. The better performance of Tree B may be explained by the lower penalty of misdiagnosing an SSM series as DTM instead of LGM. Also, Tree B produces more accurate forecasts than the original GM protocol for the series containing up to 50 observations. For longer time series, the accuracy of these two protocols is very nearly the same, regardless of the length of forecasting horizon.

3.6.6.2 Linear Growth Model (LGM)

Twelve sets of smoothing parameters α and β relevant for Holt's linear method were used to generate time series that exhibit trended behaviour according to the relationships between the method parameters and the LGM model parameters shown earlier in the chapter.

When testing the protocols on the LGM time series with low values of the α parameter, both decision trees, the $V(0)V(2)$ protocol and the original GM protocol detect the SSM model for the shorter time series (Table 3.6). For longer time series, the trees and the original GM protocol detect the DTM model. The results presented in Tables 3.6 and 3.7 confirm the theoretical expectation that the $V(0)V(2)$ protocol will identify series that exhibit strong trend; this is particularly evident for series containing more than 50 observations. The SVC protocol

Table 3.6 Percentage of LGM series correctly identified and misidentified by the protocols (LGM model, $\alpha = 0.4$ and $\beta = 0.125$)

Protocols	Time series length							
	10	20	30	40	50	75	100	150
Percentage SSM identified								
$V(0)V(2)$, both trees	96.9	88.0	68.9	52.2	40.3	24.5	17.5	11.4
Original GM	78.7	47.9	29.3	21.4	16.6	9.5	6.9	4.2
SVC	–	–	–	–	–	–	–	–
Percentage DTM identified								
Tree A	1.0	5.3	20.1	34.4	45.3	60.5	67.3	74.3
Tree B	3.1	12.0	31.1	47.8	59.7	75.6	82.6	88.6
$V(0)V(2)$	0.0	0.0	0.0	0.0	0.0	0.0	0.0	0.0
Original GM	21.2	52.1	70.7	78.6	83.4	90.5	93.1	95.8
SVC	46.8	54.8	69.8	74.6	77.6	80.6	82.0	83.8
Percentage LGM recognised								
Tree A	2.1	6.7	11.0	13.4	14.4	15.1	15.2	16.3
Tree B	0.1	0.0	0.0	0.0	0.0	0.0	0.0	0.0
$V(0)V(2)$	3.1	12.0	31.1	47.8	59.7	75.6	82.6	88.6
Original GM	0.1	0.0	0.0	0.0	0.0	0.0	0.0	0.0
SVC	53.2	45.2	30.2	25.4	22.4	19.4	18.0	16.2

recognises the LGM model for short time series only while for longer series the DTM model is identified.

For higher values of the smoothing parameter α and the same value of β parameter as in the previous table (see Table 3.7), the SVC protocol detects the LGM time series in approximately 60% of cases for all lengths of series, leading to good performance of Tree A.

Decision Tree B and the original GM protocol detect weak trended series instead of strong trended ones, as expected. It is therefore necessary to compare the forecasting performance of the protocols before drawing conclusions regarding their applicability.

From the forecasting perspective, given the GRMSE as the performance measure and taking into account only 1-step ahead forecasting horizon, the lowest forecasting error is produced by the universal application of Holt's linear method regardless of the time series length (see Table 3.8). Furthermore, Table 3.8 shows that the Decision Tree A performs better than the Tree B, while the performance of the original GM protocol depends on the series length.

Similar conclusions hold for 3-steps ahead forecasting horizon (see Fig. 3.8).

3.6.6.3 Damped Trend Model (DTM)

Eighteen sets of the smoothing parameters and the damping parameter ϕ were used to generate series exhibiting damped trend. Decision trees A and B, original GM and the SVC protocols were applied on the time series that follow the DTM model.

Table 3.7 Percentage of LGM series correctly identified and misidentified by the protocols (LGM model, $\alpha = 0.8$ and $\beta = 0.125$)

Protocols	Time series length							
	10	20	30	40	50	75	100	150
Percentage SSM identified								
$V(0)V(2)$, both trees	48.2	23.0	12.0	6.7	3.9	0.9	0.4	0.0
Original GM	22.5	6.1	1.5	0.5	0.1	0.0	0.0	0.0
SVC	–	–	–	–	–	–	–	–
Percentage DTM identified								
Tree A	18.6	28.1	34.8	36.6	39.5	39.0	39.8	39.3
Tree B	50.0	76.9	88.0	93.3	96.1	99.1	99.6	100.0
$V(0)V(2)$	0.0	0.0	0.0	0.0	0.0	0.0	0.0	0.0
Original GM	75.7	93.9	98.5	99.5	99.9	100.0	100.0	100.0
SVC	38.7	37.2	40.6	39.8	41.4	39.5	40.0	39.3
Percentage LGM recognised								
Tree A	33.3	48.9	53.2	56.7	56.6	60.1	59.9	60.7
Tree B	1.8	0.1	0.0	0.0	0.0	0.0	0.0	0.0
$V(0)V(2)$	51.8	77.0	88.0	93.3	96.1	99.1	99.6	100.0
Original GM	1.8	0.1	0.0	0.0	0.0	0.0	0.0	0.0
SVC	61.4	62.9	59.4	60.2	58.6	60.5	60.0	60.7

Table 3.8 GRMSE (1-step ahead) for the universal application of the Holt's method, $V(0)V(2)$ protocol, Decision Trees, the original GM protocol and the SVC protocol

Protocols	Time series length							
	10	20	30	40	50	75	100	150
LGM model, $\alpha = 0.4$, $\beta = 0.125$								
GRMSE 1-step ahead								
Holt's method	297.1	305.4	275.3	264.2	258.7	256.5	255.7	253.6
$V(0)V(2)$	297.7	315.6	296.1	281.6	271.4	264.0	262.0	259.4
Tree A	295.8	315.5	297.0	284.7	277.4	273.2	272.5	270.9
Tree B	296.3	315.7	298.4	286.9	279.9	275.7	274.9	273.2
Original GM	298.6	317.2	298.0	282.5	276.7	274.7	274.5	273.0
SVC	301.0	317.0	290.0	278.1	273.4	272.2	272.1	270.8
LGM model, $\alpha = 0.8$, $\beta = 0.125$								
GRMSE 1-step ahead								
Holt's method	59.0	67.4	64.7	60.6	59.5	58.0	57.9	57.7
$V(0)V(2)$	68.1	72.7	67.7	62.6	61.1	59.4	59.2	59.1
Tree A	68.0	73.5	68.3	63.0	61.6	59.9	59.8	59.6
Tree B	68.1	74.8	69.7	63.7	62.3	60.8	60.7	60.5
Original GM	69.4	77.8	70.7	63.8	62.3	60.8	60.7	60.5
SVC	68.3	74.2	68.6	63.0	61.6	59.9	59.8	59.6

The $V(0)V(2)$ protocol is not applicable since it chooses between the non-trended and trended time series regardless of the trend type. The results for four sets of the parameters are presented in Table 3.9.

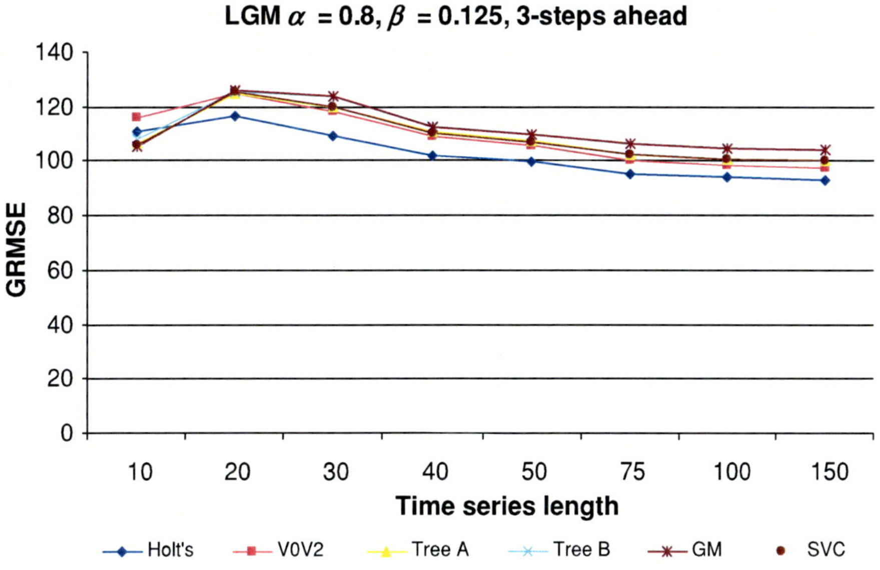

Fig. 3.8 GRMSE (3-steps ahead) for the relevant protocols (LGM model, $\alpha = 0.8$ and $\beta = 0.125$)

Table 3.9 Percentage of DTM series correctly identified by the protocols

Protocols	Time series length							
	10	20	30	40	50	75	100	150
DTM model, $\phi = 0.1$, $\alpha = 0.15$, $\beta = 0.05$								
Tree A	12.6	26.3	31.8	31.0	31.5	28.5	27.9	25.4
Tree B	36.1	67.9	83.8	92.4	96.4	99.4	99.9	100.0
Original GM	66.3	93.1	98.5	99.7	99.9	100.0	100.0	100.0
SVC	35.1	36.4	37.1	33.4	32.6	28.6	27.9	25.4
DTM model, $\phi = 0.3$, $\alpha = 0.35$, $\beta = 0.2$								
Tree A	16.5	42.1	49.1	48.3	49.8	47.6	48.3	49.7
Tree B	49.7	78.8	90.3	94.9	96.9	99.0	99.6	99.9
Original GM	71.6	90.7	94.4	96.4	97.2	99.0	99.6	99.9
SVC	29.8	46.5	50.5	48.7	49.9	47.6	48.3	49.7
DTM model, $\phi = 0.8$, $\alpha = 0.9$, $\beta = 0.4$								
Tree A	24.1	51.0	60.0	60.6	61.7	64.7	68.1	73.6
Tree B	61.7	73.2	72.1	68.5	66.6	62.5	59.3	56.5
Original GM	74.7	76.6	72.7	68.6	66.7	62.5	59.3	56.5
SVC	31.9	52.1	60.2	60.6	61.7	64.7	68.1	73.6
DTM model, $\phi = 0.9$, $\alpha = 0.65$, $\beta = 0.1$								
Tree A	16.0	29.0	32.3	32.5	33.0	28.8	28.0	26.2
Tree B	45.4	77.1	90.0	95.9	98.2	99.8	99.9	100.0
Original GM	71.9	94.7	98.6	99.5	99.8	100.0	100.0	100.0
SVC	35.5	36.0	35.8	33.9	33.6	28.8	28.0	26.2

Table 3.9 shows that Tree B and the original GM protocol selects the DTM time series in a high percentage of cases for the time series containing more than 10 observations and the low values of the smoothing parameter β regardless of the

Table 3.10 GRMSE (1-step ahead) for the universal application of the damped Holt's method, Decision Trees, the original GM protocol and the SVC protocol

Protocols	Time series length							
	10	20	30	40	50	75	100	150
DTM model, $\phi = 0.3$, $\alpha = 0.35$, $\beta = 0.2$								
GRMSE 1-step ahead								
Damped Holt's	29.4	31.6	28.3	26.0	25.6	25.3	25.1	25.0
Tree A	29.4	32.5	29.2	26.8	26.4	26.1	26.0	25.9
Tree B	29.4	31.7	28.6	26.2	25.7	25.3	25.1	25.0
Original GM	30.2	32.5	28.6	26.2	25.7	25.3	25.1	25.0
SVC	29.4	32.1	29.0	26.7	26.4	26.1	26.0	25.9
DTM model, $\phi = 0.9$, $\alpha = 0.65$, $\beta = 0.1$								
GRMSE 1-step ahead								
Damped Holt's	489.8	550.2	506.4	466.8	455.2	448.0	446.5	445.6
Tree A	490.0	556.6	512.7	472.9	460.5	452.3	450.7	450.0
Tree B	492.6	551.1	506.9	466.8	455.4	448.0	446.6	445.6
Original GM	505.2	574.1	510.5	467.4	455.3	448.0	446.6	445.6
SVC	492.8	557.4	510.3	471.5	459.9	452.2	450.7	450.0

value of the damping parameter ϕ and the smoothing parameter α. For the higher β value of 0.4, the percentage recognition by the two protocols still remains high and it is not less than 56% regardless of the series length. In those cases, the Tree A recognises the DTM model for the series containing more than 10 observations.

From the forecasting perspective, using the GRMSE as the error measure and taking into account 1-step ahead forecasting horizon, it follows that the lowest forecasting error is produced by the universal application of damped Holt's method (see Table 3.10), as expected.

Even though the Decision Tree B and the original GM protocol have selected the highest percentage of DTM time series, the difference in forecasting accuracy is almost negligible.

3.7 Real Data Analysis

The purpose of this section is to assess the forecasting performance of the protocols and decision trees for method selection as well as universal application of forecasting methods using empirical data sets gathered from different sources. As noted earlier, it is important to compare forecasting performance with universal application of damped Holt's method.

Since the focus of this chapter is the improvement of forecasting accuracy of the non-seasonal exponential smoothing methods relevant for short-term forecasting, it follows that the length of the forecasting horizon should be considered as an important factor in the empirical analysis. This issue will be addressed accordingly, using 1-, 2-, 3-, 4-, 5- and 6-step ahead horizons for the performance testing.

During the experiment with the real data, single test periods will be used and two error measures will be applied. There is no evidence that multiple test periods can help in deciding which forecasting method produces the lowest error. The method coefficients will be calibrated once, for every time series of interest. Errors are calculated using the Mean Absolute Error (MAE) and the Mean Absolute Percentage Error (MAPE).

3.7.1 Yearly Data

The entire M3-yearly data set has been used for testing the protocols and decision trees since the data are neither slow (intermittent) nor seasonal. The M3-yearly data set contains 645 time series of different lengths. The shortest time series contain only 14 observations and the longest series contain 41 observations. As the time series exhibit different lengths, the splitting rules defined in Sect. 3.6.2 were adopted.

Table 3.11 presents the classification by the protocols and Table 3.12 presents their forecasting performance using the MAPE as the error measure. Table 3.11 shows that the $V(0)V(2)$ protocol, and consequently both decision trees, select 119 time series as non-trended, while the three protocols diagnose different types of trend. Tree B favours the DTM model, while Tree A detects the DTM and LGM models almost equally. The $V(0)V(2)$ protocol, by construction, selects only the LGM model for trended series.

Table 3.12 shows a forecasting accuracy comparison between the three universal methods and the method selection protocols.

Comparing the accuracy of the universal application of forecasting methods among themselves, SES generates the most accurate forecasts for the horizons up to 3-periods ahead, while for the longer horizons damped Holt's is more accurate than SES.

It is evident that Tree B outperforms the original GM procedure for all forecasting horizons. This confirms the results of the simulation experiment discussed in the previous section of this chapter. Tree B is also more accurate than Tree A for all horizons, with Tree A performing less well than the GM procedure for a

Table 3.11 Classification of 645 yearly time series by the protocols

Protocols	Mathematical models		
	SSM	DTM	LGM
Number of series detected			
Tree A	119	271	255
Tree B	119	404	122
Orig GM	45	478	122
$V(0)V(2)$	119	0	526
SVC	0	307	338

Table 3.12 Forecasting performance (MAPE) for yearly data

Protocols	Forecasting horizon					
	1-Step	2-Steps	3-Steps	4-Steps	5-Steps	6-Steps
Universal application of forecasting methods						
SES	12.99	19.18	22.63	25.72	27.90	39.72
Holt's	13.99	22.85	27.40	31.24	36.02	45.41
Damped Holt's	13.59	19.78	22.78	25.38	26.99	38.23
Protocols						
Tree A	12.98	19.32	23.63	26.43	28.98	39.50
Tree B	**12.96**	**19.12**	**22.37**	**24.82**	**26.59**	**38.18**
Original GM	13.14	19.63	22.69	25.26	27.12	38.81
$V(0)V(2)$	12.95	19.56	23.76	26.58	28.99	36.67
SVC	13.78	21.37	25.59	29.08	33.44	44.18

Lowest MAPE values are emboldened

horizon of four periods or more. Overall, Tree B is the best performing method or protocol. This result, shown above for Mean Absolute Percentage Error (MAPE) is also confirmed for Mean Absolute Error (MAE).

The lack of a single universal method that dominates over all forecasting horizons means that Tree B's outperformance over SES and damped Holt's is greatest for different horizons: a gain of 0.63% over damped Holt's for a one-step ahead error, and a gain of 1.54% for a six-step ahead error.

3.7.2 Telecommunication Data

A data set containing non-seasonal telecommunication time series was extracted from the 3,003 M3-competition time series. The series belong to the category "other" regarding the time periods of data collection. The whole data set is homogeneous, in that the time series exhibit strong negative trend.

Table 3.13 summaries the classification results of the M3-telecommunications data set by the protocols employed in this research.

Table 3.13 shows that all the protocols have detected trended series only. The $V(0)V(2)$ protocol selects the LGM model in 100% of cases while the original GM

Table 3.13 Classification of 149 telecommunication time series by the protocols

Protocols	Mathematical models		
	SSM	DTM	LGM
Number of series detected			
Tree A	0	57	92
Tree B	0	146	3
Original GM	0	146	3
$V(0)V(2)$	0	0	149
SVC	0	57	92

variance procedure and Decision Tree B exhibited the same performance in terms of number of series detected (only 2% of time series were detected as strong trended). Furthermore, both the SVC protocol and Decision Tree A have exhibited exactly the same performance and detected 38% of time series as damped trended and 62% as strong trended.

Comparison of the universal application of single methods (see Table 3.14) shows a clear advantage to the damped Holt's method over SES for all forecast horizons. It also shows a slight gain over Holt's method (except for six-step ahead forecasts, where the improvement is more marked).

None of the protocols perform badly, because none of them select SES, for which there may be a strong penalty in forecast accuracy. There is little to choose between the protocols (see Table 3.14), with the Tree B and the GM protocol giving exactly the same result, because neither selects SES for any series. There is a small gain in accuracy of Tree B over Tree A, widening as the forecast horizon lengthens. Tree B is again the best protocol.

Unsurprisingly, these results show that if the damped Holt's method is the best method for almost all series, then there is little to be gained over universal application of that method by employing a method selection protocol. However, should any of the series have been allocated to SES by default, then the use of Tree B, or one of the other protocols, would have been of some benefit.

3.7.3 Weekly Data

Goodrich (2001) noted that in the M3-competition data set, there were no weekly series, "despite their tremendous importance to business, e.g. for materials management and inventory control", (Goodrich 2001: 561). Therefore, a weekly data set is included in this empirical study and the performance of the protocols will be analysis and reported (Table 3.16).

Table 3.14 Forecasting performance (MAPE) for telecommunications data

Protocols	Forecasting horizon					
	1-Step	2-Steps	3-Steps	4-Steps	5-Steps	6-Steps
Universal application of forecasting methods						
SES	1.42	2.60	3.80	5.02	6.29	7.52
Holt's	1.06	1.79	2.56	3.35	4.15	5.22
Damped Holt's	1.04	1.79	2.56	3.30	4.14	4.84
Protocols						
Tree A	1.05	1.80	2.59	3.36	4.37	5.43
Tree B	**1.04**	**1.79**	**2.55**	**3.29**	**4.13**	**4.80**
Original GM	**1.04**	**1.79**	**2.55**	**3.29**	**4.13**	**4.80**
$V(0)V(2)$	1.06	1.79	2.56	3.35	4.15	5.22
SVC	1.05	1.80	2.59	3.36	4.37	5.43

Lowest MAPE values are emboldened

Table 3.15 The classification of 156 weekly time series by the protocols

Protocols	Mathematical models		
	SSM	DTM	LGM
Number of series detected			
Tree A	150	3	3
Tree B	150	6	0
Original GM	104	52	0
V(0)*V*(2)	150	0	6
SVC	0	79	77

Weekly data set contains 156 non-seasonal and non-slow-moving time series. All series are the same length and they contain 125 observations each. Table 3.15 summarises the classification findings of the protocols.

In contrast to the previous dataset, this dataset is dominated by series for which SES is the best method amongst the three smoothing methods. The SVC protocol is of little relevance here, as it does not include SES in its pool of methods. The performance of the protocols and universal application of forecasting methods is presented in Table 3.16.

In this particular data set, the SES is the best performing method from 2- up to 6-periods-ahead forecasting horizon confirming that the trees have selected the non-trended model correctly. Again, none of the protocols perform badly (except the SVC), as they never or rarely select Holt's method, for which the accuracy penalty is greatest.

Unsurprisingly, these results show if Single Exponential Smoothing is the best method for almost all the series, then there is little to be gained over universal application of that method by employing a method selection protocol. However, should any of the series have been allocated to Holt's method by default, then the use of Tree B, or one of the other protocols, would have been of some benefit.

Table 3.16 Forecasting performance (MAPE) for weekly data

Protocols	Forecasting horizon					
	1-Step	2-Steps	3-Steps	4-Steps	5-Steps	6-Steps
Universal application of forecasting methods						
SES	37.76	41.47	40.41	39.80	39.68	39.06
Holt's	38.39	43.18	44.88	44.35	44.46	43.35
Damped Holt's	37.62	41.58	40.60	40.22	40.02	39.42
Protocols						
Tree A	37.77	41.58	40.91	40.06	39.83	39.26
Tree B	**37.75**	**41.47**	**40.39**	**39.79**	**39.68**	**39.06**
GM	37.79	41.67	41.31	39.97	40.38	39.58
V(0)*V*(2)	37.79	41.61	41.00	40.16	39.95	39.36
SVC	37.84	42.54	43.46	42.70	42.44	41.44

Lowest MAPE values are emboldened

3.8 Conclusions

Evidence is lacking on the comparison of decision trees for model selection with the universal application of damped Holt's method. The M3 competition shows that this method is difficult to beat. This chapter has presented results comparing decision trees with the damped Holt's method, as an ‚encompassing approach'. This serves as a base for further studies, which should include other approaches such as prediction validation and information criteria, discussed in Sect 3.3.

The simulation study showed that the Steady State Model may be detected more accurately by the *V*(0)*V*(2) protocol (and, hence, by both decision trees) than by the original GM protocol, although the performance of all protocols declines as the length of series increases. Tree B's forecasting performance is better than Tree A's in this case, because when it fails to recognise that data is non-trended, it is more likely to allocate the series to the damped Holt's method than Holt's linear method.

The simulation study also showed that the Linear Growth Model (LGM) may be detected more accurately by the *V*(0)*V*(2) protocol than the GM protocol. In this case, Tree A performs better than Tree B in the sense of recognising more LGM series. The best protocol, in terms of forecasting accuracy, is the *V*(0)*V*(2) method, as it does not misclassify trended series as damped trend. There is little to choose between Tree A and Tree B forecast accuracy, with Tree A having a slight edge. For damped trend series, the simulation study showed that the original GM protocol classified the most series correctly, with Tree B being the next best classification method. However, the difference in forecasting performance is almost negligible.

Empirical analysis of yearly data from the M3 competition shows that it is possible for a protocol to give greater accuracy than universal application of the damped Holt's method. The weekly series confirm a small gain in accuracy for Tree B. However, the telecommunications dataset shows almost identical accuracy results between Tree B and damped Holt's, as this method is diagnosed for almost all series by Tree B. The decision tree produces no accuracy benefit in this case, but neither does it yield any deterioration in performance, thus demonstrating its robustness.

The M3 yearly data also show Tree B identifying series as damped trend much more frequently than Tree A. For this dataset, Tree B gives greater forecast accuracy than both Tree A and the original GM protocol. Although the benefits are not large, they are consistent across forecasting horizons, with a general tendency for Tree B to have greater accuracy over longer horizons. For the telecommunications data, Tree B and the GM protocol give the same results, which are better than Tree A, particularly for longer horizons. The most natural explanation of this difference is the tendency of Tree A to mis-specify damped trend as linear trend. This tendency was identified in the simulation study, and it was shown that it penalises forecast accuracy. These conclusions are restricted to non-seasonal datasets. More research is needed on extending the approach to seasonal series, including a detailed analysis of the Holt-Winters method.

Overall, this study shows that Tree B is a promising approach for diagnosing trend in non–seasonal time series. It can be implemented quite straightforwardly in forecasting applications for service parts. As it is simple to understand and apply, and produces robust accuracy results, this decision tree should be considered as a potential approach to forecasting method selection along with other more sophisticated approaches.

References

Adya M, Collopy F, Armstrong JS, Kennedy M (2001) Automatic identification of time series features for rule-based forecasting. Int J Forecast 17:143–157

Akaike H (1974) A new look at statistical model identification. IEEE Trans Autom Control 19:716–723

Armstrong JS (2001) Extrapolation. In: Armstrong JS (ed) Principles of forecasting: a handbook for researchers and practitioners. Kluwer, Norwell

Assimakopoulos V, Nikolopoulos N (2000) The theta model: a decomposition approach to forecasting. Int J Forecast 16:521–530

Atanackov N (2004) Trend forecasting by constrained optimisation and method selection protocols. PhD thesis, Buckinghamshire Chilterns University College, Brunel University

Billah B, Hyndman RJ, Koehler AB (2005) Empirical information criteria for time series forecasting model selection. J Stat Comput Simul 75:830–840

Box GE, Jenkins GM (1970) Time series analysis, forecasting and control. Holden-Day, San Francisco

Boylan JE, Syntetos AA (2008) Forecasting for inventory management of service parts. In: Kobbacy KAH, Murthy DNP (eds) Complex system maintenance handbook. Springer, London

Chatfield C (1988) Apples, oranges and mean square error. Int J Forecast 4:515–518

Chatfield C (1992) A commentary on error measures. Int J Forecast 8:100–102

Collopy F, Armstrong S (1992) Rule-based forecasting: Development and validation of an expert systems approach to combining time-series extrapolations. Manage Sci 38:1394–1414

Commandeur JJF, Koopman SJ (2007) An introduction to state space time series analysis. Oxford University Press, Oxford

Durbin J, Watson GS (1951) Testing for serial correlation in least squares regression. Biometrika 38:159–177

Fildes R (1992) The evaluation of extrapolative forecasting methods. Int J Forecast 8:81–98

Fildes R, Hibon M, Makridakis S, Meade N (1998) Generalising about univariate forecasting methods. Int J Forecast 14:339–358

Fortuin L (1980) The all-time requirements of spare parts for service after sales–theoretical analysis and practical results. Int J Oper Prod Manage 1:59–69

Gardner ES Jr (1999) Note: rule-based forecasting vs damped trend exponential smoothing. Manage Sci 45:1169–1176

Gardner ES Jr (2006) Exponential smoothing: the state of the art, Part II. Int J Forecast 22:637–666

Gardner ES Jr, McKenzie E (1985) Forecasting trends in time series. Manage Sci 31:1237–1246

Gardner ES Jr, McKenzie E (1988) Model identification in exponential smoothing. J Oper Res Soc 39:863–867

Goodrich RL (1990) Applied statistical forecasting. Business Forecast Systems, Inc, Belmont

Goodrich RL (2001) Commercial software in the M3-competition. Int J Forecast 17:560–565

Hannan EJ, Quinn BG (1979) The determination of the order of an autoregression. J R Stat Soc Ser B 41:190–195

Harrison PJ (1967) Exponential smoothing and short-term sales forecasting. Manage Sci 13:821–842

Harvey AC (1984) A unified view of statistical forecasting procedures. J Forecast 3:245–283
Harvey AC (2006) Forecasting with unobserved components time series models. In: Elliott G, Granger CWJ, Timmermann A (eds) Handbook of economic forecasting, vol 1. Elsevier, Amsterdam
Holt CC (1957) Forecasting seasonals and trends by exponentially weighted moving averages. ONR memorandum, 52. Carnegie Institute of Technology, Pittsburgh, PA
Holt CC (2004a) Forecasting seasonals and trends by exponentially weighted moving averages. Int J Forecast 20:5–10
Holt CC (2004b) Author's retrospective on ‚Forecasting seasonals and trends by exponentially weighted moving averages. Int J Forecast 20:11–13
Hurvich CM, Tsai CL (1989) Regression and time series model selection in small samples. Biometrika 76:297–307
Hyndman RJ, Billah B (2003) Unmasking the Theta method. Int J Forecast 19:287–290
Hyndman RJ, Koehler AB, Ord JK, Snyder RD (2008) Forecasting with exponential smoothing. the state space approach. Springer, Berlin
Makridakis S, Hibon M (2000) The M3-competition: results, conclusions and practical concerns. Int J Forecast 16:451–476
Makridakis S, Wheelright SC, Hyndman RJ (1998) Forecasting: methods and applications, 3rd edn. Wiley, New York
Makridakis S, Assimakopoulos V, Pagourtzi E, Bougioukos N, Petropoulos F, Nikolopoulos K (2008) PYTHIA: an expert forecasting support system. In: Paper presented at the 28th international symposium on forecasting, Nice, France
Meade N (2000) Evidence for the selection of forecasting methods. J Forecast 19:515–535
Newbold P, Granger CWJ (1974) Experience with forecasting univariate time series and the combination of forecasts. J Roy Stat Soc A, Series A 137:131–165
Pegels CC (1969) Exponential forecasting: some new variations. Manage Sci 12:311–315
Roberts SA (1982) A general class of Holt–Winters type forecasting models. Manage Sci 28:808–820
Sanders N (1997) Measuring forecasts accuracy: some practical suggestions. Prod Invent Manage J (First Quarter), pp 43–46
Schwarz G (1978) Estimating the dimension of a model. Ann Stat 6:461–464
Shah C (1997) Model selection in univariate time series forecasting using discriminant analysis. Int J Forecast 13:489–500
Tashman L (2000) Out-of-sample tests of forecasting accuracy: an analysis and review. Int J Forecast 16:437–450
Tashman LJ, Kruk JM (1996) The use of protocols to select exponential smoothing procedures: a reconsideration of forecasting competitions. Int J Forecast 12:235–253
Taylor JW (2003) Exponential smoothing with a damped multiplicative trend. Int J Forecast 19:715–725
Theil H, Wage S (1964) Some observations on adaptive forecasting. Manage Sci 2:189–206
Vokurka RJ, Flores BE, Pearce SL (1996) Automatic feature identification and graphical support in rule-based forecasting: a comparison. Int J Forecast 12:495–512

Chapter 4
The Impact of Aggregation Level on Lumpy Demand Management

Emilio Bartezzaghi and Matteo Kalchschmidt

4.1 Lumpy Demand Management

In the last 20 years companies have always paid great attention on managing demand variability. Demand fluctuations are due to several reasons: quick changes in the final customer's preferences and taste are a common cause of demand variability (e.g., in the fashion industry demand for a given color can change dramatically from year to year). Marketing activities may also lead demand to suddenly change e.g., when promotional activities are conducted due to the high elasticity of demand to price. Competitors can also be a source of variability, since their behavior can influence how the demand distributes on each single company serving a specific market. The supply chain structure is also a significant cause of demand unsteadiness: the bullwhip effect (Lee et al. 1997) is a common phenomenon in different industrial contexts, leading to an increase in the variability of the demand over supply chain stages.

A vast amount of the literature has addressed the issue of designing managerial systems capable of coping with demand variability. This has been done by focusing on different leverages: from demand forecasting, aiming at increasing the capability of companies to understand variability, to production planning, trying to design efficient planning systems, capable of reacting towards sudden changes in the final demand, to inventory management, in order to manage the complex trade-off between inventory cost and service level, and so on.

E. Bartezzaghi
Department of Management, Economics and Industrial Engineering,
Politecnico di Milano, Milano, Italy
e-mail: emilio.bartezzaghi@polimi.it

M. Kalchschmidt (✉)
Department of Economics and Technology Management,
Università degli Studi di Bergamo, Bergamo, Italy
e-mail: matteo.kalchschmidt@unibg.it

N. Altay and L. A. Litteral (eds.), *Service Parts Management*,
DOI: 10.1007/978-0-85729-039-7_4,

This issue is common to all industrial contexts; however, a rather peculiar and complex situation is faced in the case of spare parts. The problem of managing spare parts demand is relevant for many reasons: first of all it influences the final product business since it affects post sale service quality. Moreover, it is a relevant business as the market is captive, thus very profitable and so firms have to pay relevant attention towards this issue. However, it is a very difficult business to cope with, since requirements are usually very dispersed over time and demand uncertainty is frequently very high.

Spare parts, in fact, often show very sporadic demand patterns for long periods of their life time. This is the case, for example, of service items that have to be stored for years as long as repair service has to be guaranteed even for products that have reached the end of their market life. Spare parts demand often tend to be highly variable and sporadic showing frequently a very peculiar pattern called *lumpy demand*.

Lumpy demand can be defined as (Bartezzaghi et al. 1999):

- variable, therefore characterized by relevant fluctuations (Wemmerlöv and Whybark 1984; Wemmerlöv 1986; Ho 1995; Syntetos and Boylan 2005);
- sporadic, as historical series are characterized by many days with no demand at all (Ward 1978; Williams 1982; Fildes and Beard 1992; Vereecke and Verstraeten 1994; Syntetos and Boylan 2005);
- nervous, thus leading to show differences between successive demand observation, so implying that cross time correlation is low (Wemmerlöv and Whybark 1984; Ho 1995; Bartezzaghi and Verganti 1995).

Managing inventories when demand is lumpy is thus a complex issue since companies have to cope with both a sporadic pattern, that usually induces high inventory investments, and highly variable order size, that make it difficult to estimate inventory levels and may affect service levels. For this reason, companies facing lumpy demand often experience both high inventory levels and unsatisfactory service levels at the same time.

Lumpiness may emerge as the consequence of different structural characteristics of the market. In particular, we may refer to the following main factors (Bartezzaghi et al. 1999):

- low number of customers in the market. Fewer customers usually induce sporadic requests for the product unit and, therefore, demand lumpiness increases;
- high heterogeneity of customers. Heterogeneous requests occur when the potential market consists of customers with considerably different sizes or buying behaviors (i.e. customers that order for very different lot sizes or with different frequencies); thus the higher the heterogeneity of customers, the higher the demand lumpiness;
- low frequency of customers requests. The higher the frequency of requests from a customer, the higher the number of different customers that ask for the unit in a given time bracket. Thus lumpiness increases as the frequency of each customer's purchase decreases;

- high variety of customers requests. Demand lumpiness increases also if each single customer has a variable reorder behavior over time. Customers may change significantly their buying behavior in specific periods of time due, for example, to promotional activities or to speculative buying;
- high correlation between customers requests. Lumpiness may occur also because customers' requests are strongly correlated with each other. Correlation, for example, may be due to imitation and managerial fads which induce similar behaviors in customers so that sudden peaks of demand may occur after periods of no requests.

Spare parts demand often shows this specific kind of variability. This is mainly due to the low frequency of requests for these items. In fact spare parts often tend to behave as "slow items" since requests for them are distributed over a long period of life. Besides, requests may change significantly between orders due to the fact that several different kinds of customers may be served by a single service unit. This is often the case when spare parts are ordered from independent units that provide the final customer with repair services. In this situation, reorder sizes are influenced by the specific reorder criteria adopted by each service provider. As a matter of fact, most of the contributions on lumpy demand management have specifically taken into consideration the spare parts case (see for example Croston 1974; Petrović et al. 1988; Cobbaert and Van Oudheusden 1996; Shibuya et al. 1998; Liu and Shi 1999; Willemain et al. 2004).

Due to its relevant impact on companies' performance, lumpy demand management has received major attention in the current literature. Specifically the literature has provided several approaches (i.e. forecasting methods and inventory models) to cope with this kind of demand. Also the specific case of spare parts has been taken into account and specific methodologies have been provided to cope with demand variability in this particular case and proposing different models to improve inventory performance (some of the works in these field are: Petrović et al. 1988; Cohen and Kleindorfer 1989; Cobbaert and Van Oudheusden 1996; David et al. 1997; Dekker et al. 1998; Shibuya et al. 1998; Liu and Shi 1999; Teunter and Fortuin 1999; Kalchschmidt et al. 2003; Syntetos and Boylan 2005).

Literature on lumpy demand management has mainly focused on methods to better evaluate demand variability (i.e. forecasting methods; e.g., Syntetos and Boylan 2005) and inventory models specific to this particular case (e.g., Teunter and Fortuin 1999). However these models usually don't address the problem of the aggregation level of data. This problem arises when the implementation of specific techniques takes place. In fact, when trying to implement forecasting and inventory models, practitioners often find out that this is much more complex than the simple design or selection of an appropriate algorithm and it involves the choice of the relevant pieces of information, the design of information systems, the control of data quality, and the definition of managerial processes. One critical issue concerning the implementation and adoption of forecasting and inventory management techniques is the choice of the appropriate level at which demand has to be managed. In particular, demand has to be defined over three dimensions:

(1) One shall define the market he/she tries to forecast; e.g., one retailer might want to forecast demand at the single store level, while a manufacturer might be interested in the demand for the overall region or country; clearly the former forecasting problem is harder to tackle than the latter;
(2) One shall define the product the demand refers to; e.g., for a given retailer it might be fairly difficult to predict the demand for a given product at the style-color-size-packaging level, whereas forecasting the total turnover for a given product category might not be that hard (Wacker and Lummus 2002);
(3) Finally one needs to define the time frame of the forecasting problem, i.e., one shall define the time bucket and the forecasting horizon; indeed forecasting demand at the day level is much more complex than forecasting total yearly demand.

The choice of the aggregation level is important since according to the specific aggregation level chosen, demand variability may show specific peculiarities and thus different techniques may apply, thus affecting forecasting and inventory performances.

In the remainder of this work we will refer to the previous three dimensions as the *level of aggregation* of the forecasting problem. The smaller the market, the more detailed the definition of the product and the smaller the time bucket, the more the forecasting problem is detailed.

This work focuses on the impact of the aggregation level of data on inventory performance and we address in particular the specific case of lumpy demand. In fact limited contributions can be found regarding how the aggregation level may influence lumpy demand.

The remainder of this paper is thus structured as follows. In the next section the level of aggregation of demand will be described and literature contributions on this issue will be summarized. Then specific research objectives and methodology will be described. In the last two sections, empirical results will be described, a proper discussion of results will be provided and future research objectives will be highlighted.

4.2 The Impact of Data Aggregation

As previously mentioned, when a forecasting problem has to be addressed, it is important to clearly state at which level of aggregation a forecast has to be provided. Typically this choice relates to the specific decision the forecast will be used for. For example, when yearly financial budget is under consideration, a company usually doesn't need a very detailed forecast: a forecast for some future months of sales at market level is going to be enough for this specific decision making process. On the contrary, if inventory decision is under consideration, probably a company will need to provide a forecast at the Stock Keeping Unit (SKU) level, for the next future days or weeks and regarding the part of the market that the specific warehouse is serving.

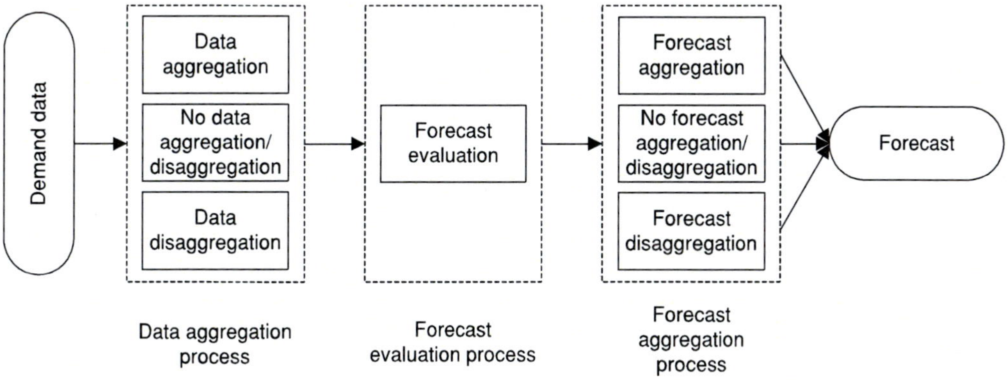

Fig. 4.1 Aggregation level options in the forecasting process

However, the level of aggregation at which the forecast has to be used (at thus provided) is not necessarily the same of the level of aggregation at which the forecast is evaluated. In particular, during the forecasting process, we face the problem of aggregation in two different situation. First, when information (i.e., past demand data) is collected, one has to choose at which level of aggregation these data should be used (we refer to this as the *data aggregation process*). Based on what companies choose here, different forecasting methods may then be selected, based on the characteristics of variability the demand shows at that level of aggregation (e.g., if monthly data is used then seasonality may be an important variability component to be considered; on the contrary if the same data is used at the yearly level seasonality becomes irrelevant, at least for the forecast evaluation process). Second, when the forecast has been evaluated it may need to be aggregated or disaggregated in order to provide the final forecast at the required aggregation level that the decision making process needs (e.g., if a market forecast is needed for budget purposes, if we evaluate forecast at customer level we then need to aggregate all these forecasts by simply summing up them). We refer to this as the *forecast aggregation process*. Figure 4.1 summarizes these different situations. In this work we will focus on the data aggregation process.

As Fig. 4.1 exemplifies, given a certain aggregation level at which a forecast is needed, one can decide to obtain this forecast though a more aggregate forecast (thus a disaggregation process is needed to provide the final forecast), through a more disaggregated process (thus an aggregation is needed) or without any aggregation or disaggregation. Forecast evaluation can be performed at different data aggregation levels. Data in fact can be aggregated before evaluating the forecast or disaggregated (e.g., in the retail industry frequently companies estimate the demand in different ways according to whether data collected on demand refer to promotional periods or not).

Based on these options companies may then structure their forecasting process differently. For example, one typical solution is when the forecast is evaluated at the same level of aggregation at which the forecast is used. In this situation, based on the aggregation level requested from the decision making process, the forecasting approach is selected according to data or information available, and no aggregation or disaggregation of the data is done. Another possibility is when the forecast is evaluated at a more detailed level compared to which the forecast is used. In this case, the forecast is evaluated based on disaggregated data and then some aggregation of these is conducted before the final forecast is provided. This is often referred to as *Bottom-Up* approach (Orcutt et al. 1968; Zellner and Tobias 2000; Weatherford et al. 2001). A typical example of this situation is when sales budget are developed: usually sales people are asked to provide their own forecast regarding their specific geographical area or regarding their specific customers. These forecasts are then aggregated all together to get to an overall market forecast. Another common situation is when the forecast is evaluated at a more aggregate level compared to which the forecast is used. In this situation, the data collection and the forecast evaluation is done at a more aggregate level compared to the one at which the forecast is then provided. This is often referred to as *Top-Down* approach (Theil 1954; Grunfeld and Griliches 1960; Lapide 1998). For example, this is often used when a weekly forecast has to be divided at the daily level for production scheduling purposes, based on some estimation of the daily seasonality.

As previously mentioned, this choice has to be done on one or more of three different dimensions:

- *Product:* companies have to choose at which level of detail of the product structure they want to evaluate forecasts. The demand can be foreseen by leveraging on very detailed data (e.g., referring to SKU) or very aggregate ones (e.g., referring to product families).
- *Market:* the demand can be foreseen by taking into consideration very precise and detailed information (e.g., the demand of each single customer) or very aggregate one (e.g., the demand at market level).
- *Time bucket:* in the end, companies have to choose whether they want to rely on detailed or aggregate time buckets: demand can be foreseen at daily level, weekly level, monthly level, etc.

The literature on demand management and forecasting has devoted some attention to the problem of choosing the proper level of aggregation. Some contributions on this issue focus on the use of aggregation to estimate seasonality curves (Dalhart 1974; Withycombe 1989; Bunn and Vassilopoulos 1993, 1999; Dekker et al. 2004). These works provided evidence that aggregating correlated time series can be helpful to better estimate seasonality since it can reduce random variability. Other works focus on the selection of the proper level of data aggregation (e.g., Chan 1993; Gonzales 1992; Weiss 1984). Some authors argue that the top-down approach (i.e., evaluating forecast at aggregate level and then dividing it at detailed level) can be helpful as it is more efficient and more accurate in times of stable demand (Theil 1954; Grunfeld and Griliches 1960; Lapide 1998). Other authors, however, reply that the bottom-up approach (i.e., building a forecast by evaluating the forecasts at detailed level and then aggregating them) is needed when there are differences across time series (Orcutt et al. 1968; Zellner and Tobias 2000; Weatherford et al. 2001). Finally, other contributions (Miller et al.

1976; Barnea and Lakonishok 1980; Fliedner 1999) take a more contingent approach and show that the choice between the aggregate and detailed approach depends on the correlation among time series. Zotteri and Kalchschmidt (2007) analytically demonstrate that in fact aggregation is preferable only under certain circumstances (i.e., high demand variability, few periods of demand, etc.).

Limited contributions can, however, be found regarding aggregation in the case of lumpy demand. Specifically, several of the mentioned contributions considered frequently the case of stationary and continuous demand. Unfortunately this is not always the case: spare parts usually show a lumpy pattern and it is not completely clear whether literature findings still hold here.

Similarly, contributions on the aggregation level selection have mainly focused on forecasting, i.e., the impact of aggregation on forecasting accuracy. Limited contributions have consider simultaneously the impact on forecasting and inventory management systems. In fact, literature on lumpy demand management has argued and, sometimes, proved that designing an integrated forecasting and inventory management system may be much more beneficial than focusing on just one of the two (Kalchschmidt et al. 2003). In this situation the forecasting method applied has to focus on estimating demand characteristics that the chosen inventory system needs to define reorder politics.

This work aims at providing a better understanding of how aggregation may influence inventory performance when demand is lumpy. In particular here attention is devoted to the case of aggregation over time (temporal aggregation). This choice is due to the fact that limited contributions can be found on this specific issue (also in the stationary and stable demand case). As a matter of fact, the vast amount of contributions on this topic usually refer to aggregation over product and over market dimensions (see previously mentioned contributions), while only limited contributions can be found regarding temporal aggregation for non-lumpy demand (some contributions can be found in Johnston and Harrison 1986; Snyder et al. 1999; Dekker et al. 2004) and very few specific to the lumpy demand case (Nikolopoulos et al. 2009). For all these reasons, in the reminder of the paper only temporal aggregation will be considered.

4.3 Objectives and Methodology

The goal of this work is to study whether temporal aggregation of lumpy demand may be beneficial in terms of impact on inventory performances. Specifically, the objectives are:

(1) Analyze the impact of temporal aggregation level in a specific situation, namely spare parts demand.
(2) Evaluate the impact of demand characteristics (e.g., lumpiness) on the choice of the proper level of aggregation.

In order to achieve these goals a simulation analysis based on real demand data has been considered. Demand data has been collected from the Spare Parts

Table 4.1 Descriptive statistics for the considered sample

	Demand		Demand interarrival		Demand size		Order frequency
	Average (units)	CV	Average (days)	CV	Average (units per order)	CV	(No. of days with nonzero demand)
Min	0.01	0.86	1.00	0.00	1.00	0.00	2
25% Ile	0.03	3.89	7.21	0.66	1.01	0.00	4
Median	0.10	6.19	18.34	0.84	1.81	0.64	9
75% Ile	0.44	8.46	40.20	1.01	4.00	1.20	28
Max	53.44	13.58	104.50	4.09	269.67	4.84	209

Management Division of a major multinational white goods manufacturer. The company provided us with daily level demand data for all its spare parts SKUs over almost 1 year period (specifically 209 working days). The company manages more than 68.000 SKUs; among these, almost 52.000 have less than two orders per year. We decided to focus only on those items that presented at least two orders over the available data set. Then we selected from 16.875 SKUs that guaranteed this requirement a sample of 1.000 SKUs chosen at random. Among these, 926 SKUs at the end were considered (some SKUs were omitted due to item specific problems, such as item recoding or termination).

This data set was divided in two samples: the first one, based on the first 105 days was allocated for fitting purposes of the selected model, while the second one (based on the remaining 104 days) was used for testing performance. Table 4.1 synthesizes some descriptive statistics on the overall sample.

As it can be noted the demand variability is quite high (median Coefficient of Variation (CV) is above 6). This is due to both variability in the demand size (median CV of demand size is 0.64) but also to the demand intermittency (median number of days with non zero demand is 9 out of 209 days of demand). Thus, coherently with our definition of lumpy demand and with previous contributions (Syntetos and Boylan 2005), we can conclude that considered data has in fact a lumpy pattern (both size is variable and demand is intermittent).

To achieve our research goals, we based our analyses on a simulation model with the following characteristics:

(1) In order to estimate inventory levels, we adopted Syntetos and Boylan (S&B) unbiased variation of Croston's method (Syntetos and Boylan 2001). We selected this forecasting approach since it is actually considered a reliable method for the case of lumpy demand compared to other known solutions (Syntetos and Boylan 2001, 2005; Altay et al. 2008). This approach reduces the bias in the estimation of the average demand in case of lumpy pattern. We refer to Syntetos and Boylan (2001) for a detailed description of the approach and to Kalchschmidt et al. (2003) and Syntetos and Boylan (2005) for comparisons with other methodologies in the case of lumpy demand. The smoothing parameters were set at 0.2 and the C parameter was set to 200 (see cited papers for details on the model). Some tests were also run with other

values for these parameters. Even if differences arise when parameters change, these do not affect significantly the results of our analyses. For this reason and for briefness sake we omitted these analyses here.

(2) The safety stock is defined according to the actual demand variability and the desired service level. Specifically we simulated different scenarios according to different average service levels i.e., 94% (the average service level the company was achieving) and 99.1% (the desired service level the company was aiming to have). We considered these two levels of performance since they represent what the company considered as reference. For briefness sake we will show the results only for the former case.

(3) The reorder model considered is an order-up-to system with daily revisions of inventory levels; backlog is allowed.

(4) Deliveries from suppliers are assumed constant and equal to 20 days for all items. The company based its own reorder politics on this specific value. We argue that according to the specific lead time suppliers provide, the impact of the aggregation level on inventory performances may change. However, we claim that the considerations we draw from our analyses are not affected explicitly from this specific assumption. We discuss this issue in deeper detail in the conclusions.

Each day of the simulation we update model parameters and evaluate inventory performance in terms of inventory levels and service level (i.e., served quantity compared to actual demand). If inventory is not enough to fulfill daily demand a backlog is accounted and demand is served as soon as inventory is available.

Simulations were run according to five different aggregation levels of demand data. Specifically we considered aggregation on a 1 day level (1d, data as it is), a 2 days level (2d, demand is aggregated between 2 subsequent days), 3 days level (3d), 10 days level (10d) and 30 days level (30d). Other intermediate aggregation levels were run but here they are omitted for sake of brevity. Since the performances of the systems under investigation are based on two objectives (service level and inventory level), in order to compare the different scenarios we run all simulations so to guarantee a 94% service level on the average of the test period. We then can directly compare inventory levels to identify the impact of the data aggregation process.

4.4 Simulation Results and Discussion

Figure 4.2 shows the average inventory level of the considered items on the testing period for different aggregation levels.

As it can be noted, the average inventory level required to guarantee a 94% average service level reduces as we aggregate demand data. The extent to which inventories benefit from aggregation is impressive (in particular when comparing the more detailed levels with the more aggregate ones) and the benefit of further aggregation tends to reduce on higher horizons. In order to verify that these

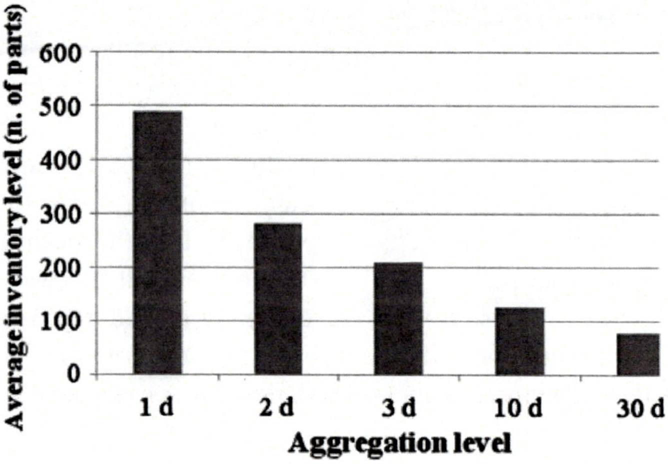

Fig. 4.2 Average inventory levels for the different temporal aggregation levels considered (average service level is 94.1%)

average results were consistent at SKUs level (and to avoid eventual bias due to few peculiar cases, e.g., high volume skus) we ran nonparametric tests on the equality of average inventories between the different simulation runs (we based our analyses on Friedman's test[1]). All tests were significant at 0.99 level, thus we can claim than on a relevant portion of our SKUs, aggregating demand improves inventory performances.

Even if on average the temporal aggregation seems to pay off, a more detailed analysis showed that this is not true for all items. Table 4.2, in fact, highlights that some items don't benefit from aggregation but, on the contrary, face a worsening of the inventory level. As we can see, among the considered SKUs, on average almost 22% show worse performance when demand is aggregated, while almost 9% on average are not affected by the aggregation level.

The identified phenomenon seems to apply differently on the items considered, thus we take a contingent approach to identify what are the key drivers that influence the optimal aggregation level. In order to identify discriminant contingent factors we ran multiple comparisons among three groups of items (namely those for which aggregation improves performance, those were aggregation is indifferent and those where aggregation lead to worse performance) for all the considered aggregation levels. *T* tests on the equality of means were run among the different groups on the following variables[2]:

- Average demand;
- Standard deviation of demand;
- Coefficient of variation of demand;
- Asymmetry of demand;

[1] The Friedman test is the nonparametric equivalent of a one-sample repeated measures design or a two-way analysis of variance with one observation per cell. Friedman tests the null hypothesis that k related variables come from the same population. For each case, the k variables are ranked from 1 to k. The test statistic is based on these ranks.

[2] For space sake we omit all statistical analyses. All contingencies have been evaluated at daily level since this was the most detailed level available.

Table 4.2 Distribution of SKUs for different aggregation levels, classified according to whether they improve performance by aggregating demand, stay the same, or worsen

	From 1 to 2 days	From 2 to 3 days	From 3 to 10 days	From 10 to 30 days
Improvement	817	529	711	517
Indifference	59	80	76	105
Worsening	50	317	139	304
Total	926	926	926	926

- Lumpiness of demand: lumpiness has been measured according to the following expression:

$$\text{Lumpiness} = \frac{CV^2}{\mu \cdot LT}$$

where CV is the coefficient of variation of demand, μ is the average demand and LT is the replenishment lead time;
- Number of days with non zero demand;
- Average size of demand;
- Standard deviation of demand size;
- Coefficient of variation of demand size;
- Average interarrival;
- Standard deviation of interarrival;
- Coefficient of variation of interarrival.

Among all, two variables seem to be constantly changing between the considered groups at the different aggregation levels: coefficient of variation of demand size (CV_s) and average interarrival between successive orders. These two measures are negatively correlated among themselves (Pearson correlation index is -0.51 with $p < 0.001$) due to the fact that both measures are affected by the number of days of non zero demand. In order to get rid of the effect of the days of actual demand, we define a standard coefficient of variation of demand size as follows:

$$CV_s^n = \frac{CV_s}{\sqrt{n}}$$

where CV_s is the coefficient of variation of demand size and n is the number of days of non zero demand. This indicator reduces the bias that CV_s has due to the number of days of actual observations.

Figure 4.3 shows the distribution of the 926 SKUs considered according to these two variables.

Based on these two variables, we divided the different SKUs in clusters. In particular, we ran first a hierarchical cluster analysis, in order to identify the proper number of groups. Specifically, we applied a hierarchical cluster analysis with between-groups linkage based on Squared Euclidean distance. Through the

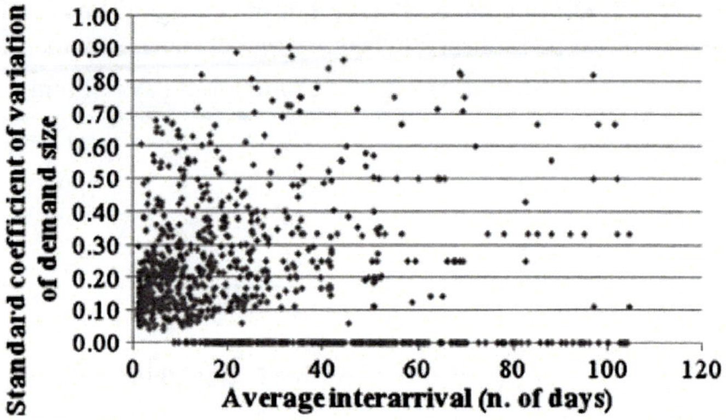

Fig. 4.3 SKUs distribution according to the standard coefficient of variation of demand size CV_s^n and the average interarrival

Table 4.3 Descriptive statistics of final clusters centroids

		Std. CV of demand size		Interarrival	
	No. of SKUs	Mean	Std. dev.	Mean	Std. dev.
HVHS	99	0.450	0.198	56.262	19.603
LVHS	237	0.011	0.038	50.798	21.020
HVS	136	0.472	0.129	15.856	8.828
LWS	454	0.154	0.076	10.364	8.281
Combined	926	0.196	0.194	26.426	24.267

analysis of the dendogram, we identified four as a proper number of clusters. A *k*-means cluster analysis with four as number of clusters led us to identify the following groups of items (Table 4.3 provides information regarding the final cluster centroids):

- High variability and highly sporadic (HVHS): these items are characterized by both high variability of demand size and sporadic pattern (on average less than 4 orders per year).
- Low variability and highly sporadic (LVHS): these items show a sporadic pattern (on average less than 4 orders per year) but demand size tends to be quite stable.
- High variability and sporadic (HVS): these items are ordered more frequently (on average more than 13 orders per year), however with highly variable quantities.
- Low variability and sporadic (LVS): these items are ordered frequently (on average more than 13 orders per year) and with very stable quantities.

The clusters obtained are also coherent to previous contributions that group SKUs according to similar variables (e.g., Syntetos and Boylan 2005).

In order to compare the different aggregation levels, we defined the Average Inventory Reduction (AIR) as the average percentage reduction of inventories between two different aggregation levels. In particular, AIR is defined as:

$$\mathrm{AIR}[i-j]=\frac{1}{k}\sum_{k}\frac{\mathrm{InvLev}\,[i]_k-\mathrm{InvLev}\,[j]_k}{\mathrm{InvLev}\,[j]_k}$$

Where InvLev $[i]_k$ is the inventory level for item k at aggregation level i.

Based on these clusters we evaluated the average inventory reduction (AIR) between the different aggregation levels within each cluster. Table 4.4 summarizes this comparison.

These results show again that improvements in inventory performance are widespread in the sample, and thus they confirm our previous evidence. Quite interestingly, however the improvements obtained through a more aggregated forecast are significant when the demand is not highly sporadic. In fact, when sporadic behavior is limited, all reductions are significant (based on pair comparisons at SKU level). Significant benefits may occur also when the demand is highly sporadic but only if the variability is limited; in fact, in most of the comparisons there is a significant reduction even if in one case a significant increase can also be seen. When the demand is highly sporadic and the demand size is highly variable, no significant improvements can be found; quite interestingly even if on average some reductions can still be found here, they are not statistically significant.

These results suggest that aggregating demand seems to be a reliable approach when demand is lumpy. However when demand sporadic behavior and variability are extremely high (i.e. HVS skus), this approach is not helpful and can in some cases lead to worse performances. This result eventually suggests that in this last situation, demand forecasting can be highly inefficient and one should design inventory management solutions based on other approaches. In our case it should also be noted that this situation affects only a limited part of the inventory problem we are addressing. In fact these items account for no more than 10% of the considered SKUs that are responsible for less than 2% of demand volumes and less than 1% of the average inventory level.

Table 4.4 Average percentage inventory reduction for each cluster between the different aggregation levels (AIR[*i-j*]: average inventory reduction with *i* days aggregation level compared to *j* days aggregation level; * $p < 0.05$, based on pair comparison of each SKU)

Demand size variability	Interarrival	
	Sporadic	Highly sporadic
High	AIR[2–1]: −51.1%*	AIR[2–1]: −20.0%
	AIR[3–2]: −33.6%*	AIR[3–2]: −10.9%
	AIR[10–3]: −47.1%*	AIR[10–3]: +4.1%
	AIR[30–10]: −21.7%*	AIR[30–10]: +9.5%*
Low	AIR[2–1]: −39.5%*	AIR[2–1]: −44.3%*
	AIR[3–2]: −23.8%*	AIR[3–2]: −12.5%
	AIR[10–3]: −37.5%*	AIR[10–3]: −19.8%*
	AIR[30–10]: −44.3%*	AIR[30–10]: +18.5%*

4.5 Conclusions

This work provides evidence that the temporal aggregation of data may be beneficial in spare parts inventory management and forecasting. The presented results show a clear effect of the aggregation of data over inventory performance, thus they emphasize the importance of paying proper attention in defining the aggregation level at which demand is managed. This consideration is coherent with previous results on this topic in the case of non-lumpy demand (see literature review for details) and provides evidence that also when demand is sporadic or lumpy, this issue has to be taken in high consideration.

A second contribution relates to the contingent analysis of the impacts of aggregation. The analyses show that even if the impact is usually significant, the characteristics of the demand significantly influence the possibility of improving inventory performance by leveraging on temporal aggregation. In particular, results provide evidence that when the demand is sporadic, impressive inventory reduction can be gained by leveraging on data aggregation. However when the demand occurrence is extremely low (in our case less than four orders in one year), leveraging on data aggregation may be effective if variability in demand size is not extremely high. On the contrary, if both sporadic nature of demand and variability of demand size are extremely high, impacts can be limited. This last situation, however, is limited to few cases in our sample (almost 10% of the considered SKUs). This result is coherent with other contributions in the field, claiming that when demand lumpiness is too high, companies should not invest too much in forecasting those patterns due to the extreme uncertainty of the situation.

This work also highlights some interesting issues that future studies should devote attention to. First of all, it would be important to define criteria that can provide companies with a clear a priori determination of the aggregation level they should adopt. In fact this work, provides some guidelines for companies willing to understand whether they should aggregate data or not. Such a contribution is important for managers since it can provide them with some guidelines to better manage their spare parts inventories.

We would also like to draw attention to some limitations of this work. First of all, we considered a specific situation in terms of data (available from one single company), thus one can doubt about the possibility to generalize these results. We argue, however, that these results are at some extent of general validity since even if the data come from a single company they represent a typical situation faced in the spare parts business. Indeed future studies should consider other data sets from other companies to verify these results. A second issue relates to the specific forecasting technique that we adopted to manage demand. This work focuses on one specific forecasting method (i.e. Syntetos and Boylan's method). It would be important to evaluate to which extent these results are method-specific and thus how the selection of the aggregation level should take the forecasting method adopted into account. We consider that some specificities of the applied method may have an impact since different methods rely on the estimation of different parameters that can be influenced heterogeneously by the aggregation level. We argue, however, that our results still constitute a relevant contribution for this topic, also due to the fact that the adopted method is considered to provide superior performances compared to others for the specific case of lumpy demand.

In the end, we would like to draw attention on the assumptions we made on suppliers lead time. As we mentioned, we assumed lead times constantly equal to 20 days for all items. Our results are for sure influenced by this assumption that we made in order to simplify analyses. We argue, however, that the overall conclusions of our work is not affected by this supposition. We claim that relaxing his assumption would be important for providing more reliable guidelines for companies and thus future works should address this topic.

References

Altay N, Rudisill F, Litteral LA (2008) Adapting Wright's modification of Holt's method to forecasting intermittent demand. Int J Prod Econ 111(2):389–408

Barnea A, Lakonishok J (1980) An analysis of the usefulness of disaggregated accounting data for forecasts of corporate performance. Decis Sci 11:17–26

Bartezzaghi E, Verganti R (1995) Managing demand uncertainty through order over planning. Int J Prod Econ 40:107–120

Bartezzaghi E, Verganti V, Zotteri G (1999) A simulation framework for forecasting uncertain lumpy demand. Int J Prod Econ 59:499–510

Bunn DW, Vassilopoulos AI (1993) Using group seasonal indices in multi-item short-term forecasting. Int J Forecast 9:517–526

Bunn DW, Vassilopoulos AI (1999) Comparison of seasonal estimation methods in multi-item short-term forecasting. Int J Forecast 15:431–443

Chan W (1993) Disaggregation of annual time-series with common seasonal patterns. J Econom 55:173–200

Cobbaert K, Van Oudheusden D (1996) Inventory models for fast moving spare parts subject to sudden death obsolescence. Int J Prod Econ 44:239–248

Cohen MA, Kleindorfer PR (1989) Near-optimal service constrained stocking policies for spare parts. Oper Res 37(1):104–117

Croston JD (1974) Stock level for slow-moving items. Oper Res Q 25(1):123–130

Dalhart G (1974) Class seasonality—a new approach. In: Proceedings of 1974 Conference of American Production and Inventory Control Society, Reprinted in Forecasting, 2nd edn. APICS, Washington, DC, pp 11–16

David I, Greenshtein E, Mehrez A (1997) A dynamic programming approach to continuous review obsolescent inventory problem. Nav Res Logist 44(8):757–774

Dekker R, Kleijn MJ, De Rooij PJ (1998) A spare parts stocking policy based on equipment criticality. Int J Prod Econ 56–57:69–77

Dekker M, Van Donselaar K, Ouwehand P (2004) How to use aggregation and combed forecasting to improve seasonal demand forecast. Int J Prod Econ 90:151–167

Fildes R, Beard C (1992) Forecasting system for production and inventory control. Int J Oper Prod Manag 12(5):4–27

Fliedner G (1999) An investigation of aggregate variable time series forecast strategies with specific subaggregate time series statistical correlation. Comput Oper Res 26:1133–1149

Gonzales P (1992) Temporal aggregation and systemic sampling in structural time-series models. J Forecast 11:271–281

Grunfeld Y, Griliches Z (1960) Is aggregation necessarily bad? Rev Econ Stat 42:1–13
Ho CJ (1995) Examining the impact of demand lumpiness on the lot-sizing performance in MRP systems. Int J Prod Res 33(9):2579–2599
Johnston FR, Harrison PJ (1986) The variance of lead-time demand. J Oper Res Soc 37:303–308
Kalchschmidt M, Zotteri G, Verganti R (2003) Inventory management in a multi-echelon spare parts supply chain. Int J Prod Econ 81–82:397–413
Lapide L (1998) New developments in business forecasting. J Bus Forecast Summer 28–29
Lee HL, Padmanabhan V, Whang SJ (1997) Information distortion in a supply chain: the bullwhip effect. Manag Sci 43(4):546–558
Liu L, Shi DH (1999) An (s, S) model for inventory with exponential lifetimes and renewal demands. Nav Res Logist 46(1):39–56
Miller JG, Berry W, Lai C-YF (1976) A comparison of alternative forecasting strategies for multi-stage production inventory systems. Decis Sci 7:714–724
Nikolopoulos K, Syntetos AA, Boylan JE, Petropoulos F, Assimakopoulos V (2009) ADIDA: an aggregate–disaggregate intermittent demand approach to forecasting. Working paper: 330/09, University of Salford, ISSN 1751-2700
Orcutt G, Watts HW, Edwards JB (1968) Data aggregation and information loss. Am Econ Rev 58:773–787
Petrović R, Šenborn A, Vujošević M (1988) A new adaptive algorithm for determination of stocks in spare parts inventory systems. Eng Costs Prod Econ 15:405–410
Shibuya T, Dohi T, Osaki S (1998) Optimal continuous review policies for spare parts provisioning with random lead times. Int J Prod Econ 55:257–271
Snyder RD, Koehler AB, Ord JK (1999) Lead-time demand for simple exponential smoothing. J Oper Res Soc 50:1079–1082
Syntetos AA, Boylan JE (2001) On the bias of intermittent demand estimates. Int J Prod Econ 71:457–466
Syntetos AA, Boylan JE (2005) The accuracy of intermittent demand estimates. Int J Forecast 21:303–314
Teunter RH, Fortuin L (1999) End-of-life service. Int J Prod Econ 59:487–497
Theil H (1954) Linear aggregation of economic relations. North-Holland, Amsterdam
Vereecke A, Verstraeten P (1994) An inventory model for an inventory consisting of lumpy items, slow movers and fast movers. Int J Prod Econ 35:379–389
Wacker JG, Lummus R (2002) Sales forecasting for strategic resource planning: practical implications and research directions. Int J Prod Oper Manag 22(9):1014–1031
Ward JB (1978) Determining reorder points when demand is lumpy. Manag Sci 24(6):623–632
Weatherford LR, Kimes SE, Scott DA (2001) Forecasting for hotel revenue management—testing aggregation against disaggregation. Cornell Hotel Restaur Adm Q August:53–64
Weiss AA (1984) Systematic sampling and temporal aggregation in time-series models. J Econom 12:547–552
Wemmerlöv U (1986) A time phased order-point system in environments with and without demand uncertainty: a comparative analysis of no-monetary performance variables. Int J Prod Res 24(2):343–358
Wemmerlöv U, Whybark DC (1984) Lot-sizing under uncertainty in a rolling schedule environment. Int J Prod Res 22(3):467–484
Willemain TR, Smart CN, Schwarz HF (2004) A new approach to forecasting intermittent demand for service parts inventories. Int J Forecast 20:375–387
Williams TM (1982) Reorder levels for lumpy demand. J Oper Res Soc 33:185–189
Withycombe R (1989) Forecasting with combined seasonal indices. Int J Forecast 5:547–552
Zellner A, Tobias J (2000) A note on aggregation, disaggregation and forecasting performance. J Forecast 19:457–469
Zotteri G, Kalchschmidt M (2007) A model for selecting the appropriate level of aggregation in forecasting processes. Int J Prod Econ 108:74–83

Chapter 5
Bayesian Forecasting of Spare Parts Using Simulation

David F. Muñoz and Diego F. Muñoz

5.1 Introduction

Forecasting demand plays a major role in many service and manufacturing organizations. Forecasts help in the scheduling of taskforce, obtaining higher service levels for the customer, and determining resource requirements among many others (Makridakis et al. 1998). Forecasting accuracy is an increasingly important objective in most firms and, in particular, plays a key role in forecasting lumpy demand.

According to many authors (e.g., Wacker and Sprague 1998; Zotteri and Kalchsdmidt 2007a), the accuracy of forecasts depends sensitively on the quantitative technique used, thus, this chapter has been motivated by an increasing need for applying and formulating new tools for demand forecasting, and in particular for the case of lumpy demand. As suggested by the work of Caniato et al. (2005) and Kalchsdmidt et al. (2006), it is necessary to propose forecasting techniques that not only take into account the time series, but also the structure of the demand-generating process (non-systematic variability). For this reason, in this chapter we illustrate how to apply simulation techniques and Bayesian statistics in a model that takes into account particular characteristics of the system under study.

In practice, a new forecasting model may become complex in the sense that analytical expressions for the point and variability estimators cannot be obtained. As an example, we can mention Croston's method (Croston 1972) proposed to forecast lumpy demands, which was revised by Rao (1973) and later again by Syntetos and Boylan (2001). This is our motivation for proposing a simulation-based methodology

D. F. Muñoz (✉)
Instituto Tecnológico Autónomo de México, Mexico, Mexico
e-mail: davidm@itam.mx

D. F. Muñoz
Stanford University, Stanford, CA, USA
e-mail: dkedmun@stanford.edu

N. Altay and L. A. Litteral (eds.), *Service Parts Management*,
DOI: 10.1007/978-0-85729-039-7_5,

to produce point and variability estimators for demand forecasting. Surprisingly, up to this day, simulation had been used to compare the performance of different techniques for demand forecasting (see Bartezzaghi et al. 1999; Zotteri and Kalchsdmidt 2007b), however the literature on using simulation as a technique for demand forecasting is still scarce. In this direction, we should mention the work of Willemain et al. (2004), where a simulation model was used to produce forecasts for intermittent demand, obtaining more accurate forecasts than do exponential smoothing and Croston's method. Although in this work the bootstrap method was applied and parameter uncertainty was not incorporated in the model.

It is worth mentioning that in classical simulation applications, people tend to fix the value of each parameter. Conversely, in forecasting applications parameter uncertainty is incorporated from sample data. The novelty of our simulation approach is that we are able to incorporate parameter uncertainty as in most forecasting techniques.

We should mention that using a Bayesian approach is particularly recommended when it is important to assess parameter uncertainty (e.g., because of short sample data), as is illustrated in de Alba and Mendoza (2007) where the authors illustrate the application of a Bayesian method for seasonal data and show that it can outperform traditional time series methods for short time series.

This chapter illustrates the development and application of a model that was used to forecast the demand for spare parts experienced by a car dealer in Mexico as well as the potential of simulation as a powerful tool for forecasting using a complex model of the system under study. Our methodology can be used as an example of how to attack forecasting problems when the complexity of the model does not allow us to obtain analytical expressions for the point and variability estimators needed for forecasting.

The organization of this chapter is as follows. In Sect. 5.2 we present a general framework to apply simulation techniques in order to produce consistent (as the number of simulation replications increases) estimators for the parameters needed for forecasting, using a complex model under a Bayesian approach. In Sect. 5.3 we develop two simple models that can be used to produce a forecast for the demand of spare parts under lumpy demand. In Sect. 5.4 we use real data from the demand of clutches at a car dealer in Mexico, and apply one of the models developed in Sect. 5.4, and the simulation techniques presented in Sect. 5.3, to illustrate how to compute a point forecast and a reorder point for the demand of clutches during the lead time of an order. Finally in Sect. 5.5 we discuss the main conclusions obtained from this research.

5.2 A Bayesian Framework for Forecasting Using Simulation

In this section, we provide a general framework to construct forecasts for a response variable from the output of simulation experiments using a complex model (in the sense that analytical expressions for the forecasting parameters are not available). We propose two main steps for forecasting using simulation (see Fig. 5.1). The first step consists in the assessment of parameter uncertainty using available data (x) and a prior density (probability in the discrete case) function $p(\theta)$, so that parameter uncertainty is assessed through a resulting posterior density function $P(\theta|X)$. In the second step we use the simulation model and the posterior density to estimate the parameters related to the forecast of the response variable (W). We now explain in detail this approach.

5.2.1 Bayesian Estimation

Although the incorporation of parameter uncertainty is extensively discussed in the literature on Bayesian statistics, we will provide a short review of the concepts and notation that are required to propose methods for forecasting using simulation.

As in most forecasting models, the output of our simulation model depends upon a vector Θ of uncertain parameters, where some of them may characterize random components of the simulation model.

We remark that a random component (also called random input) of a stochastic simulation is a sequence U_1, U_2, … of random quantities that are needed as input to the simulation. When the U_i's are assumed to be independent and identically distributed (i.i.d.), a random component is identified by the corresponding probability distribution, which is typically assumed to be a member of a parametric family. Since we wish to consider parameter uncertainty, the vector of parameters is a random variable (denoted Θ), and $\theta \in P$ denotes a particular value, where P is the parameter space.

We assume that sample data on the model parameters is available through a vector of observations $x \in \Re^n$ that satisfies a likelihood function $L(x|\theta)$. If $p(\theta)$ is a

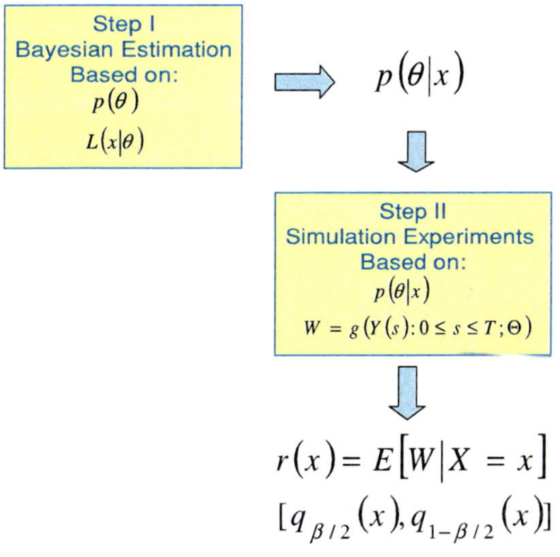

Fig. 5.1 Main steps for forecasting using simulation

prior density function for the vector of parameters Θ, then the posterior (given the data x) density function of Θ becomes

$$p(\theta|x) = \frac{p(\theta)L(x|\theta)}{\int_P p(\theta)L(x|\theta)d\theta}, \tag{1}$$

for $x \in \Re^n$ and $\theta \in P$.

Note that we are using the notation $p(\theta)$ for the prior density of Θ, and $P(\theta|X)$ for the conditional density of Θ given $[X = x]$ (although in general they are different functions), this admittedly imprecise notation will greatly simplify the exposition.

For the special case where $x = (x_1, \ldots, x_n)$ is a set of observations of a random sample $X = (X_1, X_2, \ldots, X_n)$ from a density function $f(y|\theta)$, the likelihood function takes the form of

$$L(x|\theta) = f(x_1|\theta)\,f(x_2|\theta)\ldots f(x_n|\theta) \tag{2}$$

The prior $p(\theta)$ reflects the initial uncertainty on the vector of parameters Θ, and there are essentially two points of view to proposing a prior $p(\theta)$. The first approach consists on using a non-informative prior, which is appropriate when we wish to consider a prior that does not "favor" any possible value of Θ over others, so that this is considered an "objective" point of view (see e.g., Berger et al. 2009 for further discussions). Non-informative priors have been extensively studied, and textbooks on Bayesian statistics usually provide the corresponding reference (non-informative) prior for the most commonly used distribution (see e.g., Bernardo and Smith 2000). The second approach is a "subjective" point of view, and consists on establishing a prior based on expert judgment, see e.g., Kraan and Bedford 2005 for a discussion on how to construct a prior from forecasts based on expert judgment. We should mention that a prior $p(\theta)$ need not be a proper density (in the sense that $\int_P p(\theta)d\theta = 1$. In order to have a well-defined prior $p(\theta)$, it suffices that $p(\theta) \geq 0$ for $\theta \in P$, and $\int_P p(\theta)L(x|\theta)d\theta < \infty$.

5.2.2 *Forecasting Using Simulation*

Let $Y = \{Y(s),\ s \geq 0;\ \Theta\}$ be the (possibly multivariate) stochastic process that represents the output of a simulation model corresponding to the system under study. Note that discrete-time stochastic processes can be incorporated into our framework if we let $Y(s) = Y(\lfloor s \rfloor)$, where $\lfloor s \rfloor$ denotes the integer part of s. Since we wish to forecast a response variable from a transient simulation, we let $W = g(Y(s),\ 0 \leq s \leq T;\ \Theta)$ be the response variable of interest, where T is the run length, and assume that T is a stopping time (see Asmussen 2003 for a definition) with respect to the stochastic process Y.

In general, a forecast for a response variable W is completely defined by its cumulative distribution function (c.d.f.) $F(w|x) = P[w \leq w|X = x]$. However, from a practical point of view a forecast is expressed in terms of a point forecast and an assessment of the uncertainty on the point forecast. The standard point forecast used in Bayesian statistics is the expectation

$$r(x) = E[W|X = x]. \tag{3}$$

Another performance measure that is of practical importance is the α-quantile defined by

$$q_\alpha(x) = \inf\{w : F(w|x) \geq \alpha\}, \tag{4}$$

for $0 < \alpha < 1$. In order to asses the uncertainty on the point forecast $r(x)$, α-quantiles are useful, since for $0 < \beta < 1$, a $(1 - \beta)100\%$ prediction interval in the form of $[q_{\beta/2}(x), q_{1-\beta/2}(x)]$ is usually constructed. This interval is called a $(1 - \beta)100\%$ prediction interval because $p[q_{\beta/2}(x) \leq W \leq q_1 - \beta/2(x)|X] = 1 - \beta$, provided $F(w|x)$ is continuous at $q_{\beta/2}(x)$ and $q_{1-\beta/2}(x)$.

Quantiles are also useful to compute reorder points in inventory management. In this particular application, when W is the demand during the lead time, the α-quantile $q_\alpha(x)$ can be interpreted as the reorder point for a $100\alpha\%$ (type-I) service level (see e.g., Chopra and Meindl 2004).

5.2.2.1 Forecasting Using Posterior Sampling

When analytical expressions for the forecasting parameters of Eqs. 3 and 4 cannot be obtained (or they are too complicated), simulation can be applied to estimate these parameters. In Fig. 5.2 we illustrate a first algorithm, called posterior sampling (PS), to estimate the required forecasting parameters using simulation. Under PS we sample from the posterior density $P(\theta|X)$ to obtain i.i.d. observations of the response variable W, that allow us to estimate the point forecast $r(x)$ by $\hat{r}(x)$, and the α-quantile $q_\alpha(x)$ by $\hat{q}_\alpha(x)$ (as defined in the algorithm of Fig. 5.2). As is well known, these estimators are consistent, which means that they approach the

For $i = 1$ to the number of replications m:

a. Generate (independently) θ_i by sampling from $p(\theta x)$.

b. Run (independently) a simulation experiment with $\Theta = \theta_i$ to obtain:

$$W_i = g(Y(s), 0 \leq s \leq T; \Theta).$$

End Loop

Compute:

$$\hat{r}(x) = \frac{1}{m}\sum_{i=1}^{m} W_i .$$

Sort the W_i's: $Y_1 \leq \ldots \leq Y_m$ and set $\hat{q}_\alpha(x) = Y_{\lceil m\alpha \rceil}$.

Fig. 5.2 Forecasting using posterior sampling

corresponding parameter as the number of replications m increases. However, the number of replications required to obtain estimators within a prescribed precision is problem dependent. Therefore, a good practice is to compute a measurement of the accuracy of a point estimator, so that if not acceptable, we can increase the number of replications in order to reduce the estimation error.

The standard way to assess the accuracy of a point estimator obtained from simulation experiments is to compute the halfwidth of an asymptotic confidence interval. According to the standard Central Limit Theorem (CLT), an appropriate $100(1-\beta)\%$ halfwidth for the estimator $\hat{r}(x)$ is given by

$$H_\beta[\hat{r}(x)] = z_{1-\beta/2}\frac{S(x)}{\sqrt{m}}, \tag{5}$$

where $S(x) = m^{-1/2}\sqrt{\sum_{i-1}^{m}[W_i - \hat{r}(x)]^2}$ and $z_{1-\beta/2}$ denotes the $(1-\beta/2)$-quantile of a standard normal distribution. A simple practical interpretation for a halfwidth is that the parameter $r(x)$ lies within $\hat{r}(x) \pm H_\beta[\hat{r}(x)]$ with a confidence of $100(1-\beta)\%$, and this is why a halfwidth is useful to assess the accuracy of a point estimator. Similarly, a $100(1-\beta)\%$ halfwidth for the estimator $\hat{q}_\alpha(x)$ is given by

$$H_\beta[\hat{q}_\alpha(x)] = (Y_{n_1} + Y_{n_2})/2, \tag{6}$$

where the Y_i's are defined in Fig. 5.2, and $n_1 = \left\lfloor m\alpha - z_{1-\beta/2}[m\alpha(1-\alpha)]^{1/2} \right\rfloor$, $n_2 = \left\lceil m\alpha + z_{1-\beta/2}[m\alpha(1-\alpha)]^{1/2} \right\rceil$. The asymptotic validity of this halfwidth is established in Serfling (1980).

Classical simulation techniques for transient simulation usually adopt an algorithm that is very similar to the PS algorithm, the only difference is that the value of Θ is fixed, so that sampling from $P(\theta|X)$ is no longer required. Thus, the variance of the response variable under a classical approach has the form of

$$\sigma^2(\theta) = E\left[W^2|X=x, \Theta=\theta\right] - (E[W|X=x, \Theta=\theta])^2, \tag{7}$$

where $\theta \in P$ is the fixed value of Θ. On the other hand, under our Bayesian approach, the variance of the response variable is

$$\sigma_W^2 = E\left[W^2|X=x\right] - (E[W|X=x])^2, \tag{8}$$

so that by letting $r_1(\theta)+ = E[W|X=x, \Theta=\theta]$, adding and subtracting the term $E[r_1(\Theta)|X=x]$ in (8), we can verify that

$$\sigma_W^2 = \sigma_p^2 + \sigma_s^2, \tag{9}$$

where $\sigma_p^2 = E\left[r_1^2(\Theta)|X=x\right] - (E[r_1(\Theta)|X=x])^2, \sigma_s^2 = E[\sigma^2(\Theta)|X=x]$, and $\sigma^2(\theta)$ is defined in (7). Note that $\sigma_p^2 = 0$ and $\sigma_s^2 = \sigma^2(\theta)$ under a classical approach, and this is why σ_p^2 is called the parametric variance, and σ_s^2 is called the stochastic variance.

5.2.2.2 Forecasting Using Markov Chain Monte Carlo

In order to implement the algorithm of Fig. 5.2, a valid method to generate samples from the posterior density $P(\theta|\, X)$ is required, which can be available if the family of distributions corresponding to $P(\theta|\, X)$ has been identified. However, in many situations it is very difficult to obtain an analytic expression that allows us to identify the family of distributions corresponding to the posterior density $P(\theta|\, X)$, and in this case we can apply a technique called Markov chain Monte Carlo (MCMC), which does not require an algorithm to generate samples from $P(\theta|\, X)$.

The algorithm of Fig. 5.3 is an implementation of MCMC that is called the independence sampler, because the generation of a sample from an auxiliary density $q(\theta)$ (where $q(\theta) > 0$ whenever $p(\theta) > 0$) is required. As mentioned by several authors (e.g., Asmussen and Glynn 2007), the algorithm of Fig. 5.3 performs better when $q(\theta)$ is closer to the posterior density $p(\theta|x)$.

Note that (although it is not required to compute the point estimators), we are dividing the number of replications m into b batches of length m_b. This is because we are suggesting to use the method of batch means in order to produce asymptotic confidence intervals for the point estimators. Although a CLT for the point estimators can be established (under mild assumptions), we are not trying to estimate the asymptotic variance of the point estimators, because finding consistent estimators may be a difficult task and batch means may be also required (see e.g., Song and Chih 2008). Instead we can choose a number of batches b between 5 and 20 (as suggested by Schmeiser 1982), so that under suitable

1. Generate z_0 from $q(\theta)$.
2. For $i = 1$ to the number of batches b:

   For $j = 1$ to $m_b = m/b$:

   a. Generate (independently) z_1 from $q(\theta)$, and U uniform on $(0,1)$.

   b. If $U < \dfrac{p(z_1)L(x|z_1)q(z_0)}{p(z_0)L(x|z_0)q(z_1)}$ then $z_0 \leftarrow z_1$.

   c. Run (independently) a simulation experiment with $\Theta = z_0$ to obtain:
   $$W_{ij} = g(Y(s), 0 \le s \le T; \Theta)$$

   End Loop

   Compute:
   $$\hat{r}_i = \frac{1}{m_b}\sum_{j=1}^{m_b} W_{ij}\,.$$

   Sort the W_{ij}'s: $Y_1 \le \ldots \le Y_{m_b}$ and set $\hat{q}^i_\alpha = Y_{\lceil m_b\alpha \rceil}$

   End Loop
3. Compute:
   $$\hat{r}_{MC}(x) = \frac{1}{b}\sum_{i=1}^{b}\hat{r}_i\,.$$

   Sort the W_{ij}'s: $Z_1 \le \ldots \le Z_m$ and set $\hat{q}^{MC}_\alpha(x) = Z_{\lceil m\alpha \rceil}$.

Fig. 5.3 Forecasting using Markov chain Monte Carlo

assumptions, the following $100(1-\beta)\%$ halfwidths are asymptotically (as $m \to \infty$) valid for $r(x)$ and $q_\alpha(x)$, respectively,

$$H_\beta[\hat{r}_{MC}(x)] = t_{(b-1,1-\beta/2)} \frac{S_{MC}(x)}{\sqrt{b}}, \tag{10}$$

$$H_\beta\left[\hat{q}_\alpha^{MC}(x)\right] = t_{(b-1,1-\beta/2)} \frac{S_\alpha^{MC}(x)}{\sqrt{b}}, \tag{11}$$

where $t_{(b-1,1-\beta/2)}$ denotes the $(1-\beta/2)$-quantile of a Student-t distribution with $(b-1)$ degrees of freedom, $\hat{r}_{MC}(x)$,$\hat{q}_\alpha^{MC}(x)$ are defined in Fig. 5.3, and

$$S_{MC}(x) = \sqrt{\frac{\sum_{i=1}^{b} [\hat{r}_i - \hat{r}_{MC}(x)]^2}{b-1}}, \quad S_\alpha^{MC}(x) = \sqrt{\frac{\sum_{i=1}^{b} \left[\hat{q}_\alpha^i - \overline{q}_\alpha\right]^2}{b-1}},$$

where $\hat{r}_i$,$\hat{q}_\alpha^i$ are defined in Fig. 5.3, and $\overline{q}_\alpha = b^{-1}\sum_{i=1}^{b} \hat{q}_\alpha^i$. See Muñoz (2010) for a further discussion on the validity of the confidence interval provided in Eq. 11.

5.2.2.3 Selection of *q*(*θ*) for the Independence Sampler

As mentioned before, the independence sampler of Fig. 5.3 performs better when $q(\theta)$ is closer to the posterior density $P(\theta|X)$, and this is why a procedure to select a density $q(\theta)$ can be proposed by taking into account that, under regularity conditions, a posterior density $P(\theta|X)$ satisfies a CLT (as the sample size $n \to \infty$), where the limiting distribution is (multivariate) normal. Therefore, a reasonable proposal for $q(\theta)$ is a density corresponding to a multivariate normal distribution with mean vector m_n and covariance matrix V_n, where the parameters m_n and V_n can be determined according to the particular form of $P(\theta|X)$.

The following steps to select the parameters m_n and V_n are based on Theorem 5.14 of Bernardo and Smith (2000), and can be applied when the likelihood function has the particular form of Eq. 2.

1. Let us denote $p_n(\theta) \overset{def}{=} p(\theta|x)$, where $x = (x_1, \ldots x_n)$, and set $L_n(\theta) = \log[p_n(\theta)]$.
2. Let m_n be the local minimum of $L_n(\theta)$, and compute m_n by solving $L_n'(m_n) = \nabla L_n(\theta)|_{\theta=m_n} = 0$.
3. Compute the Hessian matrix $L_n''(m_n) \overset{def}{=} \left(\frac{\partial^2 L_n(\theta)}{\partial\theta_i \partial\theta_j}\right)|_{\theta=m_n}$, which has to be positive definite for this procedure to be valid.
4. Set $V_n = [-L_n''(m_n)]^{-1}$.

A simple illustration of the application of these steps to propose a sampling distribution $q(\theta)$ for the independence sampler is provided in Sect. 5.4.

5.3 Forecasting Models for Spare Parts

In this section, we present two simple models that can be used to forecast the demand for spare parts, and will help us illustrate how the techniques described in the previous section can be applied. In both models we assume that failures occur randomly, and the only difference is how the sample data is available, as we explain below.

Under both models we have k machines, each has a critical part and failures occur independently at the same rate $\Theta \in P = (0, \infty)$. There is uncertainty on the failure rate Θ, so that $p(\theta)$ is a prior density on Θ, and for $i = 1,\ldots, k$, $N_i = \{N_i(t){:}t \geq 0; \Theta\}$ denotes the failure process for component i, which are assumed to be conditionally independent (see e.g., Chung 1974 for a definition) relative to Θ, and

$$P[N_i(t+s) - N_i(t) = j|\Theta, N_i(u), 0 \leq u \leq t] = \frac{e^{-\Theta s}(\Theta s)^j}{j!}$$
$$j = 0, 1, \ldots, t, \quad s \geq 0, \quad i = 1, \ldots, k$$

(i.e., N_i are Poisson processes with the same rate Θ).

5.3.1 A Model Using Failure Time Data

Under this model, the available sample information corresponds to the time between failures of every part. Thus, the sample data $x = (x_1, x_2,\ldots, x_n)$ comes from a random sample $X = (X_1, X_2,\ldots, X_n)$ of the exponential density

$$f(y|\theta) = \begin{cases} \theta e^{-\theta y}, & y > 0, \\ 0, & \text{otherwise}, \end{cases}$$

and it follows from (2) that the likelihood function is given by

$$L(x|\theta) = \theta^n e^{-\theta \sum_{i=1}^{n} x_i}. \tag{12}$$

An appropriate non-informative prior for the exponential density (see e.g., Bernardo and Smith 2000) is $p(\theta) = \theta^{-1}$, so that it follows from (1) and (12) that

$$p(\theta|x) = \frac{\left(\sum_{i=1}^{n} x_i\right)^n \theta^{n-1} e^{-\theta \sum_{i=1}^{n} x_i}}{(n-1)!}, \tag{13}$$

which corresponds to the $Gamma\left(n, \sum_{i=1}^{n} x_i\right)$ distribution, where, for β_1, $\beta_2 > 0$, $Gamma(\beta_1, \beta_2)$ denotes a Gamma distribution with expectation $\beta_1 \beta_2^{-1}$.

We are interested in the number of failed components during a time period of length t_0, so that the response we want to forecast is

$$W = \sum_{i=1}^{k} N_i(t_0), \tag{14}$$

given $[X = x]$.

In order to obtain an analytical expression for the point forecast $r(x)$, we can apply Proposition 1 of Muñoz and Muñoz (2008), by taking into account that $r_1(\theta) = E[W|\Theta = \theta] = kt_0\theta$, so that

$$r(x) = \int_0^\infty r_1(\theta)p(\theta|x)d\theta = kt_0 n\left(\sum_{i=1}^{n} x_i\right)^{-1}. \tag{15}$$

For the case where t_0 corresponds to the lead time for an order, the reorder point for a $100\alpha\%$ service level is given by $q_\alpha(x)$, as defined in Eq. 4. In this case, we do not have a simple analytical expression for $q_\alpha(x)$, for which using the algorithm of Fig. 5.2 would be appropriate to estimate $q_\alpha(x)$ using simulation.

Now, let us suppose that we have Q spare parts at the beginning of a period of length t_0. Two performance measures of interest in this case are the type-I service level ($100\alpha_1\%$) and the type-II service level ($100\alpha_2\%$), where

$$\alpha_1 = P[W \le Q|X = x], \quad \text{and} \quad \alpha_2 = E[\min\{1, Q/W\}|X = x]. \tag{16}$$

As we can see from Eq. 16, both service levels are particular cases of the performance measure $r(x)$ for a suitably defined response variable. Obtaining analytical expressions for these performance measures can be a difficult task, so that it would be appropriate to apply the algorithm of Fig. 5.2 to estimate α_1 and α_2.

5.3.2 A Model with Censored Data

Let us suppose that times between failures are not recorded, however for each period $i = 1, 2, \ldots, p$, we record the number k_i of machines in operation during period i, and the number of failures per machine during period i. In this case, the sample data takes the form of $x = (x_{11}, \ldots, x_{1k_1}, \ldots, x_{p1}, \ldots, x_{pk_p})$, where x_{ij} is the number of failures for the j-th machine in operation during period i, $j = 1, \ldots, k_i$; $i = 1, \ldots, p$ (in this case $n = \sum_{i=1}^{p} k_i$).

In order to simplify our notation, we assume that each period has a length of 1 time unit, which means that the failure rate is expressed in the appropriate scale (failures per time unit). Since the failure processes are assumed to be conditionally independent relative to Θ, given $[\Theta = \theta]$, the sample data x can be regarded as a random sample $X = (X_1, X_2, \ldots, X_n)$ from a Poisson distribution with a (discrete) probability function given by

$$f(y|\theta) = \frac{e^{-\theta}\theta^y}{y!}, \quad y = 0, 1, \ldots,$$

and it follows from (2) that the likelihood function is given by:

$$L(x|\theta) = \frac{e^{-\theta\sum_{i=1}^{p} k_i}\,\theta^{\sum_{i=1}^{p}\sum_{j=1}^{k_i} x_{ij}}}{\prod_{i=1}^{p}\prod_{j=1}^{k_i} x_{ij}!}. \tag{17}$$

An appropriate non-informative prior for the Poisson model (see e.g., Bernardo and Smith 2000) is $p(\theta) = \theta^{-1/2}$, so that it follows from (1) and (17) that

$$p(\theta|x) = \frac{n^{\sum_{i=1}^{p}\sum_{j=1}^{k_i} x_{ij}+1/2}\,\theta^{\sum_{i=1}^{p}\sum_{j=1}^{k_i} x_{ij}-1/2}\,e^{-n\theta}}{\Gamma\left(\sum_{i=1}^{p}\sum_{j=1}^{k_i} x_{ij} + 1/2\right)}, \tag{18}$$

which corresponds to the $Gamma\left(\sum_{i=1}^{p}\sum_{j=1}^{k_i} x_{ij} + 1/2, n\right)$ distribution.

As in the previous model, we are interested in the number of failed components during a time period of length t_0, so that the response we want to forecast is the same as in Eq. 14. By taking into account that under this model the posterior density p $p(\theta|x)$ now takes the form of Eq. 18, we can proceed as in Eq. 15 to obtain an analytical expression for the point forecast $r(x)$:

$$r(x) = \int_0^\infty r_1(\theta)p(\theta|x)d\theta = n^{-1}kt_0\left(\sum_{i=1}^{p}\sum_{j=1}^{k_i} x_{ij} + 1/2\right). \tag{19}$$

Also in this case, the use of the algorithm of Fig. 5.2 would be helpful to estimate a reorder point $q_\alpha(x)$, or a service level as defined in (15).

We would like to mention that in the two simple models presented we do not require the use of the algorithm of Fig. 5.3, because we can recognize the family of distributions corresponding to the posterior density (in both cases is the Gamma family). However, if we do not use the objective (non-informative) prior in these models, it may be the case that the family of distributions corresponding to the posterior density $p(\theta|x)$ may not be easily recognized, and the algorithm of Fig. 5.3 will be useful, as we will illustrate in the next section.

5.4 An Application to Forecasting Spare Parts for a Car Dealer

In this section we apply the model that uses censored data to forecast the demand of clutches (from a particular model), that is experienced by a car dealer. We will illustrate the use of the methods described in previous sections using demand data obtained from a car and spare parts dealer in Mexico.

5.4.1 Using PS to Compute Reorder Points and Service Levels

Car model A is the ongoing model of a very successful line, which continues to be produced and offered to this day. Table 5.1 presents the data available for this model, which include car and clutch demand for every month of 2008. To apply the model with censored data, we assume that customers that acquire their cars at a particular dealer also purchase spare parts from it. Thus, the number k_i of cars in operation at period i is equal to the amount of model A cars sold by the dealer up to that period of time, so that $k_i = k_{i-1} + S_i$, where S_i is the amount of cars that were sold in period i. As we can see from Table 5.1, model A began the year 2008 with 337 cars in operation (according to sales from previous years).

We are interested in forecasting the demand W, as given in Eq. 14, when $t_0 = 0.5$ and $k = 500$, since we are assuming that the lead time for an order of clutches is approximately 15 days, and there are 500 cars in operation during the forecasting period.

Using (19) we computed the point forecast for the demand of clutches during the lead time, and obtained $r(x) = 1.82701$. Then, we applied the algorithm of

Table 5.1 Car sales and clutch demand for Model A in 2008

Month (i)	Car Sales (S_i)	Cars in Operation (k_i)	Clutch Demand $\left(\sum_{j=1}^{k_i} x_{ij}\right)$
January (1)	4	341	3
February (2)	1	342	3
March (3)	6	348	2
April (4)	9	357	3
May (5)	6	363	3
June (6)	15	378	3
July (7)	7	385	3
August (8)	2	387	1
September (9)	8	395	4
October (10)	16	411	3
November (11)	20	431	3
December (12)	15	446	2
Total	109	4584	32

Fig. 5.2 in order to estimate $q_{0.9}(x)$, the reorder point for a 90% service level. The results for the estimation of $r(x)$ and $q_{0.9}(x)$ using $m = 10^6$ replications are summarized in Table 5.2. In the case of $r(x)$ the halfwidth indicates that $\hat{r}(x)$ has an error that is below 0.00228, although we already know that $r(x)$ and $\hat{r}(x)$ are equal within 4 decimal places.

In the case of the estimation of $q_{0.9}(x)$, we see from Table 5.2 that the halfwidth is 0, which is not surprising if we take into account that in our application W is a discrete random variable, and its c.d.f. is piecewise constant. Note also that $P[W \leq q_\alpha(x)|X = x]$ is not necessarily equal to α (as is the case when W is a continuous random variable), for which it would be of interest to estimate the true service level corresponding to $q_{0.9}(x)$.

In Table 5.3 we report the estimated type-I service level (cumulative probability) for different values of Q, obtained from applying the PS algorithm with $m = 10^6$ replications. From this table we see that the true service level for $q_{0.9}(x) = 4$ is approximately 95.74%.

As we can observe from Tables 5.2 and 5.3, the number of replications $m = 10^6$ was large enough to obtain accurate estimators for all the estimated parameters, which may not be true when the number of replications is small. To illustrate how the number of replications affect coverage and halfwidths, we repeated the estimation experiment $M = 1000$ times for different values of m, and computed the empirical coverage, average halfwidth and standard deviation, mean square error and bias for each set of replications. Since theoretical (true) values of $r(x)$ and $q_{0.9}(x)$ are required for these computations, we used the values of $r(x) = 1.82701$ and $q_{0.9}(x) = 4$ obtained before. The results of these experiments are presented in Table 5.4.

Table 5.2 Estimated performance measures using PS with $m = 10^6$ replications

Performance measure	Simulation estimate	Halfwidth
$r(x)$	$\hat{r}(x) = 1.82695$	0.00228
$q_{0.9}(x)$	$\hat{q}_{0.9}(x) = 4$	0

Table 5.3 Estimated type-I service levels for different values of Q using PS with $m = 10^6$

Initial inventory (Q)	Estimated type-I service level ($\hat{\alpha}_1$)	Halfwidth
1	0.460923	4.09E-04
2	0.722068	3.30E-04
3	0.881953	1.71E-04
4	0.95742	6.71E-05
5	0.986648	2.17E-05
6	0.996307	6.05E-06
7	0.999127	1.43E-06
8	0.999789	3.47E-07
9	0.999956	7.24E-08
10	0.999989	1.81E-08
11	0.999999	1.64E-09
12	1	0.00E+00

Table 5.4 Performance of the PS algorithm from *M=1000* experiments and different numbers of replications *m*

Posterior sampling	Estimated parameter	Empirical coverage	Halfwidth		Mean square error	Bias
			Average	St. dev.		
$m = 100$	$r(x)$	0.878	0.2269	0.0187	0.0211	0.0064
	$q_{0.9}(x)$	0.679	0.5585	0.2467	0.3400	−0.3340
$m = 400$	$r(x)$	0.898	0.1139	0.0045	0.0047	0.0022
	$q_{0.9}(x)$	0.869	0.336	0.2348	0.1310	−0.1310
$m = 1600$	$r(x)$	0.904	0.0570	0.0011	0.0012	0.0003
	$q_{0.9}(x)$	0.987	0.1295	0.2191	0.0130	−0.0130

As can be observed from the results presented in Table 5.4, the empirical coverage approaches the nominal 90% as the number of replications increases, and for $m = 1600$ we already obtain an acceptable coverage. Also note that for $m = 1600$ we obtained an over-coverage in the estimation of $q_{0.9}(x)$, which is explained by the fact that the response variable W is discrete. It is worth mentioning that increasing the number of replications also increase the accuracy of the point estimation, and consistently reduce the halfwidth, mean square error and bias.

5.4.2 Using MCMC to Incorporate a Subjective Prior

As explained in Sect. 5.3.2, when using the non-informative (Gamma) prior for the model with censored data, we were able to identify that the posterior distribution is also Gamma, for which we can apply the PS algorithm for the estimation of $r(x)$ and $q_\alpha(x)$. In this section we illustrate how MCMC can be applied when using a subjective prior for which we cannot identify the family of distributions corresponding to the posterior density.

We considered a uniform prior on $[a, b]$, where the values of a and b can be supplied by the user. An analytical expression for the posterior density using this prior is not easy to obtain, so that the use of the MCMC algorithm of Fig. 5.3 is justified. For the implementation of this algorithm we considered a sampling distribution $q(\theta)$ that was obtained using the procedure suggested in Sect. 5.2.2.3, as we explain below.

Since $p(\theta) = 1$, for $a < \theta < b$, it follows from (1) and (14) that:

$$L_n(\theta) = \log[p(\theta|x)] = \sum_{i=1}^{p}\sum_{j=1}^{k_i} x_{ij}\log(\theta) - \theta\sum_{i=1}^{p} k_i - \log\left(\prod_{i=1}^{p}\prod_{j=1}^{k_i} x_{ij}!\right),$$

for $a < \theta < b$, so that by solving $L_n'(m_n) = 0$ we obtain the mean

$$m_n = \frac{\sum_{i=1}^{p}\sum_{j=1}^{k_i} x_{ij}}{\sum_{i=1}^{p} k_i}. \tag{20}$$

On the other hand, since $L_n''(\theta) = -\theta^{-2}\sum_{i=1}^{p}\sum_{j=1}^{k_i} x_{ij}$, we obtain the variance

$$\sigma^2 = \left[-L_n''(m_n)\right]^{-1} = \frac{m_n^2}{\sum_{i=1}^{p}\sum_{j=1}^{k_i} x_{ij}}. \tag{21}$$

Finally, by considering that $p(\theta|x)$ is zero when $\theta < a$ or $\theta > b$, we selected $q(\theta)$ as the density corresponding to a truncated $N(m_n, \sigma^2)$ distribution on $[a, b]$, where m_n and σ^2 are defined in (20) and (21), respectively.

Using the sampling distribution $q(\theta)$ defined before, we implemented the method described in Fig. 5.3, to forecast the demand W as defined in (14). The parameters of the prior were set at $a = 0$ and $b = 0.02$ for $t_0 = 0.5$ and $k = 500$. The results for the estimation of $r(x)$ and $q_{0.9}(x)$ using $m = 10^6$ replications with this method are summarized in Table 5.5.

We may see from Tables 5.2 and 5.5 that the estimations obtained with a uniform (subjective) prior and the non-informative (objective) prior are extremely similar (up to the second decimal place in the case of $r(x)$, and exactly the same for $q_{0.9}(x)$). This result is not surprising if we note from Table 5.1 that our sample size $\left(\sum_{i=1}^{p} k_i = 4584\right)$ is large, and as is well known, in this case the posterior distribution is dominated by the data.

Another interesting consequence of possessing a large sample data is that parameter uncertainty is reduced, and the variance σ_W^2, as defined in (9), is dominated by the stochastic variance σ_s^2. To illustrate this, note that our model satisfies $r_1(\theta) = \sigma^2(\theta) = kt_0\theta$, so that, under the objective (Gamma) prior, we can verify that $\sigma_p^2 = (kt_0)^2\beta_1\beta_2^{-2}$, and $\sigma_s^2 = kt_0\beta_1\beta_2^{-1}$, where $\beta_1 = \sum_{i=1}^{p}\sum_{j=1}^{k_i} x_{ij}$, and $\beta_2 = \sum_{i=1}^{p} k_i$. These results allow us to verify that, for the data of Table 5.1, σ_s^2 accounts for 94.82% of the variance σ_W^2.

To provide an example of how these Bayesian methods perform differently according to the available information, we simulated a different data set for our same forecasting estimation. For the new data, the number k of machines in operation is small, whereas, the failure rate is rather high. Our simulated data is shown in Table 5.6, and was obtained by considered a fixed number of machines

Table 5.5 Estimated performance measures using MCMC with $m = 10^6$ replications

Performance measure	Simulation estimate	Halfwidth
$r(x)$	$\hat{r}_{MC}(x) = 1.86013$	0.00254
$q_{0.9}(x)$	$\hat{q}_{0.9}^{MC}(x) = 4$	0

Table 5.6 Simulated data for 10 machines and a rate $\theta = 1$

Month (i)	Machines in operation (k_i)	Failures $\left(\sum_{j=1}^{k_i} x_{ij}\right)$
January (1)	10	8
February (2)	10	13
March (3)	10	13
April (4)	10	14
May (5)	10	9
June (6)	10	14
July (7)	10	12
August (8)	10	8
September (9)	10	11
October (10)	10	13
November (11)	10	13
December (12)	10	16
Total	120	144

in operation $k_i = 10$, $i = 1, \ldots, 12$, with failures distributed according to a Poisson distribution with rate $\theta = 1$ failures/month. The estimation of $r(x)$ and $q_{0.9}(x)$ using $m = 10^6$ replications, $k = 10$ and $t_0 = 1$ are summarized in Table 5.7 for both the PS and the MCMC method. For the PS algorithm we considered the objective (Gamma) prior, whereas for the MCMC method we considered the uniform distribution on [0.8, 1.2].

In Table 5.7 we summarized the results for the estimation of $r(x)$ and $q_{0.9}(x)$ using different priors. In the case of PS, we are working with a non-informative prior, which is used to assume very little prior knowledge about parameter Θ. On the other hand, for the MCMC case, we considered a uniform prior on [0.8, 1.2] in order to provide more information on Θ. As we see from Table 5.7, the influence of the prior on the estimated parameters is far greater than that observed with the data of Table 5.1. Furthermore, in Fig. 5.4 we present plots of the c.d.f. of W obtained under the two different priors, where we can more easily recognize the difference. Note that the distribution of W under the objective (non-informative) prior exhibits a larger variance.

5.5 Conclusions and Directions for Future Research

In this chapter, we show how a simulation model can be used in conjunction with Bayesian techniques to estimate point and variability parameters needed for solving forecasting problems related to spare parts inventory. The advantage of using simulation as a forecasting tool lies in the fact that even very complex models that incorporate detailed information about the system under study may be analyzed to produce reliable forecasts.

Table 5.7 Estimated performance measures using PS and MCMC with $m = 10^6$ replications

Method	Simulation estimate	Halfwidth
PS	$\hat{r}(x) = 12.0449$	0.00594
	$\hat{q}_{0.9}(x) = 17$	0
MCMC	$\hat{r}_{MC}(x) = 11.2499$	0.00529
	$\hat{q}_{0.9}^{MC}(x) = 16$	0

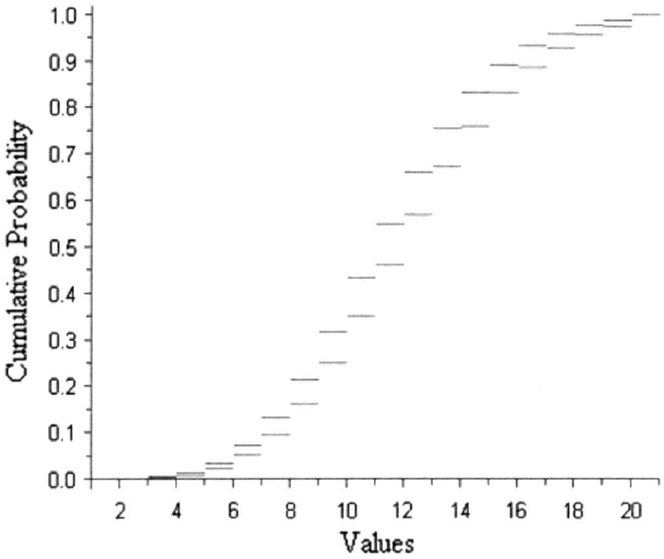

Fig. 5.4 Cumulative empirical distribution of the response variable using simulated failure data

This chapter presents a Bayesian framework for forecasting using two different methods. The first method, posterior sampling, requires a valid algorithm to generate samples from the posterior density function $p(\theta|x)$, whereas the second one, Markov chain Monte Carlo, can be implemented when the family of distributions corresponding to $p(\theta|x)$ has not been identified. We illustrate the potential applications of these methods by proposing two simple models, a model with failure time data and a model with censored data, which can be applied under a Bayesian framework to forecast the demand of spare parts.

We applied the model for censored data, using real data from a car and spare parts dealer in Mexico, and illustrated the application of both the PS and MCMC methodologies. In the case of the PS method, a non-informative prior was used to establish the posterior distribution, whereas, for MCMC, a uniform (subjective) prior was considered. The results of applying these methods allow us to arrive to several interesting conclusions.

Firstly, the accuracy of the point estimators, as well as the halfwidth, bias and mean square errors, are highly dependent of the number of replications m of the simulation experiments. Thus, it is very important, from a practitioner's point of view, to establish the number of replications m according to the desired accuracy, and the computation of the corresponding halfwidth is important to assess the accuracy of the point estimators obtained using simulation.

Secondly, the results using real data show that having a large sample size let the posterior distribution dominated by the data. Thus, the results obtained with an objective prior are very similar to the ones obtained with a more informative prior, since the priors have little influence on the posterior distribution. To illustrate this phenomenon, another set of data was simulated using a smaller sample with a higher failure rate. In this case, the influence of the prior is made evident and, thus, the results are considerably different.

Finally, the relevance of the methodologies illustrated in this chapter depends on the ability of the proposed model to imitate the real system, which is closed related to an adequate selection of the parameters for which sample data is available as well as the random components that incorporate the stochastic uncertainty. In this direction, the models proposed in Sect. 5.3 can be modified to resemble more accurately the system under study (e.g., assuming a different family of distributions for the time between failures). Likewise, this framework could also prove useful in forecasting problems related to supply chain management (see e.g., Kalchschmidt et al. 2006).

Acknowledgments This research has received support from the Asociación Mexicana de Cultura A.C. Jaime Galindo and Jorge Luquin have also participated. They both shared their knowledge on the auto-parts sector. As such, the authors want to express their most sincere gratitude.

References

Asmussen S (2003) Applied probability and queues. Springer, New York

Asmussen S, Glynn P (2007) Stochastic simulation algorithms and analysis. Springer, New York

Bartezzaghi E, Verganti R, Zotteri G (1999) A simulation framework for forecasting uncertain lumpy demand. Int J Prod Econ 59:499–510

Berger JO, Bernardo JM, Sun D (2009) The formal definition of priors. Ann Stat 37:905–938

Bernardo JM, Smith AFM (2000) Bayesian theory. Wiley, Chichester

Caniato F, Kalchschmidt M, Ronchi E, Veganti R, Zotteri G (2005) Clustering customers to forecast demand. Prod Plan Control 16:32–43

Chopra S, Meindl P (2004) Supply chain management, 2nd edn. Prentice Hall, New Jersey

Chung KL (1974) A course in probability theory, 2nd edn. Academic Press, San Diego

Croston JD (1972) Forecasting and stock control for intermittent demands. Oper Res Q 23:289–303

de Alba E, Mendoza M (2007) Bayesian forecasting methods for short time series. Foresight 8:41–44

Kalchschmidt M, Verganti R, Zotteri G (2006) Forecasting demand from heterogeneous customers. Int J Oper Prod Manag 26:619–638

Makridakis S, Wheelwright SC, Hyndman RJ (1998) Forecasting: methods and applications, 3rd edn. Wiley, New York

Muñoz DF (2010) On the validity of the batch quantile method in Markov chains. Oper Res Lett 38:222–226

Muñoz DF, Muñoz DG (2008) A Bayesian framework for the incorporations of priors and sample data in simulation experiments. Open Oper Res J 2:44–51

Rao AV (1973) A comment on forecasting and stock control for intermediate demands. Oper Res Q 24:639–640

Schmeiser B (1982) Batch size effects in the analysis of simulation output. Oper Res 30:556–568

Serfling RJ (1980) Approximation theorems of mathematical statistics. Wiley, New York

Song TW, Chih M (2008) Implementable mse-optimal dynamic partial-overlapping batch means estimators for steady-state simulations. In: Mason SJ, Hill RR, Mönch L, Rose O, Jefferson T, Fowler JW (eds) Proceedings of the 2008 winter simulation conference, IEEE, New Jersey, 426–435

Syntetos AA, Boylan JE (2001) On the bias of intermittent demand estimates. Int J Prod Econ 71:457–466

Wacker JG, Sprague LG (1998) Forecasting accuracy: comparing the relative effectiveness of practices between seven developed countries. J Oper Manag 16:271–290

Willemain TR, Smart CN, Schwarz HF (2004) A new approach to forecasting intermittent demand for service parts inventories. Int J Forecast 20:375–387

Zotteri G, Kalchsdmidt M (2007a) Forecasting practices: empirical evidence and a framework for research. Int J Prod Econ 108:84–99

Zotteri G, Kalchsdmidt M (2007b) A model for selecting the appropriate level of aggregation in forecasting processes. Int J Prod Econ 108:74–83

Chapter 6
A Review of Bootstrapping for Spare Parts Forecasting

Marilyn Smith and M. Zied Babai

6.1 Introduction

Inventory forecasting for spare parts constitutes a very important operational issue in many industries, such as automotive and aerospace. It has been reported that the stock value for spare parts may account for up to 60% of the total stock value in any industrial setting (Johnston et al. 2003). Since demand for spare parts arises from corrective and preventive maintenance related activities, generally these items are characterized by a highly variable demand size, with many periods when there is no demand at all. Such patterns are often called sporadic or intermittent.

Intermittent demand patterns are built from constituent elements (demand sizes and intervals) and they are very difficult to forecast. Croston's method (Croston 1972) and its variants (in conjunction with an appropriate distribution) have been reported to offer tangible benefits to stockists forecasting intermittent demand. Nevertheless, there are certainly some restrictions regarding the degree of lumpiness that may be dealt with effectively by any parametric distribution. In addition to the average inter-demand interval, the coefficient of variation of demand sizes has been shown in the literature to be very important from a forecasting perspective (Syntetos et al. 2005). However, as the data become more erratic, the true demand size distribution may not comply with any standard theoretical distribution. This challenges the effectiveness of any parametric approach. When SKUs exhibit a lumpy demand pattern one could argue that only non-parametric bootstrapping approaches (that do not rely upon any underlying distributional assumption) may provide opportunities for further improvements in this area.

M. Smith (✉)
Winthrop University, Rock Hill, SC, USA
e-mail: smithm@winthrop.edu

M. Z. Babai
BEM Bordeaux Management School, Bordeaux, France
e-mail: Mohamed-zied.babai@bem.edu

N. Altay and L. A. Litteral (eds.), *Service Parts Management*,
DOI: 10.1007/978-0-85729-039-7_6,

Non-parametric bootstrapping approaches rely upon sampling randomly individual observations from the demand history to build a histogram of the lead-time demand distribution. Since the seminal work of Efron (1979), there has been considerable research literature that deals with bootstrapping approaches for forecasting intermittent demand items (e.g., Snyder 2002; Willemain et al. 2004; Porras and Dekker 2008). The work that has received most attention is that published by Willemain et al. (2004), in which the authors claimed improvements in forecasting accuracy achieved over parametric approaches. However, this work has been criticized in terms of its methodological arrangements and experimental structure. This also has motivated a number of calls to academics and practitioners to broaden the empirical knowledge-base in this area (e.g., Gardner and Koehler 2005). The objective of this chapter is: (i) to review the literature on bootstrapping methodologies, focusing in particular on those related to spare parts inventory forecasting; and (ii) to present the algorithmic steps associated with their implementation in practice. The latter may serve as an information repository for future research and empirical comparisons in this area.

The remainder of this chapter is organized as follows. In Sect. 6.2, a brief description of the research on the origins of the bootstrapping approach is provided. Section 6.3 presents a detailed review of bootstrapping, also positioning it in the context of inventory forecasting for spare parts. Finally, some natural extensions for further work in this area are given in Sect. 6.4. Details related to the implementation of the bootstrapping methods discussed in this chapter are presented in the Appendix.

6.2 Research Background on the Bootstrapping Approach

6.2.1 Efron's Introduction of the Bootstrap

During the 1977 Rietz Lecture at the Seattle joint statistical meetings, Professor Bradley Efron introduced the bootstrap technique to estimate the sampling distribution of an observed sample. Later, the lecture was published in *The Annals of Statistics* (Efron 1979). The paper presented the mathematical development for bootstrapping as an extension of the jackknife method, which had first been introduced by Quenouille for estimating bias, and then developed further by Tukey for estimating variance, both of which were described by Miller (1974). Professor Efron was asked in an interview if he had an application in mind when he developed the bootstrap, but he indicated that he had been trying to determine why the jackknife did not always give dependable variance estimates (Aczel 1995). His study of the jackknife led him to the bootstrap, and while he modestly stated in the abstract of the landmark work that the bootstrap worked "satisfactorily", he demonstrated bootstrap's superiority over the jackknife in estimating the sample median, variance of a sample mean, and estimating error rates in a linear discrimination problem. The key to the bootstrap was that it did not just use data from a population (n), but it took a sample from the set with the mass of 1/n, and replaced it in a set to form a new sample. The sample and replacement steps, or making bootstrap samples, are repeated for a number of replications. These bootstrap samples can then be analyzed for the distribution. Using direct calculations is too time consuming for most problems, but Efron presented a short example with $n = 13$, using values of zero or one for the variables. If a Taylor series expansion is used to evaluate the bootstrap samples, then the results are the same as the jackknife. More information on the algorithmic steps related to the application of Efron's method is given in Appendix A.1. Hence with modern computing technology (discussed later), the preferred method to generate the bootstrap samples is to use a Monte Carlo simulation, and 50 replications were run for each trial. In the example used, Efron was careful to note that the bootstrap provided frequency statements, not likelihood statements.

While Efron published several articles, monographs, and books related to the bootstrap, a limited number are reviewed here for their specific application to spare parts inventory forecasting. These articles capture the major statistical models developed from the bootstrap. In 1983, Efron and Gong presented the bootstrap, jackknife, and cross-validation from a more "relaxed" perspective. That is, the paper is more descriptive of the bootstrap technique, with limited mathematical equations and proofs. The paper introduces a nonparametric example of LSAT (law school admission test) scores and undergraduate GPA (grade point average). In this problem 100 replications were run, and then 200 replications, but the differences in the results were too small to be noted. If the reader is somewhat rusty in mathematical and statistical theory, it may be better to start with this 1983 paper, and then move to the original 1979 paper.

Later Efron (1987) extended the original bootstrap concept with bias-corrected confidence intervals, and further improved the confidence intervals to wider problems. The law school student data introduced earlier was again used to demonstrate the confidence intervals. Also, it was noted that while runs of 100 gave good results for the standard error estimates, trial and error indicated that 1,000 bootstrap replications are needed for determining confidence intervals. In 1990, Efron published bootstrap methods for bias estimates and confidence limits that generate the bootstrap samples the same way, but require fewer computations than the original models. After the bootstrap sample is generated, a probability vector (P) is generated where the ith component is the proportion of bootstrap sample equaling x_i. The law school data were used again to illustrate the method, using the sample correlation for P.

As an aside, Efron (2000) noted that the name bootstrap was taken from the fables of Baron Muenchausen (also spelled Munchhausen and Munchausen). These fables were published in the late 1700's, and with centuries of time and multiple translations, as well as fables published under the Baron's name which he did not write, an exact citation is difficult. However, in all the fables, the Baron has extraordinary powers, such as our modern Superman, but also the phenomenal tool making ability of Angus MacGyver from the 1980's television series. Professor

Roger Johnson (2009) has found a reference where the Baron falls a couple of miles from the clouds (another tall tale), into a hole nine fathoms deep. The Baron reports "Looking down, I observed that I had on a pair of boots with exceptionally sturdy straps. Grasping them firmly, I pulled myself with all my might. Soon I had hoist myself to the top and stepped out on terra firma without further ado." The Baron's adventures were entertaining with no moral to the story, but the moral to this story is to illustrate the power of the bootstrap statistical technique based on the origins of its name.

6.2.2 The Bootstrap and the Computer

The use of the computer has been integral to the bootstrap from the beginning, with the bootstrap samples in the landmark paper being generated by Monte Carlo simulation using a 370/168 computer. Efron reported the Stanford University computer time for those runs cost $4.00. As his work progressed, computer technology progressed, and by 1991 problems using up to 400 bootstrap samples were being run on a personal computer (Efron and Tibshirani 1991). In 1994, Thomas Willemain (1994) published a short paper describing how to use Lotus 1-2-3 to construct a basic spreadsheet for a bootstrap problem. The steps are well documented, so the procedure could easily be adapted to Excel.

On October 26, 1999, *Business Wire* reported that Smart Software, Inc. was releasing its new Intermittent Demand Forecasting for spare parts and other intermittent demand ("Smart Software brings…"). Smart and Willemain (2000) described the software, the bootstrapping methodology it uses and gave an example problem in the June 2000 *Performance Advantage.* The pending patent for the "system and method for forecasting intermittent demand" used by the software was granted to Smart Software, Inc. March 20, 2001 (Smart Software 2001) (A later section provides more information about the bootstrap model used in the patent). Today, Smart Software promises that SmartForecasts will "typically reduce standing inventory by as much as 15–20% in the first year, increase parts availability 10–20% and more, and reduce the need for associated costs of emergency transshipments" (Smart Software "Intermittent Demand…").

In 2002b, Smart Software's SmartForecasts was listed in a review of 52 forecasting packages. However, the review only listed support, price, and functionalities, such as data entry, file export, graphic capabilities, and the statistical methods used, but bootstrap was not listed as a choice for statistical technique and would have only been captured in "Other". The review did not provide any caparison of performance (Elikai et al. 2002). Sanders and Manrodt (2003) surveyed practitioners regarding their use and satisfaction of marketing forecast software, and their study could be used as a model for an updated study of inventory forecasting software use, features, performance, and user satisfaction.

6.3 Bootstrapping Methods for Spare Parts Forecasting

6.3.1 Foundational Work

The intermittent demand pattern for spare parts means that traditional inventory models, such as those with the economic order quantity (EOQ) or materials requirements planning (MRP) as their foundations, are not applicable, since EOQ models require a constant demand or a normal distribution, and MRP models require fixed lot sizes. Most papers on spare parts inventory or intermittent demand begin with a cursory introduction about the need for different models. These models assume spare parts usually have periods of zero demand, then a part breaks on some random interval, the part must be replaced, and the research issue is to be able to predict the demand for the replacement parts. That is, the random time between demands for spare parts and the number demanded do not tend to follow patterns for parametric methods, which require a readily distinguishable statistical distribution and meaningful measures of mean, standard deviations, etc. Kennedy et al. (2002) began their literature review with a more extensive discussion of the unique characteristics of maintenance inventories that would support the need for non-parametric models. The special data related to these parts, includes supplier reliability, stock-out objectives, inventory turn goals, age-based replacement of working parts, spare parts obsolescence, and repairable items, as well as some special cases such as emergency ordering.

Croston's 1972 landmark work on intermittent demand uses parametric statistics, but it serves as the benchmark for later non-parametric work using the bootstrap. He extended the exponential smoothing method by adding variables for the size of the non-zero demands when they occur and an estimated inter-arrival time between non-zero demands that follow a Bernoulli process. It should be noted that he assumed the demand variables and the time between demands were independent. He used an example with 180 observations, with demand occurring on average every six review periods, and the average demand 3.5 units with a standard deviation of 1.0 unit. The Croston model prevented stockouts 95% of the time, while the exponential smoothing model resulted in stockouts in 20% of the periods.

Bookbinder and Lordahl (1989) developed a model using the bootstrap to estimate the lead-time distribution (LTD) and determine re-order points. They then compared their bootstrap model results to inventory models based on normal distributions using simulated populations of varying shapes. The calculations were done on a TRS (Model III) microcomputer, with only samples of $n \leq 30$ considered. Seven different probability density functions were used: uniform, truncated normal, log normal, two-point, positively skewed bimodal, symmetric bimodal, and positively skewed normal. The algorithmic steps for the implementation of Bookbinder and Lordahl's method are given in Appendix A.2. The researchers presented a comprehensive table of results of the acceptability of service at .8, .9, and .95, which captured all seven tested distributions, different sample sizes and varying coefficients of variance. Their results show when the bootstrap performed better, when the normal better, when both were acceptable, and when neither was acceptable. Generally, the normal only performed better than or as well as the bootstrap for the log normal and truncated normal distributions. The bootstrap tended to dominate in the case of the two point distribution, and negative skew binomial, which would better model the demand for spare parts. Since the data was simulated, they could compare the known optimal costs to the costs from the bootstrap model and the costs based on the normal theories. The costs results were also presented for all the combinations of the service level results. The bootstrap estimates gave better cost results for the distributions with some type of bimodal data, as well as slightly better cost results for the uniform data. The authors also performed a Newman–Keuls test for an analysis of variance (ANOVA) for the bootstrap and normal procedures. The results showed that the bootstrap was much less sensitive to the lead time distribution.

Snyder (2002) used data from an automobile company to illustrate four models that used a parametric bootstrap for demand forecasting of slow and fast moving inventory items. Like other researchers, he considered exponential smoothing and Croston's method. In his Monte Carlo simulation, he needed estimates of mean, bias, and variance, so he used least squares estimates in their place making the model a parametric bootstrap. He also reviewed the Croston method and identified problems with how seed values are selected for the exponentially weighted mean averages, and developed a modified version called MCROST. MCROST was then modified for a log transformation of non-zero demands, giving the log-space adaptation (LOG). Lastly, he experimented with a modification of Croston's model that used variances instead of mean absolute deviation and a second smoothing variable was introduced to define how variable changes over time (AVAR). Details related to the implementation of the MCROST, the LOG and the AVAR methods are presented in Appendix A.3. The simulation used 10,000 replications with the key performance measure being the order-up-to level (OUL) that represented the ideal level of stock that achieved a 95% fill rate target while being as low as possible. For the slow moving item, which would most closely represent the typical demand for spare parts, the OUL's were very close.

6.3.2 SmartForecasts

Before the development of SmartForecasts, Willemain et al. (1994) tested and extended Croston's work. They used simulated data, as well as real world data to violate Croston's assumptions, and also compared Croston's model to exponential smoothing when some of his assumptions were violated. First, they gathered intermittent demand data from four organizations, in four very different industries, where the companies indicated these data were particularly challenging to forecast. Based on these data, they identified areas where Croston's assumptions may not apply. In the first scenario, the normal distribution for the demand was changed from normal to lognormal. While Croston had assumed no relationship between demand interval and demand size, Willemain et al. (1994) tested for both positive and negative cross correlations. Croston had also only considered the case of demand sizes to be independent, but Willemain et al. (1994) investigated both positive and negative correlations. In addition, they tested the scenarios for both positive and negative correlations between the demand intervals.

In their Monte Carlo comparison of Croston and exponential smoothing, Willemain et al. (1994) measured accuracy by mean absolute percentage error (MAPE), median absolute percentage error, root mean square error, and mean absolute deviation. For all four scenarios, Croston's model gave more accurate results for the true demand on all four measures than exponential smoothing. Further, on average, it was also more accurate than exponential smoothing for the industrial data.

At the 2002 APICS International Conference, Charles Smart, President of Smart Software, Inc presented more details about the SmartForecasts mentioned above (This presentation is also available as a white paper on the Smart Software web site, www.smartcorp.com). The research to develop the software was a joint effort of Smart Software and Rensselaer Polytechnic Institute through an Innovative Research Grant from the National Science Foundation (Thomas Willemain, the co-developer, is a Professor of Decision Sciences and Engineering Systems at Rensselaer, as well as Vice President of Smart Software). The developers studied 28,000 data series from nine companies in the US and Europe. The research confirmed that exponential smoothing and Croston's method were effective for forecasting demand per period for intermittent items, but both techniques were lacking in forecasting the entire distribution, especially when service level requirements are imposed. The paper gives an example of intermittent demand data for 24 periods and demonstrates how the bootstrap method can be used to estimate demand per period and the requirements to meet customer service levels. With an example lead time of 3 months, random demands for 3 months are taken to get an estimated lead time demand (bootstrap), and the process is repeated on the computer for thousands of times to get an estimated lead time demand distribution.

Smart's (Smart 2002a) conference presentation provided impressive results for SmartForecasts. Two organizations had tested the software using their own data. A warehouse with 12,000 SKUs got almost 100% accuracy using the software, first at a 95% service level, and then at a 98% service level. An aircraft maintenance operation used their data on 6,000 SKUs to estimate that if they had used the SmartForecasts, they could have saved \$3 million per year in inventory holding costs. The software had already helped NSK Corporation save \$1 million in their aftermarket business unit and increase on-time delivery above 98%, and NSK predicted future savings of about \$3 million.

Willemain et al. (2004) provided more details about the mathematical model used by SmartForecasts, as well as statistical comparison of its performance to exponential smoothing and Croston's method. Again, the key improvements in the model are the lead time distribution based on bootstrapping and the method of assessing the accuracy of the forecast. With respect to the use of the bootstrap, this model expands the inventory applications of Bookbinder and Lordahl (1989) and Wang and Rao (1992), since neither of these applications considered the special case of intermittent demand. The paper presents general information about the nine organizations used to develop the model, as well summary statistics for their respective demand data.

The bootstrap was modified in three ways to better model the intermittent inventory data with autocorrelation, frequent repeated values, and relatively short series. First, positive and negative autocorrelation are added, since demand can run in "streaks". These autocorrelations are modeled using a two state, first order Markov process to get zero and non-zero demands. The next step is to generate numerical values for the nonzero forecasts. If only the bootstrap replacement values were used, then only past values could occur in the future. The model then uses a patented "jittering" process to allow more variation around the nonzero values. The example is given that a 7 may be replaced with a 7 or 6 or 10. The authors report that the jittering was shown to improve the accuracy, especially for small sample sizes. More information on the algorithmic steps for the implementation of Willemain et al.'s method is given in Appendix A.4.

The lead time distribution for the model was evaluated and compared to exponential smoothing and Croston's method. Results were examined on the upper tail as a measure of the ability of the model to satisfy the availability requirement and on the lower tail as a measure of holding costs. While the bootstrap method was shown to be the most accurate forecasting method, all the methods were least well calibrated in the tails of the distributions. The authors concluded their paper with three significant suggestions for further work. First, they recognized that there are problems with the jittering step, and it will not work for items that are sold in different size cases. However, in the paper it is not clear exactly how one determines the standard normal random deviate for this step. The second issue is nonstationarity, or the fact that the demand distributions for the items may be changing due to seasonality, life cycle stage of the product, or some other complication or trend. Finally, the authors acknowledge that they began the research with the idea that the intermittent demand followed some kind of Poisson process. While they were unable to find an acceptable Poisson based model, they still consider this an avenue for future research.

Michael Lawrence, editor of the *International Journal of Forecasting,* published a special commentary following the article described above. He noted the unique aspect, perhaps the first ever, of publishing a patented technique in an academic journal. One of the reviewers had expressed a concern about other researchers being able to use and extend these contributions, and then publish their comparisons with patent protected methods. In their response to these reviewer comments, Willemain et al. (2005) stressed the importance of researchers being able to patent their work and gain the potential economic benefits of the work. Further, they note that discussions of patented algorithms do not result in legal patent infringement. Finally, they offer that "it is easy to arrange a nominal licensing fee for researchers doing non-competitive work" (Lawrence 2004).

More discussion ensued about the work described above. Gardner and Koehler (2005) expressed reservations about the work because Willemain et al. had not used some later published research for the simple exponential smoothing model and Croston's model. The comparison to the exponential smoothing method used an estimated standard deviation based on work from 1959, but this estimate had been improved by later researchers, including Snyder et al. (1999). Willemain et al. (2004) assumed a normal distribution for the lead time demand in the exponential smoothing model, but Gardner and Koehler suggested a bootstrap lead time demand based on the work of Snyder et al. (2002). Gardner and Koehler also noted published research that improved upon the original Croston model, but Willemain et al. had only compared their work to the original Croston. In their response, Willemain et al. acknowledge the use of the early versions of exponential smoothing and Croston's method, and they encourage researchers to continue this stream of research by comparing the changes proposed by Gardner and Koehler to the bootstrap method. However, they note that there should be more research to determine how these proposed changes to exponential smoothing and Croston would compare when their new accuracy metric was applied.

The patent granted to Smart Software, Inc. for the "System and Method for Forecasting Intermittent Demand" provides more information than the articles mentioned above. Some of the patent's references are provided in this chapter, but the patent references more book chapters and monographs, as well as more broadly related topics such as forecasting accuracy and statistical smoothing methods, which are outside the scope of this work.

The patent includes five high level flow diagrams (Figs. 6.1–6.5) of the computer system, software, method of forecasting, selection of demand values, and the method of forecasting using a subseries method. Figure 6.6 in the patent, the experimental demand data, is the same as Table 6.2 in Willemain et al. (2004). The patent includes subseries methods (normal and log normal) which approximate the distribution by using the sums of overlapping distributions. This component of the patent is not featured in Willemain et al. (2004). However, the patent reports that the subseries forecasts were more accurate than exponential smoothing and Croston, and can be computed faster than their bootstrap. Figure 6.7 is the same as Table 6.3 in Willemain et al. (2004), except Fig. 6.7 also includes accuracy results for the subseries models. Figures 6.8, 6.9 and 6.10 show the forecasting accuracy of exponential smoothing, Croston's method, bootstrap, as well as the subseries normal and subseries log normal, as measured by a table of mean log likelihood ratios, chi-square values, and a graph of mean log likelihood values, respectively.

The patent gives a small example to aid understanding the concepts, with intermittent demand for 12 months presented in Tablc 6.1. Table 6.2 lists examples of five replications for possible demand and lead time distributions over months 13–16 (Table 6.3 is the same as Table 6.1). Table 6.4 gives examples of four different 12 month strings of zero and non-zero (1) data generated from Markov transition probabilities based on the LTD in Table 6.1. Table 6.5 illustrates how bootstrap converts the non-zero values (the 1's) in Table 4 to integers.

Then Table 6.6 presents the demand values which were determined by jiggering the bootstrap demand values from Table 6.5. The jittering process is presented in general terms, but detailed enough to more clearly explain the specific data.

6.3.3 *Other Significant Work on Bootstrapping and Spare Parts Inventory*

Hua et al. (2006) developed an integrated forecasting method (IFM) method to predict auto correlated nonzero demands and presented two methods of assessing forecast methods. Their forecasting method uses a first order Markov process, first estimating the nonzero demand, and then the lead time distribution. In their algorithm, a user specified a variable to adjust for nonzero demand that may have some special explanation. Once the nonzero demands have been identified, they used a bootstrap to estimate the lead time distribution. Their research also included the development of two new forecasting performance assessment measures. One measure is the error ratio of nonzero demand judgments (ERNJ) over time, which would apply to binary forecasts, and then used the average of the average of the ERJN (AERNJ) for each forecasting method. They also developed a mean absolute percentage error of lead time distribution (MAPELTD). The new methods and measures were compared with exponential smoothing, Croston's method, and a modified bootstrap (MB), which is the same as the bootstrap method described by Willemain et al. (2004), but without their patented jittering technique. Based on 40 kinds of spare parts data from a petrochemical enterprise in China, the AERNJ performed significantly better than the MB over an 8-month period. Also, IFM was better at forecasting occurrences of nonzero demand than MB. When the IFM was compared to the three other forecasting methods mentioned above, it was found to better forecast LTD.

In Hua et al.'s (2006) examples used to demonstrate their model, they included plant and equipment overhaul plans to illustrate the special times when nonzero demands could be expected. Also, they suggest that current information technology systems make it possible to incorporate other information about explanatory variables into the model. In a related paper about forecasting in general, Bunn and Wright (1991) suggested the development of bootstrap models that incorporate judgments. They identify four general, common issues that must be addressed within all classes of model building which incorporate judgment. These issues, which would also seem to apply to forecasting the intermittent demand for spare parts, are: variable selection, model specification (in terms of specifying the relationships between variables), parameter estimation, and data analysis. A recent work by Syntetos et al. ("The effects of integrating management judgment...," 2009b) presents a model that integrates management judgment into the forecasted intermittent demand for end products. Their work used data on slow moving items in an international pharmaceutical company, which uses a commercial forecasting software package. The model for forecast adjustments and assessing forecast performance incorporated concepts previously developed and tested by the research team. The adjusted demand forecasts were better than the system forecast for 61% of the SKUs studied, which suggests potential for expanding the model to study judgment adjustments for spare parts forecasts. The comprehensive literature review in Syntetos et al. (2009a) provides examples of related types of forecasting models that have included judgment in the forecasts.

In 2008, Porras and Dekker published the results of a major study of re-order points for spare parts inventory. The study used data covering a 5-year period from a petrochemical company in the Netherlands with 60 plants, totaling 14,383 spare parts. The current method is based in the SAP, which does not capture and accommodate for the special case of the intermittent demand The parts were classified into criticality classes, demand classes (based on demand pattern and demand level), and price classes. The classes were combined, so that an item's three digit code indicated its criticality, demand, and price classes. First, classes were optimized, and then items within classes were optimized. The simulation model used an ex-ante approach, where once the distribution had been fitted, and then a different distribution was used for testing. It also used an ex-post approach where the same data set was used for fitting and testing. The performances of both approaches were tested. The researchers tested four methods for determining LTD: normal, Poisson, Willemain's method, and a new empirical method. Their new empirical method samples demands over blocks of time equal to the lead time length (capturing implicitly underlying auto-correlation structures). Details related to the implementation of this bootstrapping method are presented in Appendix A.5. The optimization model determines the lot size (Q) and the smallest reorder point that satisfies a 100% fill rate. Then the model determines cost savings using the holding cost and ordering costs of the current model and the model being tested.

Porras and Dekker (2008) gleaned several interesting findings from their study. When they used the ex-post approach, all of the tested models outperformed the current approach. The normal LTD performed best overall, but Willemain's model and their empirical model were close. Their empirical model produced cost savings of 1.05 million euros and Willemain's model saved .96 million euros over the current model. The results for the ex-ante were similar. In their conclusion, they reiterated the need for including high demands for preventative maintenance in spare parts inventory models.

Varghese and Rossetti (2008) developed a model for use on the spare parts used by the Naval Aviation Maintenance Program of the US Navy. The model is a Markov Chain demand occurrence (MC) and auto regressive (ARTA) to any demand amount (IDF), and it is a parametric bootstrap (PB); that is MC-ARTA-IDF-PB). They compared their model to the work of Croston (1972) and Syntetos (2001) on mean square error, mean absolute deviation, and mean absolute percentage error. With a bootstrap of 1,000 replications, they did not find any significant differences between the three forecasting methods, leading them to conclude that more work is needed on their estimation algorithm and to experiment with using their model for differing inventory policies.

Teunter and Duncan (2009) used UK Royal Air Force data for 5,000 items over a 6-year period to compare forecasting methods for spare parts. The objective of the system is to minimize stock levels, while meeting a service level constraint. They compared six methods of forecasting demand. A zero forecast was the benchmark, since traditional inventory models do not incorporate a forecast. The traditional forecasting techniques of moving average and exponential smoothing were used. They also considered Croston's original model for intermittent demands and the variation to Croston's model that was developed by Syntetos and Boylan (2005). Bookbinder and Lordahl's (1989) bootstrapping method was also included. To compare the first five methods, they used the relative geometric root mean square error (RGRMSE), which calculates the performance of one method compared to another as the ratio of the geometric mean of the squared errors. That is, they calculated the average mean square error for each item, and then calculated the geometric average of these numbers over the entire data set. When they used the RGRMSE, the zero forecast gave the best results, but the authors concluded that forecast performance cannot be measured strictly on error measures.

When Teunter and Duncan (2009) compared service levels, they used normal and lognormal for lead time demand, except for the bootstrap method, which generated its own lead time distribution. The six methods were compared for service level and a combination of service level and average inventory level. The zero forecast performed the worst. The bootstrap and the two Croston type methods performed the best, with the results being very close. However, initially all the methods gave results that were significantly below the ideal target. When the researchers realized that an order in a period is triggered by a demand in that period, they adjusted the lead time distribution to lognormal for the bootstrap and Croston based methods. The improvement in service level accuracy was significant.

6.4 Future Research

Based on the review above, a few ideas for future work surface. First, an empirical study of the commercial forecasting software packages that are available would be timely. Such a study could follow Elikai et al. (2002), but also include capture information about whether bootstrapping is used, and specifically whether the forecasting techniques are applicable to intermittent demand. Other work, similar to that of Sanders and Manrodt (2003) could examine the user satisfaction and performance of commercially available software for forecasting of spare part demand. A study like this was done by McNeill et al. (2007) with AMR Research, but it does not appear to be in the public domain.

In their response to a reviewer comment regarding the ability of academics to make comparisons of their work to the patented SmartForecasts (Lawrence 2004), Willemain et al. offered to make their patented SmartForecasts available to academic researchers for a "nominal licensing fee" to encourage future scholarly research in this area. So far, it does not appear that anyone has accepted the offer, even though it could help researchers understand and extend the patented "jittering" step. In addition, Willemain et al. (2004) suggested work on examining a Poisson based model for spare parts forecasting.

As noted above, the research to develop SmartForecasts used 28,000 inventory items. Hence, the researchers decided against individual item analysis and pooled data across items, which impacts error measures. However, the SmartForecasts patent notes that the forecasting program is designed to run on any computing hardware. So as hardware has improved, and continues to improve, then it is assumed that the feasibility of performing comparisons and error analysis on per item basis would increase.

Hsa et al. used plans for plant overhauls (scheduled preventative maintenance) in their model, and other researchers have suggested models that can incorporate other types of judgments (Bunn and Wright 1991; Syntetos et al. 2009b). As Hsa et al. noted, while management information technology improves and the integration abilities increase, this should become easier. Also, Willemain et al. (2004) had the considerations of seasonality, or other issues, as possible extensions of their work.

Appendix

A.1 Efron's Bootstrapping Method

Efron's bootstrapping method works according to the following steps:

1. Obtain historical demand data in chosen time buckets (e.g., days, weeks, months);
2. Express the lead-time L as an integer multiple of the time bucket;
3. Sample randomly with replacement L demands from the demand history;
4. Sum the sampled values to get one predicted value of Lead Time Demand (LTD);
5. Repeat steps 3–4 many times;
6. Sort and use the resulting distribution of LTD values.

A.2 Bookbinder and Lordahl's Bootstrapping Method

Bookbinder and Lordahl's method works according to the following steps:

1. Obtain historical demand data in chosen time buckets (e.g., days, weeks, months);
2. Express the lead-time L as an integer multiple of the time bucket;

3. Sample randomly with replacement L demands from the demand history;
4. Sum the sampled values to get one predicted value of Lead Time Demand (LTD);
5. Repeat steps 3–4 many times;
6. Calculate the mean and standard deviation of the resulting LTD values;
7. A theoretical density function of the LTD values is obtained by considering the empirical mean and standard deviation.

A.3 Snyder's Bootstrapping Methods

A.3.1 Modified Croston (MCROST)

1. Obtain historical demand data in chosen time buckets (e.g., days, weeks, months);
2. Express the lead-time L as an integer multiple of the time bucket;
3. From the demand history, determine least squares estimates of the parameters α, μ and σ;
4. Generate binary values $(x_1), (x_2) \ldots x_L$ for the indicator variable x_t from a Bernoulli distribution with probability p;
5. Use Monte Carlo random number generation methods to obtain values of the errors $(\varepsilon_1), (\varepsilon_2) \ldots \varepsilon_L$ generated from a normal distribution with mean zero and variance $x_t\sigma^2$;
6. Generate realizations $(y_1), (y_2) \ldots y_L$ of future series values by using the equations: $(y_t) = x_t\mu_{t-1} + \varepsilon_t$, $\mu_t = \mu_{t-1} + \alpha\varepsilon_t$ *and* $\mu_0 = \mu$;
7. Sum the L values of y_t;
8. Repeat steps 4–7 many times;
9. Sort and use the resulting distribution of LTD values.

A.3.2 Log-Space Adaptation (LOG)

The steps are the same as in the MCROST method except for the step 6, where the smoothing equations are modified to not allow for negative values as follows:

$$y_t = x_t\mu_{t-1} + \varepsilon_t, y_t^+ = x_t \exp(y_t), y_t = \begin{cases} \log(y_t^+) & \textit{if } x_t = 1 \\ \textit{arbitrary} & \textit{if } x_t = 0 \end{cases}, \varepsilon_t = x_t(y_t - \mu_{t-1}),$$

$$\mu_t = \mu_{t-1} + \alpha\varepsilon_t \quad \text{and} \quad \mu_0 = \mu$$

A.3.3 Adaptive Variance Version (AVAR)

The steps are the same as in the MCROST method except for the steps 5 and 6, where the equations are modified as follows:

$$\varepsilon_t \sim NID\left(0, \sigma_{t-1}^2\right), y_t = x_t\mu_{t-1} + \varepsilon_t, y_t^+ = x_t \exp(y_t), \mu_t = \mu_{t-1} + \alpha\varepsilon_t \quad \text{and} \quad \mu_0 = \mu$$

A second smoothing parameter β is also used in the following equation to define how the variability changes over time.

$$\sigma_t^2 = \sigma_{t-1}^2 + \beta x_t\left(\varepsilon_t^2 - \sigma_{t-1}^2\right), \sigma_0^2 = \sigma^2$$

A.4 Willemain's Bootstrapping Method

Willemain's method works according to the following steps:

1. Obtain historical demand data in chosen time buckets (e.g., days, weeks, months);
2. Estimate transition probabilities for two-state (zero vs. non-zero) Markov model;
3. Conditional on last observed demand, use Markov model to generate a sequence of zero/non-zero values over forecast horizon;
4. Replace every non-zero state marker with a numerical value sampled at random, with replacement, from the set of observed non-zero demands;
5. 'Jitter' the non-zero demand values—this is effectively an ad-hoc procedure designed to allow greater variation than that already observed. The process enables the sampling of demand size values that have not been observed in the demand history

   If X^* is the historical demand value selected at random:
   Generate a realization of a standard normal random deviate Z
   $$\text{Jittered value} = 1 + \text{INT}(X^* + Z.\sqrt{X^*})$$
   If Jittered value ≤ 0 Then: Jittered value $= X^*$

6. Sum the forecast values over the horizon to get one predicted value of Lead Time Demand (LTD);
7. Repeat steps 3–6 many times;
8. Sort and use the resulting distribution of LTD values.

A.5 Porras and Dekker's Bootstrapping Method

Porras and Dekker's method works according to the following steps:

1. Obtain historical demand data in chosen time buckets (e.g., days, weeks, months);
2. Express the lead-time L as an integer multiple of the time bucket
3. Consider the L successive demands from the demand history starting at the first time bucket
4. Sum the sampled values to get one predicted value of Lead Time Demand (LTD)
5. Repeat step 3–4 by starting at the second, third,… time bucket
6. Sort and use the resulting distribution of LTD values.

References

Aczel (1995) Interview with Bradley Efron. Irwin/McGraw-Hill Learning Aids http://mhhe.com/business/opsci/bstat/efron.mhtml. Accessed 8 June 2009

Bookbinder J, Lordahl A (1989) Estimation of inventory re-order levels using the bootstrap statistical procedure. IIE Trans 21(4):302–312

Bunn D, Wright G (1991) Interaction of judgemental and statistical forecasting methods: issues and analysis. Manag Sci 37(5):501–518

Croston J (1972) Forecasting and stock control for intermittent demands. Oper Res Q 23(3):289–303

Efron B (1979) Bootstrap methods: another look at the jackknife. Ann Stat 7(1):1–26

Efron B (1987) Better bootstrap confidence intervals. J Am Stat Assoc 82(397):171–185

Efron B (1990) More efficient bootstrap computations. J Am Stat Assoc 85(409):79–89

Efron B (2000) The bootstrap and modern statistics. J Am Stat Assoc 95(452):1293–1295

Efron B, Gong G (1983) A leisurely look at the bootstrap, the jackknife, and cross validation. Am Stat 37(1):36–48

Efron B, Tibshirani R (1991) Statistical data analysis in the computer age. Science 253(5018):390–395

Elikai F, Badaranathi R, Howe V (2002) A review of 52 forecasting software packages. J Bus Forecast Summer 2002:19–27

Gardner E, Koehler A (2005) Comments on a patented bootstrapping method for forecasting intermittent demand. Int J Forecast 21:617–618

http://www.amrresearch.com/Content/View.aspx?compURI=tcm%3a7-34107&title=Service+Parts+Planning+and+Optimization+Landscape%3a+Saving+Millions+Through+Inventory+Reductions+and+Increased+Service+Levels. Accessed 15 June 2009

Hua Z, Zhang B, Yang J, Tan D (2006) A new approach of forecasting intermittent demand for spare parts inventories in the process industries. J Oper Res Soc. doi:10.1057/palgrave.jors.2602119

Johnson R (2009) Statistical quotes and humor. http://tigger.uic.edu/~slsclove/stathumr.htm. Accessed 2 June 2009

Johnston FR, Boylan JE, Shale EA (2003) An examination of the size of orders from customers, their characterization and the implications for inventory control of slow moving items. J Oper Res Soc 54:833–837

Kennedy W, Patterson J, Fredendall L (2002) An overview of recent literature on spare parts inventories. Int J Prod Econ 76(2):201–215

Lawrence M (2004) Commentary on: a new approach to forecasting intermittent demand for service parts inventories. Int J Forecast 20:389–390

McNeill W, Fontanella J, Ruggles K (2007) Service parts planning and optimization landscape: saving millions through inventory reductions and increased service levels. http://www.amrresearch.com/Content/View.aspx?compURI=tcm%3a7-34107&title=Service+Parts+Planning+and+Optimization+Landscape%3a+Saving+Millions+Through+Inventory+Reductions+and+Increased+Service+Levels. Accessed 15 June 2009

Miller R (1974) The jackknife—a review. Biometrika 61:1–15

http://www.smartcorp.com/pdf/Intermittent_Demand_Forecasting_WhitePaper.pdf. Accessed 27 May 2009

Porras E, Dekker R (2008) An inventory control system for spare parts at a refinery: an empirical comparison of different re-order point methods. Eur J Oper Res. doi:10.1016/j.ejor.2006.11.008

Sanders N, Manrodt K (2003) Forecasting software in practice: use, satisfaction, and performance. Interfaces 33(5):90–93

Smart C (2002) Accurate intermittent demand/inventory forecasting: new technologies and dramatic results. In: 2002 international conference proceedings, American Production and Inventory Control Society

Smart C (2002) Accurate intermittent demand forecasting for inventory planning: new technologies and dramatic results. http://www.smartcorp.com/pdf/Intermittent_Demand_Forecasting_WhitePaper.pdf. Accessed 27 May 2009

Smart Software (Oct 26, 1999) Smart Software brings new forecasting technologies to market. Business Wire: 0406

Smart Software (2001) U.S. patent no. 6,205,431 B1. US Patent and Trademark Office, Washington, DC

Smart Software. Intermittent Demand Planning and Service Parts Forecasting. http://www.smartcorp.com/intermittent_demand_planning.asp Accessed 27 May 2009

Smart C, Willemain T (2000) A new way to forecast intermittent demand. Perform Adv June 2000:64–68

Snyder R (2002) Forecasting sales of slow and fast moving inventories. Eur J Oper Res 140: 684–699

Snyder R, Koehler A, Ord J (1999) Lead time demand for simple exponential smoothing: an adjustment factor for standard deviation. J Oper Res Soc 50:1079–1082

Snyder R, Koehler A, Ord J (2002) Forecasting for inventory control with exponential smoothing. J Forecast 18:5–18

Syntetos A (2001) Forecasting of intermittent demand. PhD Dissertation. Brussels University

Syntetos A, Boylan J (2005) The accuracy of intermittent demand estimates. Int J Forecast 21:303–314

Syntetos AA, Boylan JE, Croston JD (2005) On the categorization of demand patterns. J Oper Res Soc 56:495–503

Syntetos A, Boylan J, Disney S (2009a) Forecasting for inventory planning: a 50 year review. J Oper Res Soc 60:S149–S160

Syntetos A, Nikolopoulos K, Boylan J, Fildes R, Goodwin P (2009b) The effects of integrating management judgment into intermittent demand forecasts. Int J Prod Econ 118:72–81

Teunter R, Duncan L (2009) Forecasting intermittent demand: a comparative study. J Oper Res Soc 60:321–329

Varghese V, Rossetti M (2008) A parametric bootstrapping approach to forecast intermittent demand. In: Proceedings of the 2008 industrial engineering research conference, pp 857–862

Wang M, Rao S (1992) Estimating reorder points and other management science applications by bootstrap procedure. Eur J Oper Res 56:332–342

Willemain T (1994) Bootstrap on a shoestring: resampling using spreadsheets. Am Stat 48(1):40

Willemain T, Smart C, Shockor J, DeSautels P (1994) Forecasting intermittent demand in manufacturing: a comparative evaluation of Croston's method. Int J Forecast 10(4):529–538

Willemain T, Smart C, Schwartz H (2004) A new approach to forecasting intermittent demand for service parts inventories. Int J Forecast 20:375–387

Willemain T, Smart C, Schwartz H (2005) Author's response to Koehler and Gardner. Int J Forecast 21:619–620

Chapter 7
A New Inventory Model for Aircraft Spares

Michael MacDonnell and Ben Clegg

7.1 Introduction

This paper presents an optimization model for the planning of spare parts levels used in the support of aircraft operations. In particular, the model addresses the problem of planning rotable inventory: serialized items that are maintained and restocked rather than discarded upon failure. This special problem can be characterized as an operational situation where inventory levels do not change over the planning period (typically a year) and the flow of inventory occurs in a closed loop, from operation (installed on an aircraft), to the repair cycle, to spares inventory and returning to operation. Thus inventory items classed in this way will survive for the lifetime of the parent aircraft fleet. Changes in inventory are made by buying and selling overhauled items in the case of a mature fleet. The trigger for inventory activity in this operational situation is the failure of an item and its removal from service: this then prompts a repair activity, which will later result in replenishment. Thus demand is based purely on forecast failures, so there is no need for ordering decisions in the short term. Inventory changes are reviewed and planned in the medium term. Historically, inventory levels tend to creep up in response to epidemics of failure. Given the pressure to maintain operational reliability, it is usually considered acceptable to high levels of rotable stocks, even though utilization of these stocks may be poor.

Treatment of the rotable problem in the literature is sparse: much of the prior work addresses the forecasting of demand for consumable spares, even when the operational context describes repairable assemblies. Much of the literature

M. MacDonnell (✉)
School of Business, University College Dublin, Dublin 4, Ireland
e-mail: michael@ucd.ie

B. Clegg
Aston Business School, Aston University, Birmingham B4 7ET, England
e-mail: b.t.clegg@aston.ac.uk

N. Altay and L. A. Litteral (eds.), *Service Parts Management*,
DOI: 10.1007/978-0-85729-039-7_7,

addresses the problem at the item level, whereas the present work optimizes at a system level. Given the stochastic nature of demand, the problem is re-stated to give an objective of a service level (SL) target for an entire pool of inventory, rather than applying a service level to each line item. In this manner it is possible to skew inventory holdings by cost: the objective function changes from "exceed target SL for each part" to "exceed target SL for the system of parts, at minimum cost". One method in the literature, also observed in practice—Marginal Analysis—addresses the problem at the system level, but is not an optimization.

A large-scale optimization model is developed and formulated for solution as a binary integer linear program. This model is solved for a range of sensitivity values and re-formulated for multiple operating scenarios, reflecting changing operational criteria, such as increased fleet size, reduced repair times and varying target SLs.

Current practice from the field and the literature is replicated and compared with the new solution. The linear programming optimization is shown to produce superior results to current practice: inventory investment can be reduced by 20% or more without reducing SL. This is achieved through a coordinated increase in availability of lower-cost parts with a reduction in holdings of higher-cost parts.

While the new solution is computationally intensive, its benefits exceed implementation effort and the model may be applied to other operational settings where expensive spares are held.

7.2 Literature Review and Problem Description

This paper looks at published model specifications and experience with specific reference to the aircraft rotable inventory problem. Thus the focus is narrowed to the rotable problem only, and does not consider the cases of consumable items, which are better understood in mainstream practice. There are two main possible approaches to classifying the rotable scheduling problem: planning parts at the individual level, and planning for systems of parts where demand can be combined and considered together. The latter is preferable since it more closely reflects the reality of stochastic demand and is seen to give better results.

Demand for spares may arise in several ways (Ghobbar and Friend 2003a): due to hard time constraints (for example, a landing gear assembly must be changed after 500 flights), on condition (an item is inspected against a defined standard, e.g., tyre tread depth) and condition monitoring (real time diagnosis of performance, e.g., brake pad wear). Rotables can generally be considered as arising for maintenance on condition, meaning that their performance is observed to be deficient upon inspection, or often in operation. However, it is best to consider rotables as arising for removal through condition monitoring, since their failure will usually be observed during operation, so the removal does not typically result from a planned inspection. Ghobbar's analysis gives a detailed statistical analysis of parts demand for maintenance items but there is no optimization involved.

Most companies owning rotable inventory follow practices recommended by the aircraft OEMs, chiefly Airbus and Boeing; these policies are typically limited to individual line item treatments. Airbus advise that 99% of their listed spare parts are rotables and recommends a simple calculation to derive the mean number of expected demand events of a given part in a year (Rutledge 1997):

$$E = FH \times n \times N \times (1/MTBUR \times 365) \times TAT$$

In the above equation, E = expected demand, FH = average flight hours per aircraft, n = number of units on an aircraft, N = number of aircraft in the fleet, MTBUR = mean time between unscheduled removals and TAT = repair turn around time. Note that E is proportional to repair time, TAT: E is the number of failures multiplied by the proportion of the time that a failed part is unavailable due to being in the repair process. Thus, for example, if the number of failures is 10 and the proportion of time that a spare is in the repair cycle is one-tenth of a year (36 days), then the demand for the inventory calculation is 1.

Using the above calculation for initial provisioning (spares purchase when a fleet is first commissioned), the mean inventory demand figure is applied to a Poisson distribution to give an inventory count that meets a stated probability of demand being satisfied. Thus, if there is a need to meet 90% of requests for the above item, which has a mean expected demand of 1 (when TAT is taken into account), then the number of parts giving a probability of over 90% with a mean of 1 is 2. If 2 parts are held in stock there is a 92% chance that all requests are met, if the average demand over a long time is 1.

Several measures are proposed to reduce the cost of initial provisioning (Panayiotou 1998): price reduction (not a long-term measure), improved reliability, reduced shop processing time and cost-optimized planning. The last item involves some phasing of the provision of expensive spares, so that not all spares are bought at the outset but are introduced as the fleet accumulates flying time. This assumes that failures follow the Poisson distribution and needs to be managed carefully. The cost-optimized planning principle also introduces reduced SL targets for non-essential items (Haupt 2001). Thus, while the required SL for items of essentiality code 1 ("no-go") is maintained at 94–96%, values for items with dependent essentiality ("go-if", i.e., systems with redundancy) are assigned 85–92% and non-essential ("go") items are required to be provided 70–80% of the time. Thus average values of 95, 89 and 75% are prescribed, as compared with 95, 93 and 90% in common use.

Of an airline's spares inventory, rotables or line replaceable units account for 25% by quantity and 90% by value (Haupt 2001).

The line item-level calculations are developed further in the Models chapter later as the Poisson model (Model 1).

While the above treats items at the individual level, there is also scope for considering the combined demand for parts. Haas and Verrijdt (1997) discusses a model for a multi-indentured assembly, an engine, which comprises three levels of indenture. Several stages of supply, or echelons, are also modeled. The model consists of an aggregation of Poisson forecasts for individual part demand to meet an overall SL target. However, there is no account of cost in the model, other than aggregation: since an engine is effectively a hierarchical grouping of modules, there is not much scope for cost optimization.

Adams (2004) compares item-level and system-level approaches to aircraft rotable optimization, concluding that item-level forecasting is the least risky but will over-provide spares. Meanwhile Marginal Analysis combines parts, takes account of costs and can be modeled for multi-echelon scenarios. However, while Marginal Analysis gives good results, they are not optimal. A genetic algorithm approach is also tested—this is a complex approach, which may improve on Marginal Analysis but is not necessarily optimal.

The Marginal Analysis method was first developed by Sherbrooke (1968) in a military setting, in the Multi-Echelon Technique for Recoverable Item Control (METRIC) model. It is interesting to note that the US Air Force investment in recoverable items (rotables) is reported at $5bn in 1967. The METRIC model addresses overall optimality of spares stocks and the balanced distribution of spares in a network with two echelons, or levels of supply: bases and depots. Bases are the locations from which aircraft operate (as in the supply chain schematic Fig. 2.1) while depots are central inventory locations, usually with comprehensive repair capabilities. The model represents failed requests for parts as back-orders, so that a failed request survives until it is filled. The model is of type (s − 1, s), i.e., a replacement is ordered when a part is taken from stock.

The approach adopted in this study differs from the METRIC model because:

(a) demand is presented here in mean terms, not recurring, so that behavior over a planning period is represented by a SL;
(b) back-orders are not tolerated in commercial aviation—some action must be taken to satisfy the demand, usually borrowing a part or expediting delivery in the supply chain;
(c) Sherbrooke acknowledges that the marginal contribution of increasing part numbers should be concave—marginal contribution should reduce as the number of parts increases—but this is not the case and so the Marginal Analysis approach is flawed.

The METRIC model is improved on with MOD-METRIC and VARI-METRIC versions (Sherbrooke 1986), with better forecasting of expected back-order rates. The "best" results, closest to optimal, are derived by simulation and the new versions of the METRIC model are shown to be closer to those that are presumed optimal.

General Electric Rotable Services claim massive savings in inventory through the use of Marginal Analysis—see Fig. 7.4. Where current practice achieves 77% SL with $50M in inventory, the same performance is claimed with $28M in inventory, a 44% reduction, through the use of Logistechs k2s (knowledge to spares) solution, in which GE holds a stake. Figure 7.4 also shows that, for the existing $50M in inventory, the operator could increase SL from 77 to 90%

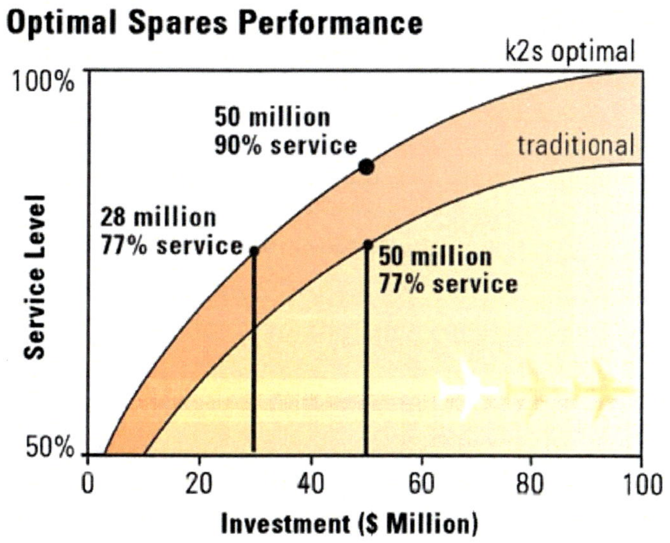

Fig. 7.1 Cost/SL gain using marginal analysis (GE Engine Services 2002)

through optimal planning. The system provides demand forecasting, optimization and simulation to customer including Air Canada and America West (Logistechs 2006) (Fig. 7.1).

Studies have been performed on demand prediction for aircraft spares (Ghobbar 2004) and the best inventory policies for determining optimal stock levels for spares (Friend et al. 2001; Ghobbar and Friend 2003b) but they generally treat individual line items so are not explored further here, since it is intended to consider only pooled inventory here, i.e., the combined cost and performance of many parts together.

It is claimed that cost reductions of 30% can be achieved by pooling spares among airlines (Kilpi and Vepsalainen 2004) but there is a trade-off in short-term SL, since it will be necessary to incorporate a delay of typically 12 h to allow for provision of pooled spares. This work once again considers individual line items.

A genetic algorithm model is used to find optimal inventory levels for multiple locations (Lee et al. 2007) but is again confined to a single line item.

A simulation model is used to determine a re-stocking priority order for multiple bases holding rotables (Lye and Chan 2007), but this does not treat groups of parts as a system, focusing on the airline network and its distribution of demand.

In the above instances, there are models of varying sophistication, dealing with the problem of multiple locations. Marginal Analysis seems to be the sole model for grouping parts together with the aim of achieving a desired SL while arranging the inventory selection so as to minimize cost.

Computer World (2005) gives a detailed account of Southwest Airline's supply chain optimization project, which uses a range of mathematical programming solutions to plan inventory levels for its fleet of 385 Boeing 737 aircraft, with an average utilization time of 12 h per day. Mathematical programming is applied to expensive, slow-moving critical parts, but the details of the model are not disclosed. An optimization-based heuristic, Constrained Marginal Analysis, is used for faster-moving parts: this recognizes the problem that Marginal Analysis has with parts with infrequent demand, namely marginal contribution that does not diminish continually. Through its supply chain optimization project, which included reducing inventory in the supply chain as well as in stores, Southwest cut its 2003 budget for rotable purchase from $26 to $14M, identified $25M of excess inventory and avoided repair costs of $2M, while increasing SL from 92 to 95%.

A model has been proposed using a small example for illustration (MacDonnell and Clegg 2007), which involves using a linear programming model to pick an optimum selection set of inventory levels for a connected set of parts. The resulting solution should meet a target SL at the least overall cost. It is predicted that this can be achieved on a larger scale by increasing stocks of cheap parts while reducing stocks of expensive parts. The consequential equivalent SLs for the individual inventory items will therefore deviate from the target SL but the global SL is maintained. An issue that remains to be addressed is the different essentiality levels of parts in the same inventory, as they have different SL requirements and should contribute differently to a connected solution. A version of this model has been tested with a large MRO (maintenance, repair and overhaul service provider) SR Technics (Armac 2007), showing potential for a 25% reduction in capital investment in inventory based on a 4-month trial reviewing new purchase requests. As part of the study, the company showed that a 2-day reduction in component repair cycle time would enable a reduction of $7.5M in their UK inventory.

7.3 Formulation and Implementation of the New Solution

Table 7.1 shows the values of a Poisson distribution for a range of mean values. This is used as the basis for planning inventory levels at the line-item level in current industry practice. Thus, if a given part number fails an average of 3 times a year, then a stock of 6 spares is needed to ensure that at least 95% of requests for spares are met. Note that demand is scaled down to allow for the return of stock: for example, if a removed item is missing from stock (in the repair cycle) for one-tenth of a year, then it is available for nine tenths of the year, so demand is scaled down by a factor of ten. In such a case, the demand figures 2, 3, 4 and 5 in Table 7.1 would equate to failure rates of 20, 30, 40 and 50.

The conventional approach described above is limited as follows:

Table 7.1 Sample cumulative Poisson distribution values

Expected value, x											
		1	2	3	4	5	6	7	8	9	10
Mean	2	0.41	0.68	0.86	0.95	0.98	1	1	1	1	1
	3	0.20	0.42	0.65	0.82	0.92	0.97	0.99	1	1	1
	4	0.09	0.24	0.43	0.63	0.79	0.89	0.95	0.98	0.99	1
	5	0.04	0.12	0.27	0.44	0.62	0.76	0.87	0.93	0.97	0.99

- rounding of quantities means that the target SL is often exceeded;
- there is no reference to cost;
- the real objective is to attain a SL for the combined system of parts.

The new model is stated as follows: for the combined total number of demand events, ensure that the SL is met. For instance, if the total number of events is 5,000 and target SL 95%, ensure that 4,750 requests are filled. Finding the solution that achieves this at the least cost will increase stock performance for lower-value parts and allow requests for high-value parts to fail relatively more often. This problem is solved as a binary integer linear programming formulation as follows:

- variables are created for part number quantities in a defined range, such that X1,1, X1,2, X1,3,... denote quantities of 1, 2, 3,... for part number 1;
- exactly one value from Xn,1, Xn,2, Xn,3,... is chosen (set to 1);
- the combination of values for all Xn items exceeds the SL by satisfying minimum stock levels to meet total demand;
- the objective function is to minimize the combined value of the chosen stock levels.
- A trivial linear programming formulation is shown in Fig. 7.2 above. Looking at the binary constraint, there are two part numbers with five possible values each, such that X1a represents a quantity of 1 for part number 1, X1b is quantity two

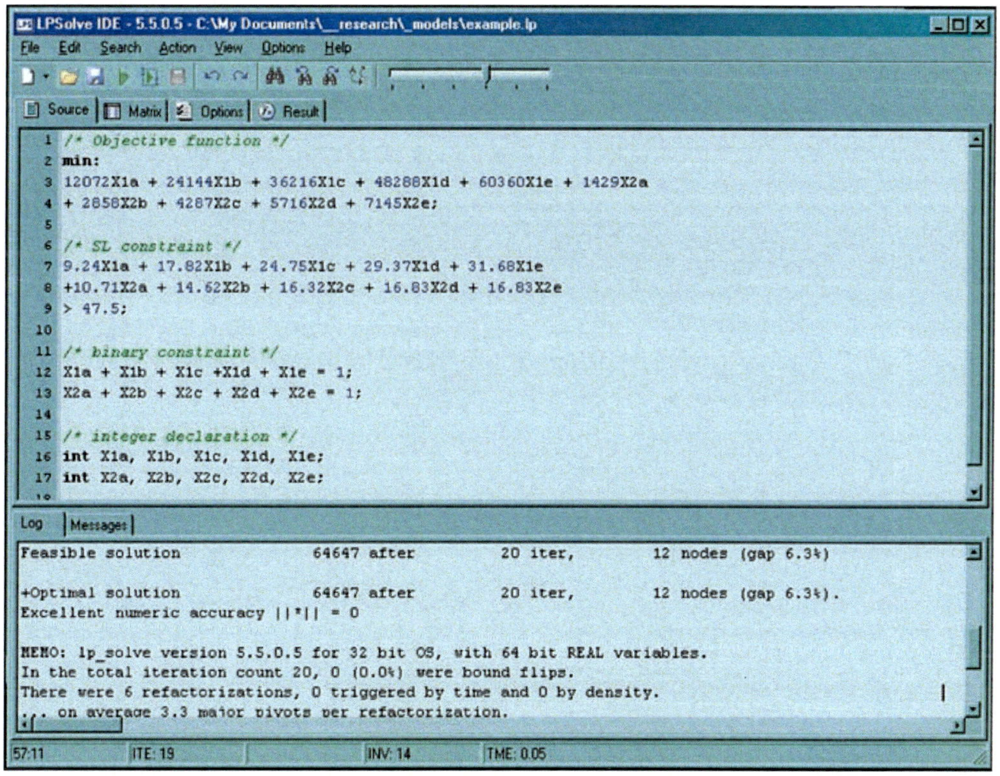

Fig. 7.2 Linear programming formulation

for part number 1, and so on. The binary constraint and integer declaration require that X1 and X2 will have exactly one quantity variable selected.

- The SL constraint in Fig. 7.2 requires that the combined number of satisfied requests for inventory for the two parts will exceed 47.5, which is the total number of expected demand events multiplied by the target SL.
- The objective function in Fig. 7.2 minimizes the total cost of the solution that satisfies the constraints above.

Figure 7.3 shows that the solution calls for a quantity of 5 of part number 1 and 3 of part number 2. Note that, while part 1 is more expensive ($12,072 compared with $1,429 from Fig. 7.2), it also has a higher fill rate, so a quantity of 5 of part 1 gives 31.68 fills, compared with 16.83 for 5 of part 2 (SL constraint in Fig. 7.2).

The total cost is $64,647 and the SL attained is derived from the number of fills = 31.68 + 16.32 = 48. If the target fill rate of 47.5 represents 95% SL, then the SL attained can be calculated as: 0.95 * (48/47.5) = 96%, or 48 out of 50.

Several variants of the conventional and linear programming models are tested to give a rigorous assessment; these are tested for a range of scenarios derived from a common data set. In all, five models were tested on five scenarios, giving 25 formulations and sets of results, which are presented in the next section.

The data set used is a sample provided for Boeing 737 fleet support by a large maintenance organization. The actual inventory holdings are also presented for

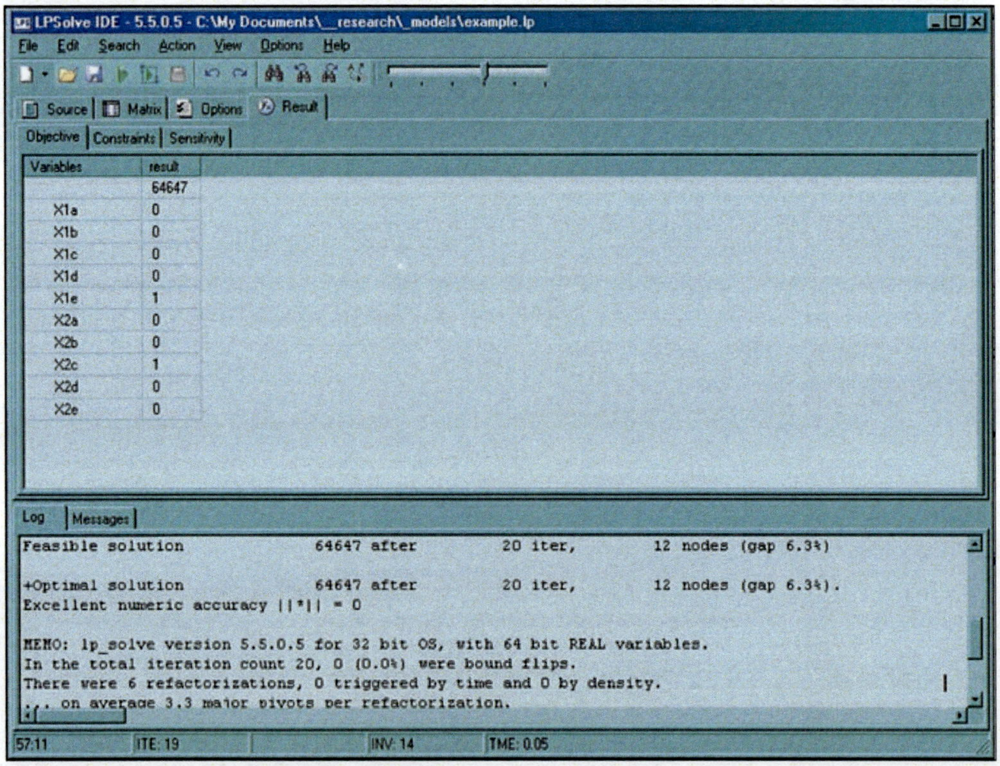

Fig. 7.3 Linear programming solution

comparison. The sample contains 300 inventory line items, which is about one tenth of a typical inventory package for an aircraft family.

7.4 Results and Analysis

Table 7.2 presents results for five models, each tested against a common group of five data sets, numbered 1–5 for each model. The five data sets represent different operational scenarios, such as larger fleet size and shorter repair time, to observe

Table 7.2 Results of multiple model tests for rotable inventory planning

Run	SL (%)	Total cost ($M)	Actual cost ($M)	Total inventory count	Average item value ($)
Actual	89	32.9		2325	14164
Poisson					
P1	96	15.3	17.6	995	15362
P2	94	14.0	19.0	930	15006
P3	96	13.2	19.8	865	15220
P4	93	25.2		1649	15294
P5	94	21.0		1338	15713
Marginal analysis					
M1	96	11.1	21.8	1050	10597
M2	95	10.6	22.4	1006	10509
M3	96	9.5	23.4	916	10395
M4	99	25.2		2060	12227
M5	97	18.4		1623	11364
Cost-wise skewed					
C1	95	11.8	21.1	979	12063
C2	95	11.8	21.0	982	12178
C3	95	10.6	22.3	820	12984
C4	95	23.0		1634	14083
C5	95	18.4		1389	13260
LP					
L1	94	10.3	22.6	1019	10106
L2	93	9.8	23.2	972	10035
L3	94	8.6	24.3	882	9807
L4	95	17.5		1785	9786
L5	95	14.6		1513	9665
LP3					
L3-1	94	9.7	23.2	1006	9662
L3-2	91	8.7	24.2	932	9381
L3-3	94	8.3	24.6	854	9709
L3-4	94	16.4		1734	9469
L3-5	91	12.2		1381	8858

the effect of changing operational conditions and to add rigor to the evaluation. The five scenarios are as follows:

(1) Base—operating conditions obtained from an actual fleet support case and pertaining to the data set used
(2) Fewer—Airbus recommends lower SL values for non-essential parts than Boeing does—{95, 89, 75%} against {95, 93, 90%}, so these values are used as an alternative scenario.
(3) Faster—repair times are given as 20, 28 or 38 days in the test data set used. An alternative scenario reduces this time by 5 days.
(4) Bigger—operators often question the increase in spare stock utilization that would result from a greater fleet utilization, either by increasing flight activity or pooling demand with another operator. This scenario assesses a doubling in utilization.
(5) Best—this scenario combines Airbus recommended SLs, reduced repair times and higher utilization.

The five models shown include conventional practice (Poisson), a published system-level heuristic (Marginal Analysis), a simple heuristic sorting parts into cost bands (Cost-wise skewed) and two linear programming (LP) models. LP models differ in their treatment of parts with different target SLs. As explained earlier, there are typically three SL values used to reflect three levels of essentiality of parts. The LP model above comprises a single, large formulation where demand for items with lower essentiality is scaled down relative to demand for the parts with highest SL. This is an approximation but has the advantage of being a single, large model so could provide a more efficient solution. The LP3 model solves three separate formulations for different target SLs so removing the approximation introduced in the LP model by scaling demand for parts with lower SL.

With reference to the columns in Table 7.2, *Run* lists the solution sets generated, *SL* shows the SL attained, *Total Cost* is the extended cost of the solution, *Actual-cost* is the improvement over the current inventory holding, *Total inventory count* is the number of parts prescribed and *Average item value* is the total cost divided by the inventory count. The values for *Actual-cost* left empty reflect scenarios 4 and 5, which represent doubling the size of the supported fleet, so comparison with the actual inventory holding is not meaningful.

Looking at the results for each model, the five rows represent different scenarios, so that P1 to P5 use the same methodology (industry standard Poisson analysis) on the five different variations of operational conditions described.

Turning to the results, the first finding is that the Poisson model over-stocks: for P1, where the target SLs are 95, 93 and 90%, the overall achieved is 96%. This is because the process operates at the part level to exceed the target, giving an excessive combined result. The Marginal Analysis model also overshoots substantially, the Cost-Wise Skewed model reaches the highest SL, 95%, and the two LP models have lower aggregate results, meaning that they are more precise.

Comparing overall results for the different models, it is striking that the actual observed stock performance gives low SL but high cost—this is thought to result from a practice over many years of over-ordering expensive items following a low stock event. Interviewing managers in the organization, the root cause appears to be poor management of repair times and the fact the management priority is seen to be stock availability, not total inventory cost. Also, there is a sense that low-cost items are less important and thus receive less attention, although clearly their importance is the same as that of more costly items.

The Poisson model is the most expensive for all scenarios, while LP3 is the least expensive. Marginal Analysis gives good results for the first three scenarios but over-stocks for the larger utilization levels, which is a recognized flaw in the model, referred to as the need for a concave distribution, or diminishing marginal return, which does not occur for higher demand rates.

The Cost-Wise Skewed model, a simple spreadsheet model that groups parts by cost and lowers SL for the highest-cost items, gives better results than current practice, so would be useful as an improvement. However, the LP models gives significantly better results and should be recommended in all cases as superior. The LP3 model, which solves separate formulations for different SL values, is significantly superior to the LP model that combines SL using scaled-down demand for non-essential items.

As well as better targeting of SLs, as can be seen from the overall SL values in Table 7.2, it can be seen that, for each solution that gives a lower total cost than its predecessor, the average item value is lower. Thus, while the total number of parts tends to increase, they are less expensive parts. This is consistent with the aim of reaching target inventory performance at the lowest cost. Therefore, having a higher inventory count is not a negative: if average cost is lower, then the total cost is also lower. Simply put, the aim is to ensure that, if a given number of demand failures are expected to occur, then it is preferable that the stock-outs be for the most expensive parts.

Having examined the aggregate results, which promise a potential cost reduction of up to 40% comparing different models, it is interesting to examine the distribution of the results. Current practice performs a line-by-line set of calculations, so there is no relative analysis in relation to cost. The other models take cost into account, so it is expected that their results prescribed high stock levels for low-cost parts with stock levels (and consequent SLs) falling in proportion to cost for the items with higher value.

Figure 7.4, which shows part numbers in order of increasing value, confirms that low-value parts have high SL performance as they are highly stocked, while high-value items have decreasing stock levels and SL performance. The curves plotted are for 5 scenarios tested using model LP3. The noise in the curves is attributed to rounding of small stock numbers.

It can be seen from Fig. 7.4 that there are two dominant gradients in the results, with a steep slope for the rightmost portion of approximately one-fifth of the data range. Fitting lines to the data gives two slopes for the first 80% and top 20% by

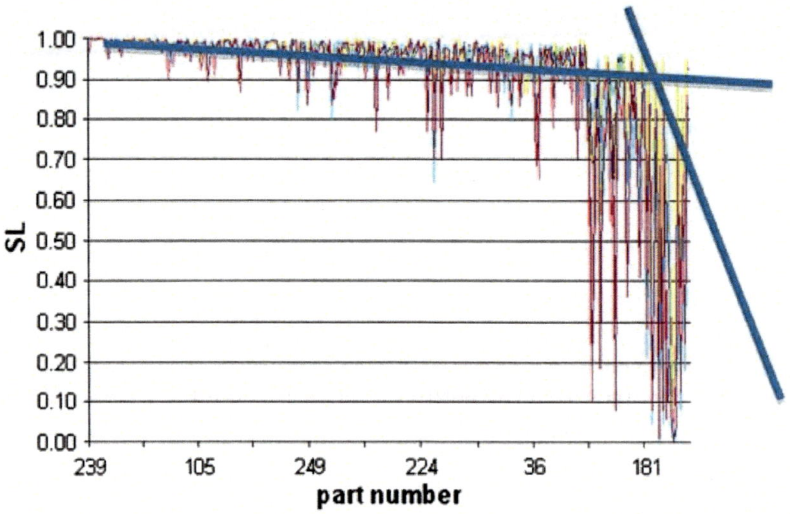

Fig. 7.4 SL performance for parts ranked by value, 5 cases for LP3 model, with two gradient lines

value of the inventory list. The tangent of the left slope is approximated as:$(0.9945 - 0.9482)/0.8 = 0.058,$ giving anangle of 3.3° below horizontal.

The right-hand slope is approximated as:

$$(0.9482 - 0.4917)/0.2 = 2.285, \quad \text{which is thetangent of } 66.4^{\circ}.$$

This method of deriving gradients could be used to provide an approximation tool for planning a large inventory set without performing a linear programming analysis and offer the potential for a real-time heuristic planning a scenario analysis tool.

7.5 Conclusions and Recommendations for Improved Practice

This study shows that there is a clear benefit in planning rotable inventory stocks by using a system-wide cost-oriented approach. This can achieve very significant cost savings when compared with current practice.

The data set tested represents approximately one tenth of the inventory list held in support of a large Boeing 737 fleet. The actual value of the inventory holding for the part number selection tested is over $30M, so a total estimate of up to $300M is given to this operational scenario. The potential to reduce this investment by up to 40% gives strong motivation for any large operator to move beyond current practice and implement an optimization model such as that presented here.

The model shown here has been implemented as an enterprise application and tested on full sets of fleet data. This data and the results are not disclosed for commercial reasons, but inventory value reductions of 20–40% have been predicted in several instances, without loss of performance (SL).

Without implementing the large-scale, complex and intensive LP solution, it is possible to apply gradients derived from test results to give a scaled prioritization of inventory holdings in order of cost. For example, it is feasible to prescribe that the lowest-value parts seek 99.45% SL, with the highest-value in the band of 80% of part numbers having an SL target of 94.82%, with a slope of 0.058 for part numbers between these limits, and a similar graduation of parts in the top 20% band with limits of 94.82 and 49.17% and a slope of 2.285. The results of both the LP solution and the proposed gradient heuristic are easily checked by calculating the total number of demand fills by each line item, and the resulting global SL achieved.

In conclusion, current practice, which mainly employs the Poisson calculation, is far from optimal and there are significant benefits to using a system-wide LP solution for planning aircraft rotable inventory levels. Where it is not feasible to formulate and run the LP model, the gradient heuristic will give a good approximation of the LP results.

References

Adams C (2004) Inventory optimisation techniques, system vs item level inventory analysis. In: IEEE: Reliability and maintainability symposium, Los Angeles, CA, January 2004, Vol XVIII–679, pp 55–60 ISBN 0-7803-8215-3

Armac (2007) http://armacsystems.com/downloads/SR%20Technics%20Case%20Study.pdf

Computer World (2005) Southwest supply chain optimization project. Computer World. http://www.cwhonors.org/Search/his_4a_detail.asp?id=4916. Accessed Jul 2008

Friend C, Swift A, Ghobbar AA (2001) A predictive cost model in lot-sizing methodology, with specific reference to aircraft parts inventory: an appraisal. Prod Inventory Manag J 42(3/4):24–33

Ghobbar A (2004) Forecasting intermittent demand for aircraft spare parts: a comparative evaluation of methods. J Aircr 41(3):665–673

Ghobbar A, Friend C (2003a) Evaluation of forecasting methods for intermittent parts demand in the field of aviation: a predictive model. Comput Oper Res 30:2097–2114

Ghobbar A, Friend C (2003b) Comparison of lot-sizing methods in aircraft repairable component inventory systems. J Aircr 40(2):378–383

Haas H, Verrijdt J (1997) Target setting for the departments in an aircraft repairable item system. Eur J Oper Res 99:596–602

Haupt M (2001) Cost reduction for initial spares investment. Airbus Fast 27:11–14

Kilpi J, Vepsalainen A (2004) Pooling of spare components between airlines. J Air Transp Manag 10(2):137–146

Lee L, Chew E, Teng S, Chen Y (2007) Multi-objective simulation-based evolutionary algorithm for an aircraft spare parts allocation problem. Eur J Oper Res 189:476–491

Logistechs (2006) Expect results. Logistechs. http://www.rotable.com. Accessed 2007

Lye K, Chan L (2007) A virtual warehouse simulation tool for aerospace rotables management. In: IEEE aerospace conference. Big sky, MT, 3–10 March 2007, pp 1–7 ISSN 1095-323X

MacDonnell M, Clegg B (2007) Designing a support system for aerospace maintenance supply chains. J Manuf Technol Manag 18(2):139–152

Panayiotou O (1998) Common, reliable and punctual: the path to lower spares costs. Airbus Fast 23:12–19

Rutledge J (1997) Spare parts = cost benefit management. Airbus Fast 21:25–29

Sherbrooke C (1968) METRIC—a multi-echelon technique for recoverable inventory control. Oper Res 16(1):122–141

Sherbrooke C (1986) VARI-METRIC: improved approximations for multi-indenture, multi-echelon availability models. Oper Res 34(2):311–319

Chapter 8
Forecasting and Inventory Management for Spare Parts: An Installed Base Approach

Stefan Minner

8.1 Introduction

As customers are more demanding with respect to after sales operations and service level agreements put challenging availability targets on equipment uptime, the provision and deployment of service parts becomes of focal interest for many original equipment manufacturers. High demand variability and uncertainty driven by different phases of a product's and its critical components life-cycles make spare parts demand forecasting and safety inventory management a major challenge. Boone et al. (2008) identify service parts demand forecasting as the unanimously agreed challenge in service parts management in their Delphi study with senior service parts managers. Parts management of a non-stationary demand process over the phases of series production, end-of-production (EOP), end-of-service (EOS), and end-of-life (EOL) is even further complicated by heterogeneous customers and different ages of equipment in use due to different points in time of purchases and component replacements.

Mainstream stochastic inventory management approaches for spare parts make use of distributional assumptions for demands. In reality, distributions and their parameters are hardly known and need to be estimated. Practical suggestions recommend the use of time series based forecasting approaches and derive safety stock requirements from observed forecast errors. Boylan and Syntetos (2008) give a recent excellent overview and classification in the area of service parts demand forecasting. These ideas are supported by many Advanced Planning Systems and commercial service parts supply chain solutions, e.g., SAP Service Parts Planning. According to the literature, e.g., Dickersbach (2007), the majority of methods offered by these systems are standard time series forecasting methods including

S. Minner (✉)
University of Vienna, Brünner Straße 72, 1210 Vienna, Austria
e-mail: stefan.minner@univie.ac.at

N. Altay and L. A. Litteral (eds.), *Service Parts Management*,
DOI: 10.1007/978-0-85729-039-7_8,

extensions to intermittent demand patterns. The life-cycle dynamics are incorporated by specifying certain phase-in and phase-out patterns.

In this paper we pursue a causal demand modeling approach that combines theoretical models from reliability pham (2003) and inventory theory Silver et al. (1998) to derive improved service parts demand forecasts. Information technology, machine monitoring tools, and detailed maintenance data from customers provide additional knowledge on age and status of products and systems in use, and about customer maintenance and replacement policies. This data, in the following generally referred to as installed base information, can be used to improve spare parts demand models compared to the application of basic time series methods often found in practice. Song and Zipkin (1996) present a state-dependent inventory management approach which is related to the idea to base service parts demands on the state of current installed base. Ihde et al. (1999) propose the use of continuous reliability models to estimate the market potential of service parts demand. Schomber (2007) provides an overview and a simulation study to quantify the value of age-based information. Jalil et al. (2009) investigate the value of installed base data quality in the context of IBM's service parts network. Their focus is especially on geographical aspects of stock and customer locations.

We present an integrated life-cycle demand modeling and inventory management framework that exploits different kinds of available information about installed base. The focus is especially on low volume, high value products and components, e.g., complex engines or medical equipment. The implementation of the framework is illustrated by a stylized service parts system that consists of the following model components. For every point in time, the state of the system from the perspective of a service parts provider is represented by the number of products in use, differentiated by age category. Sales of original parts follow some life-cycle pattern under uncertainty. The products and components respectively have a random life-time and failure characteristic with given probability distributions, and the customers follow some well known maintenance and replacement policies, e.g., age replacement or block replacement. This information is used to parameterize the required demand distributions for different inventory management decisions and is dynamically updated. In situations of closed loop inventory systems where product returns, disassembly, and remanufacturing can be used as an alternative source for service parts requirements, the proposed methodology is capable to further strengthen the estimation of returns and their dependency on past sales. Spengler and Schröter (2003), e.g., use a system dynamics approach to model the complex interdependencies in such a system. As illustrated in Inderfurth and Mukherjee (2008), a stochastic dynamic programming decision support framework requires this information to dynamically optimize manufacturing and remanufacturing decisions.

Based on a small simulation experiment we illustrate the benefits of this accurate but more complex approach over simple time series based forecasting techniques and forecast error driven safety stock setting approaches for different life-cycle patterns of products and different phases of a product's life-cycle. Section 8.2 gives an overview on main ingredients of the modeling framework with respect to demand life-cycle, inventory management, and maintenance policies. In Sect. 8.3 we present the installed base forecasting and inventory management concept which is further illustrated and benchmarked against simple-minded time series smoothing forecasting approaches in Sect. 8.4. Conclusions and an outline for future research and refinements of the concept are given in Sect. 8.5.

8.2 Foundations of Dynamic Service Parts Management

Dynamic service parts management explicitly addresses the different phases during a product's life cycle. The first phase from the perspective of service parts management is from product introduction and therefore ramp-up of parts manufacturing to end of sales and EOP. After sales operations then take place until the EOS, the most critical phase for the provision of service parts if they are no longer produced together with parts for new products. From EOS until the EOL, there might still be products in use, but without any obligation to provide service parts. The following subsections sketch modeling basics for the different phases that are used to present the installed based demand modeling framework in Sect. 8.3.

8.2.1 Spare Parts Forecasting and the Product Life Cycle

Let D_t denote the original equipment sales in period t for the product under consideration. As a modeling example, the demands might follow some theoretical demand distribution where mean demand follows a suggested life-cycle pattern function. In the literature, two popular functional representations are

1. Albach–Brockhoff formula (with parameters a, b, c)

$$D_t = at^b e^{-ct} \tag{8.1}$$

2. Logistic growth function (with parameters a, b, c) where S_t denotes cumulative sales until period t

$$S_t = \frac{a}{1 + e^{b-ct}}. \tag{8.2}$$

For modeling uncertain product demands, the above functional representation needs to be extended by a measure of demand variability, or following some stochastic process (see, e.g., Beichelt 2006). Independence of demands between periods is not a necessary assumption for the following modeling and can be supplemented by some autocorrelation function, e.g., as it is typical for automotive sales, see Stephan et al. (2009).

8.2.2 Spare Parts Inventory Management

Incorporating life-cycle dynamics, the foremost difference of strategic service parts inventory management is a non-stationary demand process. In the following, we use service parts and spare parts interchangeably. Because of different manufacturing options in the production and after sales phase, and potentially different service level requirements for service parts, not only the demand process but also other parameters require stochastic dynamic modeling for inventory management, though the available literature and main results are predominantly for stationary inventory models (see, e.g., Silver et al. 1998).

Kennedy et al. (2002) give an overview on spare parts inventory management. Important ingredients for service parts inventory management are customer specific service level requirements from service level agreements, e.g., product availability often represented by non-stockout probability in planning frameworks. Further, inventory management tasks can be classified into short term operational replenishment operations where, according to some replenishment concepts and rules, service parts are reordered from a single or multiple suppliers (e.g., Minner 2003) and medium term strategic tasks like volume contracting and inventory placement (Minner 2000). The latter comprises where to stock service parts (centrally or decentrally) and how much to stock (strategic safety stock levels). In contrast to operational replenishment planning, the strategic planning of inventories uses aggregate forecasts for service parts demands and is the special focus of the following installed base modeling framework. Sherbrooke (2004) and Muckstadt (2005) are respective state-of-the-art monographs for multi-echelon techniques in inventory management for service parts. For the one-time service parts provision at the time of end-of-production, Teunter (1998) provides an overview and several research contributions to inventory modeling.

8.2.3 Reliability, Maintenance, and Repair Policies

Explaining service parts demands by a causal method requires knowledge about the interaction of component reliability, maintenance policies, and repair and replacement behavior of product users. An important initial driver for service parts demands is component failure. However, not every failure generates a service part demand if the component can be repaired, the component will be replaced by a substitute or competitor product, or if the failure yields to the end-of-use of the entire product. Secondly, the maintenance and replacement policy influences service parts demands. For example, under age replacement, a component is replaced upon failure, but at the latest after a maximum age has been reached. Under block replacement, a failed component is either repaired or replaced by a new one, for an overview, see, e.g., Dohi et al. (2003). The knowledge about customers' maintenance and replacement policies therefore can yield a significant improvement and this knowledge can be utilized in order to parameterize the component demand models presented in the following.

8.3 Installed Base Demand and Return Forecasts

The following presents a state-dependent forecasting and service parts demand planning framework with age-based installed base information. The state of the system in a given period is specified by the composition of the number of products in use in that period being detailed by age.

8.3.1 Product Level Dynamics

We follow a discrete time approach and for ease of presentation assume that states and dynamics are recorded at the end of the respective time periods $t = 1, 2, \ldots$, EOS. The installed base dynamics are driven by new product sales D_t entering the base and end-of-use products leaving the base. The sequence of events assumed for the following presentation of state dynamics is: demands occur, maintenance and replacement of components is performed, end-of-use of products is recorded, and finally, the installed base record is updated.

Let B_t and B_{it} denote the number of products in total and with age i at the end of period t, respectively. For notational convenience, let $B_{0,t-1} = D_t$. Further, let E_{it} denote the number of end-of-use products that leave the system with age i at the end of period t prior to installed base recording. Then, the dynamics for the age-specific number of products is

$$B_{it} = B_{i-1,t-1} - E_{it} \quad t = 1, \ldots, \text{EOS}; \; i = 1, \ldots, t. \tag{8.3}$$

Table 8.1 shows these dynamics over time and per age category.

Table 8.1 Installed base dynamics

	B_{1t}	B_{2t}	…	B_{it}
$t = 1$	$D_1 - E_{11}$			
$t = 2$	$D_2 - E_{12}$	$D_1 - E_{11} - E_{22}$		
$t = 3$	$D_3 - E_{13}$	$D_2 - E_{12} - E_{23}$		
…			…	
t	$D_t - E_{1t}$	$D_{t-1} - E_{,t-1} - E_{2,t}$		$D_{t-i+1} - \sum_{j=1}^{i} E_{j,\, t-i+j}$

Let p_i denote the probability that a product of age i is taken out of use in an arbitrary period. p_1 denotes the probability that the sale of a period is no longer in use in the next period. Assuming independence of end-of-use within and between age categories, each component failure follows a Bernoulli process and the end-of-use by age category follows a binomial distribution with parameters $B_{i-1,t-1}$ and p_i.

$$P(E_{it} = e) = \binom{B_{i-1,\,t-1}}{e} p_i^e (1 - p_i)^{B_{i-1,\,t-1}-e} \quad e = 0, \ldots, B_{i-1,\,t-1}. \tag{8.4}$$

The total number of end-of-use products in period t is the sum of the age-specific quantities for all ages up to t,

$$E_t = \sum_{i=1}^{t} E_{it}. \tag{8.5}$$

The distribution of total end-of-use E_t given the installed base at the end of period $t - 1$ is the sum of t (independent) binomially distributed random variables. The convolution can numerically be determined recursively as follows.

$$P(E_k = e) = \sum_{j=0}^{e} P(E_{k-1} = j)P(E_{kt} = e - j) \quad k = 2, \ldots, t \quad \text{and} \quad E_1 = E_{1t}. \tag{8.6}$$

Starting with the distribution of products with age 1, E_{1t}, the probability that e products of age k and less are taken out of use is given by all possible combinations of j products with maximum age $k - 1$ and $e - j$ products of age k. This is recursively repeated until the distribution of the number with age $k = t$ has been obtained.

8.3.2 Component Level Dynamics

Each product might consist of multiple critical components. At the component level, the dynamics are driven by sales, ageing, end-of-use-process, and component replacement. For notational convenience, we assume bill-of-material coefficients being equal to one between component and product level. Further, it is assumed that there will be no multiple failures for the same product within a period, though this restricting assumption can easily be relaxed to general failure characteristics. As the age of the product in use and its built-in components might differ due to maintenance and replacement, we will detail components in the following by age of the associated product and age of the respective component or part.

Let C_{ijt} define the number of components of age i in a product of age j $(i \leq j)$ at the end of period t. Further, let q_i denote the probability that a component of age i has to be replaced. M_{ijt} denotes the number of replaced components of age i in products of age j in period t that are added to component age category 1 and leave component age category i. Then, the component state dynamics are given by

$$C_{ijt} = \begin{cases} D_t - E_{1t} & i = 1,\ j = 1 \\ \sum_{k=1}^{j} M_{kjt} & i = 1,\ j = 2, \ldots, t \\ C_{i-1,\,j-1,\,t-1} - M_{ijt} - E_{ijt} & j = 2, \ldots, t,\ i = 2, \ldots, j \end{cases}. \tag{8.7}$$

In the first row, the number of components in products of age 1 is equal to demand less first-period end-of-use. In the second row, the number of components of age 1 in products with an age j is equal to all component failures from products of age j in period t. The last row shows component dynamics where the number of components equals the number in the corresponding category at the end of the previous period less the number of component failures and of components within products that were terminated. Then, the total number of replaced components is

$$M_t = \sum_{j=1}^{t} \sum_{i=1}^{j} M_{ijt}. \tag{8.8}$$

Assuming independence as done for the analysis on the product level, M_{ijt} is binomially distributed with parameters q_i and $C_{i-1,\,j-1,\,t-1}$.

$$P(M_{ijt} = m) = \binom{C_{i-1,\,j-1,\,t-1}}{m} q_i^m (1 - q_i)^{C_{i-1,\,j-1,\,t-1} - m} \quad m = 0, \ldots, C_{i-1,\,j-1,\,t-1}. \tag{8.9}$$

For the service parts demand in a period driven by failures of components within that periods sales, we find

$$P(M_{11t} = m) = \sum_{d=m}^{\infty} P(D_t = d) q_1^m (1 - q_1)^{d-m}. \tag{8.10}$$

Using (8.9) and (8.10) in (8.8) and applying the same convolution logic as presented in the previous subsection, the probability distribution of total component demands can be determined.

8.3.3 Derivation of Future Installed Base Distributions

As a result of the above state analysis, we can derive the probability distributions of future service parts requirements given the installed base information at the end of period t. Note that this information can be used in a manifold way, to give a single value demand forecast (e.g., mean, median or mode demand), give a confidence interval, or directly use this distribution for inventory planning and contracting purposes.

The future demand distribution can be developed using convolutions of individual age category demands. For the age-specific installed base volumes of the next period we find

$$P(B_{i+1,\,t+1} = b|B_{it}) = \begin{cases} \sum_{d=0}^{\infty} P(D_{t+1} = d)P(E_{i+1,\,t+1} = B_{it} + d - b) & i = t,\; t \leq \text{EOP} \\ P(E_{i+1,\,t+1} = B_{it} - b) & i < t \text{ or } t > \text{EOP} \end{cases} \tag{8.11}$$

The determination of the probability distribution of next period's installed base differs for periods before EOP and after EOP. Before EOP, the probability that b units are in use, given that currently the base consists of B_{it} is given by all combinations of new sales D_{t+1} and end-of-use E_{t+1} that yield a new base of b. After EOP with end-of-use only, it is determined by the probability that end-of-use products equal the difference in installed base.

For the determination of the probability distribution of service parts demands in the next period, we use (8.8). The convolution can be performed sequentially as described above for total end-of-use in (8.5) using (8.10) and (8.9).

8.4 Numerical Study

In order to illustrate the benefits of the above installed base dynamics analysis, we use a simple comparison with exponential smoothing based methods in the following. The proposed installed-base approach is benchmarked against simple first-order exponential smoothing (though not being appropriate under a life-cycle pattern driven random demand model from a theoretical perspective), the naïve forecast that next period service parts demands will equal the observed service parts demand from the current period, and an experience driven smoothing forecast that updates service parts demand observed in a certain period t of the life-cycle over the repetitions of the simulation.

8.4.1 Experimental Design

The experimental design and the required parameters are set as follows.

8.4.1.1 Demand and Life-Cycle Function

We use two life-cycle settings with a short and a long production cycle. The parameters that represent dynamic mean demand development are LC-1 (life-cycle

Type 1) with $a = 1{,}000$, $b = 2$, $c = 1$, EOP $= 5$, and EOS is either 10 or 15. In the second set LC-2 (life-cycle Type 2) we use $a = 150$, $b = 2$, $c = 0.4$, EOP $= 10$, and EOS either 15 or 20. Figure 8.1 shows the dynamic development of expected demands over time.

Demand uncertainty is represented by assuming that period demands are independently Gamma-distributed with the mean according to the life-cycle pattern and the coefficient of variation cv $= \sigma/\mu$ selected from $\{0.2, 0.4, 0.8\}$ and assumed to be independent of time.

8.4.1.2 End-of-Use and Component Failure Characteristics

We consider two kinds of component replacement and end-of-use probability characteristics, constant (age-independent) and increasing with age. The chosen values are $p_t = 0.1$ and $q_t \in \{0.2, 0.1\}$ for all $t = 1, \ldots,$ EOS and $p_t = 1/\max\{1, 10 - t\}$ and $q_t \in \{1/\max\{1, 5 - t\}, 1/\max\{1, 10 - t\}\}$ for all $t = 1, \ldots,$ EOS.

In total, we have 48 different parameter configurations; each of them is replicated 500 times.

8.4.2 Illustrative Results

Figure 8.2 shows an example of a single simulation run and illustrates original product sales, service parts demand, and total installed base volume for each period until EOS.

Table 8.2 shows the installed base specified by age of the built-in components at the end of period 10, taken from one single simulation example.

In the following we benchmark the installed-base driven inventory planning approach against three simple smoothing based methods.

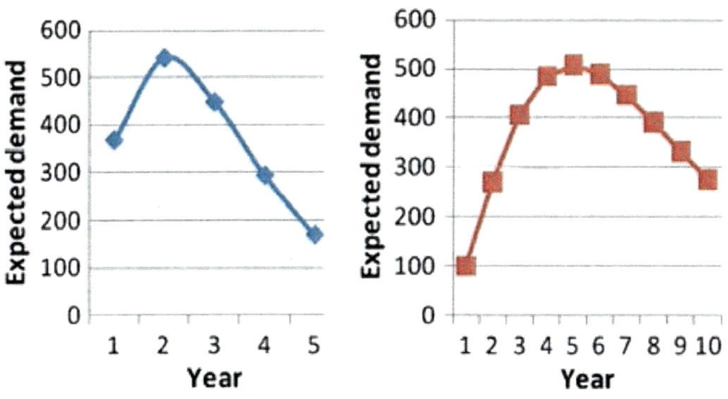

Fig. 8.1 Life cycle patterns LC-1 (*left*) and LC-2 (*right*)

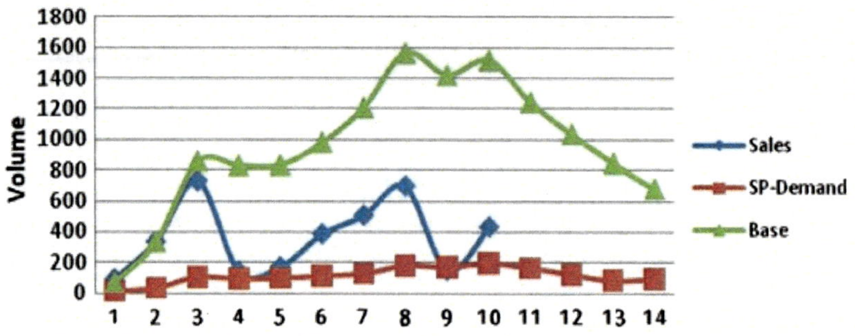

Fig. 8.2 Simulation example: sales, service parts demand, and installed base

Table 8.2 Installed base detailed by age of built-in components

Product age	Component age									
	1	2	3	4	5	6	7	8	9	10
1	350									
2	17	89								
3	50	35	297							
4	28	24	17	139						
5	18	13	11	6	89					
6	5	6	5	4	2	30				
7	6	5	2	2	5	4	18			
8	21	18	17	8	13	8	8	67		
9	13	7	8	7	0	4	3	3	20	
10	1	1	1	2	0	1	0	4	0	5

1. First-order exponential smoothing with smoothing constant $\alpha = 0.3$, chosen at the upper range of typical text-book recommendations. The starting value for smoothing is set equal to first period service parts demand.
2. Naïve forecast by using the last observation as the forecast for the next period (note that this is a special case of 1 with $\alpha = 1$).
3. Exponential smoothing driven learning: Here a specific forecast for each period (phase) of the life-cycle is smoothed over the 500 replications of the simulation to imitate long experience of service parts demand over previous products having the same life-cycle characteristics. The used smoothing constant is set equal to $\alpha = 0.1$ at the lower range of recommended parameters to prevent too volatile estimates in the learning environment.

For each method, the forecast error is dynamically monitored using the mean squared error MSE which is then used to update the required inventory level by using the currently available value of MSE instead of the variance of a theoretical demand distribution.

The required strategic target inventory level for the next period then is set equal to the derived service parts forecast plus a safety factor k times the estimated standard deviation of service parts demands. The required safety factor was set to $k = 1.282$. As we consider a strategic inventory approach with annual inventory targets, short term, operational lead times are negligible. For the installed base method, we use the derived service parts demand distribution and set the required inventory level such that a non-stockout probability of 90% is guaranteed.

Table 8.3 shows the average inventory increase of the respective smoothing method over the installed base approach. On average over all considered problem instances, the information on installed base and its incorporation into tactical target inventory planning offers an inventory reduction potential of 16% (over the learning approach) to 50% (over simple first-order exponential smoothing). The other rows in Table 8.3 show the respective increases of inventory levels detailed by design parameters, that is for all instances having the same characteristics or parameter as indicated in the first column.

As an inherent shortcoming from the exponential smoothing based methods, the ex-ante learning capability lags behind the life-cycle development until end-of-production and in the after-production phase, fails to properly adapt for end-of-use and component failures that drive service parts demands. As a consequence, these approaches do not meet the required service target early and overachieve the target later. In order to prevent this happening, knowledge about the life-cycle has to be incorporated. A detailed comparison for each varied parameter characteristic shows that especially life-cycles with irregular pattern and early end-of-production (LC-1) and late EOS have the largest impact on inventory improvement using installed base information. The coefficient of demand variation and failure characteristics do not show a clear indication of differences in improvements.

For all forecasting methods, Fig. 8.3 shows the average inventories over all 500 replications of the instance with life-cycle type 2, EOS = 15, cv = 0.4, $p_t = 0.1$, and $q_t = 0.2$ detailed by the period within the life-cycle.

During the early production cycle, all smoothing methods adapt too slowly to the service parts demands whereas in the after-sales phase with increasing replacements and end-of-use, they do not adapt fast enough to the decreasing number of components in use.

Table 8.3 Average relative inventory differences: time series versus installed base

	Exp. smoothing (%)	Naive (%)	Learning (%)
LC-1	64	43	22
LC-2	36	17	9
EOS = EOP + 5	39	23	13
EOS = EOP + 10	66	37	16
cv = 0.2	49	29	13
cv = 0.4	44	26	24
cv = 0.8	56	34	9
$p = 0.1$, $q = 0.2$	27	17	12
$p = 0.1$, $q = 0.1$	39	24	15
$p = 1/(10 - t)$, $q = 1/(5 - t)$	76	45	19
$p = 1/(10 - t)$, $q = 1/(10 - t)$	59	35	16
All instances	50	30	16

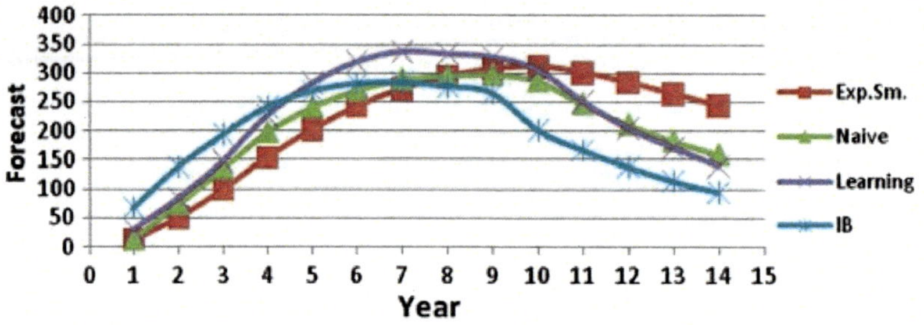

Fig. 8.3 Development of average inventory levels for different methods

8.5 Summary and Conclusions

We have presented a framework for using information about product and component installed-base, component age distributions according to previous maintenance and replacement that is provided by recent information technology and equipment monitoring tools. We illustrated the value of this knowledge for medium term inventory forecasts, e.g., in negotiations and volume contracting with component suppliers after end-of-production. Compared to (admittedly not appropriate but widely used for dynamic service parts demand forecasting) simple time series methods for predicting service parts demands, the framework offers a substantial inventory reduction potential, especially for longer life-cycles. The approach offers a framework to unlock the value of information in service parts management by transforming the knowledge about sales, maintenance and replacement characteristics and the current age-structure of components in the product base into a statistical model for mid-term inventory planning.

The presented framework can be adapted and extended into different directions. Most obvious, a comparison with more sophisticated time series methods is required to assess the full benefits of the proposed framework. In addition, other error measures than MSE can be utilized when comparing the different forecasting methods. The exemplary analysis can further be refined by introducing different customer classes with individual product use and component replacement distributions. As an example, consider customers replacing certain components under a maintenance contract and age replacement. Several assumptions about the availability and observability of information are quite strong, especially about the end-of-use characteristics and customer loyalty to OEM parts when components need to be replaced. Depending on the industry and product under consideration, the available data and its quality might differ. Some or all of the parameters that were assumed to be given need to be estimated too, e.g., within a Bayesian framework. Another fruitful application is to extend the methodology to multi-step forecasts, e.g., when the EOL procurement decision needs to be taken (e.g., Teunter and Fortuin 1999). However, then the determination of the total service parts demand distribution over several future periods is somewhat more complicated as service parts requirements of consecutive periods are correlated (through the dependence of base dynamics and component dynamics). As a venue for computational simplifications, instead of determining the exact distributions of a sum of binomial distributions, for sufficiently large base volumes, we could approximate the binomial by normal distributions where then, the sum of the normal distributions is again normally distributed.

References

Beichelt F (2006) Stochastic processes in science engineering and finance. Chapman & Hall, Boca Raton

Boone CA, Craighead CW, Hanna JB (2008) Critical challenges of inventory management in service parts supply: A Delphi study. Oper Manag Res 1:31–39

Boylan JE, Syntetos AA (2008) Forecasting for inventory management of service parts. In: Kobbacy KAH, Murthy DNP (eds) Complex system maintenance handbook. Springer, Berlin, pp 479–508

Dickersbach JT (2007) Service parts planning with my SAP SCM. Springer, Berlin

Dohi T, Kaio N, Osaki S (2003) Preventive maintenance models: replacement, repair, ordering, and inspection. In: Pham H (ed) Handbook of reliability engineering. Springer, London, pp 349–366

Ihde GB, Merkel H, Henning R (1999) Ersatzteillogistik. 3. Aufl. Huss-Verlag, München

Inderfurth K, Mukherjee K (2008) Decision support for spare parts acquisition in post product life cycle. Cent Eur J Oper Res 16:17–42

Jalil MN, Zuidwijk RA, Fleischmann M, van Nunen JAEE (2009) Spare parts logistics and installed base information. ERIM working paper ERS-2009-002-LIS, Erasmus University Rotterdam, The Netherlands

Kennedy WJ, Patterson JW, Fredendall LD (2002) An overview of recent literature on spare parts inventories. Int J Prod Econ 76:201–215

Minner S (2000) Strategic safety stocks in supply chains. Springer, Berlin

Minner S (2003) Multiple supplier inventory models in supply chain management: A Review. Int J Prod Econ 81–82:265–279

Muckstadt JA (2005) Analysis and algorithms for service parts supply chains. Springer, Berlin

Pham H (2003) Handbook of reliability engineering. Springer, London

Schomber G (2007) Forecasting for service parts management—a comparison of causal and time series based methods. Master Thesis, University of Mannheim

Sherbrooke CC (2004) Optimal inventory modeling of systems: multi-echelon techniques, 2nd edn. Wiley, New York

Silver EA, Pyke DF, Peterson R (1998) Inventory management and production planning and scheduling, 3rd edn. Wiley, New York

Song JS, Zipkin PH (1996) Evaluation of base-stock policies in multiechelon inventory systems with state-dependent demands. Part II: state-dependent depot policies. Nav Res Logist 43:381–396

Spengler T, Schröter M (2003) Strategic management of spare parts in closed-loop supply chains—a system dynamics approach. Interfaces 33(6):7–17

Stephan H, Gschwind T, Minner S (2009) A multi-stage stochastic dynamic programming approach for capacity planning under uncertainty. Working paper, University of Vienna

Teunter RH (1998) Inventory control of service parts in the final phase. PhD Thesis, University of Groningen

Teunter RH, Fortuin L (1999) End-of-life service. Int J Prod Econ 59:487–497

Chapter 9
A Decision Making Framework for Managing Maintenance Spare Parts in Case of Lumpy Demand: Action Research in the Avionic Sector

M. Macchi, L. Fumagalli, R. Pinto and S. Cavalieri

9.1 Introduction

Traditionally, maintenance has been considered as a core in-house activity. Only a few years ago outsourcing maintenance has become a common practice (Taracki et al. 2006). According to Harland et al. (2003), "outsourcing involves the use of specialists to provide competences, technologies and resources". Three fundamental factors nowadays bring firms to use outsourcing in maintenance: (i) cost reduction, since providers are more experienced, thus they can provide better service efficiency; (ii) more skills, since using external resources is a way to rapidly get new skills via qualified people and specific instruments; (iii) a highest service level, since this is normally bound to penalties established during the contracting phase measured by clear Key Performance Indicators (KPIs) to which the provider is duly responsible.

A well known contractual instrument to this concern is the Service Level Agreement (SLA). SLA allows the customer to evaluate a provider's performances and pay the service accordingly. The basis of this contract clause is the definition of a Service Level (SL) to be granted by the provider and, then, a no-claim bonus malus

---

M. Macchi (✉) · L. Fumagalli
Dipartimento di Ingegneria Gestionale, Politecnico di Milano,
Piazza leonardo da Vinci 32, 20133 Milan, Italy
e-mail: marco.macchi@polimi.it

L. Fumagalli
e-mail: luca1.fumagalli@polimi.it

R. Pinto · S. Cavalieri
Dipartimento di Ingegneria Industriale, Università degli Studi di Bergamo,
Viale marconi 5, 24044 Dalmine, Bergamo, Italy
e-mail: roberto.pinto@unibg.it

S. Cavalieri
e-mail: sergio.cavalieri@unibg.it

N. Altay and L. A. Litteral (eds.), *Service Parts Management*,
DOI: 10.1007/978-0-85729-039-7_9,

Table 9.1 Example of processes involved in outsourcing contracts (adapted from Kumar et al. 2004)

Candidate processes	Main activities
Operative maintenance	Includes all the activities at the operational level, such as the equipment repair, intervention preparation, intervention reporting, and so forth
Maintenance management	Includes all the activities at the management level, such as the work order requests management, work order assignment, operative data storage, resources management, and so forth
Contractual management	Includes all the activities needed to transfer administrative information, contracts and documents between customers and provider
Data management	Includes all the activities related to the acquisition and transfer of data from the equipment in a customer's site to the provider's service workshop, for their further analysis in maintenance engineering, condition monitoring, and so forth
Spare parts management	Includes the acquisition and storage of spare parts and analysis on data regarding spare parts logistics
Maintenance engineering	Includes all the activities aimed at improving the performances of the maintenance contract by leveraging on preventive (cyclic and condition based) maintenance and proactive maintenance
KPI monitoring	Includes the monitoring of maintenance activities, at different level of analysis (e.g., KPI at operative, tactical, strategic level)
Training activities	Includes the activities offered to the customer's maintenance personnel in order to enable the transfer of information and knowhow required by the maintenance interventions

system to manage differences between the SL and the effective outcomes, measured by a KPIs' dashboard covering the main aspects included in the outsourcing scope.

A critical issue that may arise when discussing about maintenance outsourcing regards the identification of the processes that can be effectively outsourced. In order to provide a quick reference, Table 9.1—adapted from Kumar et al. (2004)—presents, with no claim of exhaustiveness, a range of candidate processes for outsourcing.

In this chapter we consider the case of an Original Equipment Manufacturer (OEM) contractually committed—in its role of service provider—to execute four of the above mentioned processes:

Operative Maintenance: the OEM performs repair activities on pieces of equipment sent from the customer to a centralized service workshop.

Maintenance Management: the OEM manages work order requests for the repair activities coming from different customers' sites.

Contractual Management: all interventions are reported and any document required in the contract is duly completed to be available for customer's audits.

Spare Parts Management: whenever a defect is detected at the service workshop, the component must be replaced; the OEM is responsible for the availability of the proper components at the repairing premises.

Focusing in particular on the *Spare Parts Management* process, a structured procedure supporting the management of a service contract from the OEM standpoint is presented. Such a procedure tackles three main decisions:

1. to detain or not stocks of a spare item in the OEM's service workshop;
2. to define the right level of inventories of a spare item, given it is detained;
3. to keep in stock full assemblies or sub-assemblies, in case of complex bill of materials.

Although there is a huge body of academic literature devoted to provide a valid and pragmatic answer to these questions, in the experience of the authors quite few industrial companies seem to apply them rigorously and consistently. Two main challenges arise, making difficult the utilization of models and methods normally known from literature: first of all, each piece of equipment, subject to the repair activities and replacement with spare items, runs in different operating conditions/mission profiles, i.e. depending upon the diverse locations of the customers worldwide; secondly, there is a partial or total lack of visibility on how each equipment is actually utilized on field.

In such a context, the adoption of a *decision making procedure* has been preferred to specific analytical approaches. At each step of the procedure, simple tools and algorithms have been devised to progressively solve the spare parts management problem. The proposal of such a procedure is an evolution of a series of works on this subject carried out by the same authors as a result of collaborations with companies operating in different industrial sectors. The positive feedbacks gained from the industrial partners support the authors' assumption that an adequate combination of simple tools—already available in literature and collected for their use in a structured procedure—can lead to a robust decision even in a complex industrial environment.

The problem-solving orientation of the research and the strong collaborative and participative pattern of relationship with the industrial counterpart—operating in the avionic sector—enforced the idea to use Action Research as the most suitable research methodology for this specific study. Scope, objectives, content and main decisions have been carried out with a strong involvement of the management of the company, thus being quite confident that its main findings are likely to be of high relevance to at least a section of the practitioner community (e.g. the immediate industrial partner and other companies operating in the same industry).

The decision making framework, derived from a recent published contribution of the same authors (Cavalieri et al. 2008), is described in Sect. 9.2. By using this framework also as a taxonomy, an in-depth review is presented in the same section providing a quick summary of the main models and methods for managing maintenance spare parts available in literature. In Sect. 9.2 motivations for using the action research as a research methodology are provided. The action research in the avionic sector is then described in Sect. 9.4: the full demonstration of the decision making framework is provided by presenting, in context of the action research, the use of models and methods selected from literature as promising "tools" for factual and quantitative assessment. Section 9.5 draws some conclusions by discussing the managerial implications related to the adoption of the framework; besides, some recommendations for future researches are outlined.

9.2 Spare Parts Management: A Brief Literature Review

Spare parts management is undoubtedly not a novel topic in the academic and industrial world. Several approaches, coming mainly from logistics, operations management and operations research communities, have been proposed in the past. Given the plethora of contributions, there have been many systematic overviews and comprehensive surveys, among which it is worthwhile to cite the works of Guide and Srivastava (1997) and Kennedy et al. (2002).

A visible fall-out of such an intensive strive has been the development and commercial proposal of numerous vertical IT solutions, often hyped to the potential industrial customers as the real panacea for their day-by-day issues in managing their spare parts inventory. Not neglecting the substantial added value provided by some of these solutions, what seems really missing in the industrial practice is the capability to follow a sound and consistent logical procedure in tackling the spare part management problem. This motivated the same authors to propose in Cavalieri et al. (2008) a stepwise decision making path in order to orienteer an industrial manager on how to pragmatically handle the management of maintenance spare parts in an industrial company. The framework is organized into five sequential steps: (i) Part coding, (ii) Part classification, (iii) Part demand forecasting, (iv) Stock management policy selection, (v) Policy test and validation. A summary of the objectives of each step is provided in Table 9.2.

Since in the performed action research the part coding resulted as a non-critical step, the remainder of this chapter will specifically address the other four phases.

9.2.1 Part Classification

The identification of criticality of spare items is the primary output of part classification. This is normally obtained by considering different aspects of the parts and the considered environment. It is worth noticing that there are several definitions of criticality; hence it is useful to delineate criticality in a meaningful way for the purposes of the chapter.

According to Dekker et al. (1998), criticality is the level of importance of a piece of equipment for sustaining production in a safe and efficient way. A classification of level of equipment criticality can be used to put evidence on the critical spare parts which deserve more attention and are to be kept in stock to sustain production. Along with Huiskonen (2001), the criticality of a spare item is related to the consequences caused by the failure of a part on the process in case a replacement is not readily available. Hence, it could be named as *process criticality*. Therefore, a spare part can be considered critical for the process when it causes either long lasting or frequent stoppages (see also Schultz 2004), while no alternative production facility is available in order to guarantee the production continuity (see also Gajpal et al. 1994).

Table 9.2 The five decision making steps of the framework (adapted from Cavalieri et al. 2008)

Steps	Objectives and contents
Part coding	A specific spare item coding system has to provide a prompt understanding of the technical features of the item, the equipment tree it refers to, the involved supplier (especially for specific and on-design parts) and, for stocked items, their physical location in the warehouse. A part coding system is mandatory in order to make decisions properly, since it realizes a rationalized set of data on which decisions can rely upon
Part classification	A proper classification of spare items is needed because of the high variety of materials used for maintenance and repair purposes; their technical and economical features can be highly different. A proper classification system should give fundamental information for establishing the criticality and, as a consequence, the strategies for developing the logistics for different classes of maintenance spare parts
Part demand forecasting	Special forecasting techniques are required for some types of spare items. Neglecting consumables, a common feature of many spare items is their relative low level of consumption, due to breakdown or preventive maintenance. Sometimes, time intervals between requests span over several years. Moreover, the consumption rate of a spare part is highly dependent on the number of pieces of equipment where the part is installed, as well as its intrinsic level of reliability
Stock management policy selection	A stock management policy customized upon each class of spare items is required; it might range from no-stock and on-demand policies to the traditional EOQ-RL (Economic Order Quantity - Reorder Level) approach
Policy test and validation	Test and validation of the results achieved applying the above mentioned steps are to be accomplished and refinement may be applied when necessary. This can be obtained by what-if analysis carried out in different scenarios of consumption and supply/repair of the spare items

Beside aspects concerning the consequences on the process, other aspects are related to the possibilities to control the logistics in a given service supply chain setting. To this concern, the same Huiskonen (2001) suggests a so called *control criticality*, including aspects more related to criticality in managing the spare parts logistics as: (i) the availability of spare part suppliers, (ii) the supply lead-times, (iii) the presence of standard versus user-defined features of a spare item, (iv) the purchasing and inventory holding costs, (v) the demand patterns (either easing or not the predictability of the maintenance events). Other authors mention aspects related to the control criticality as the case of Haffar (1995), Dhillon (2002), Mukhopadhyay et al. (2003), Syntetos (2001) and Ghobbar and Friend (2002).

Part classification methods represent a support in considering all the aspects needed to characterize both the *process* and the *control criticality*. A part classification may adopt both quantitative methods, implying the adoption of drivers

Table 9.3 Short review of methods for part classification

Types of methods	Main references
Quantitative methods based on traditional Pareto approaches, single-driver (only one driver used)	Haffar (1995): ABC analysis of logistics performances. This analysis typically enables to focus on a specific aspect, like for example, in the case of turnover rate, identifying slow moving spares or even low rotating items
Quantitative methods based on traditional Pareto approaches, multiple-drivers (more drivers combined)	Schultz (2004): ABC analysis of MTTF (Mean Time To Failure) and MDT (Mean Down Time). This analysis enables to identify the spare items which deserve more attention since they cause a process criticality, either due to frequent failures (low MTTF) or long lasting stoppages (high MDT) or both
	Dhillon (2002): ABC analysis of annual demand and annual purchasing cost. This analysis enables to identify the spare items which deserve attention because of their control criticality, having major concern to their contribution in the annual maintenance budget
Quantitative methods based on the analysis of the demand patterns	Mukhopadhyay et al. (2003): Analysis of the demand patterns based on measurements of the moving rates for each period. This analysis leads to identify the Fast (F), Slow (S), Non Moving (N) spare items, based on number of the parts consumed for each period (FSN approach)
	Syntetos (2001) and Ghobbar and Friend (2002): Analysis of the demand patterns based on the measurements of the average time between two consecutive orders of the same part and the variation of the demand size. This analysis leads to four classes of demand patterns, based on combinations of the average time between orders and variation of the demand size: smooth, intermittent, erratic and lumpy demand
	In both cases, the demand patterns measure aspects of control criticality
Qualitative methods based on consultation with maintenance experts	Mukhopadhyay et al. (2003): Analysis of the subjective judgment of how Vital (V), Desirable (D), Essential (E) is considered a spare item for a system (VED analysis on the process criticality)
	Gajpal et al. (1994): VED analysis applied to the scores assigned by means of consultation of experts through an Analytic Hierarchic Process (AHP) (for details on AHP, see Saaty 1988 and Saaty 1990)
	Due to the high flexibility, the two methods can be adopted both to measure aspects of process and control criticality

based on a numerical value, and qualitative methods, assigning criticality based on a rough judgment or on subjective scoring methods. Table 9.3 provides an overview of the most cited methods in literature.

The identification of criticality of spare items is a relevant step towards the identification of the development strategies of maintenance spare part logistics.

9.2.2 Part Demand Forecasting

Spare parts demand time series can show diversified patterns, depending upon the type of part considered and the specific industry. In most cases, spare parts demand is characterized by a sporadic behavior, which implies a large proportion of zero values (i.e., periods in which there is no demand at all) and a great variability of demand size, when it occurs. The consumption rate is not stationary; hence the demand statistical properties are not independent by the time.

Describing the patterns of sporadic demand for forecasting purposes, the terms erratic, intermittent and lumpy are often used as synonyms. A better specification of these terms can be made by introducing two explicit measures of the demand patterns (Syntetos 2001):

1. The average time between two consecutive orders of the same part, evaluated through the Average Demand Interval (ADI) coefficient.
2. The variation of the demand size, evaluated through the square of the Coefficient of Variation (CV).

Depending on the values of these two indicators demand patterns can be classified into four categories:

1. Smooth demand, which occurs randomly with few or none periods with no demand and with modest variation in the demand size.
2. Intermittent demand, which appears randomly with many time periods having no demand, but without a substantial variation in demand size.
3. Erratic demand, which is highly variable in the demand size and presents few or none periods with no demand.
4. Lumpy demand, with many periods having no demand and high variability in the demand size.

Table 9.4 provides an outline of the most common methods for forecasting spare parts. In particular, two primary classes of techniques are:

- Reliability based forecasting (RBF), to be used when the installed base, that is the number of current installations and their own technical operating conditions, is known.
- Time series forecasting (TSF), suitable when the only available data are related to the time series of the spare items consumption or repair records, while no information about the reliability of the installed base is retrievable.

Table 9.4 Short review of methods for part demand forecasting

Types of methods	Main references
RBF methods based on data banks expressly devoted to collect the failure rates of different typologies of items	Petrovic and Petrovic (1992), Birolini (2004) and Tucci and Bettini (2006): Reliability forecasts, based on a priori knowledge of the conditions of use, required performance and duty cycle of an item, and supported by a computer aided access to data banks for ease consultation
RBF methods based on the life data analysis of different typologies of items	Lawless (2003), Birolini (2004) and Murthy et al. (2004): Reliability forecasts, based on the statistical analysis of a history of failures registered in a Computer Maintenance Management System (CMMS) or as outcome of reliability tests. The Weibull analysis is one of the most commonly used methods: it consists of a data fitting procedure which aims at finding the Weibull distribution that best fits the records of the registered item failures
TSF methods based on the analysis of the orders issued for a spare part (either the repair orders of existent items or the supply orders of new ones)	Croston (1972): proposes a forecasting method to deal with intermittent demand items. It takes into account both demand size and inter-arrival time between demands
	Syntetos and Boylan (2001) and Levén and Segerstedt (2004): Syntetos and Boylan showed that Croston's method was positively biased and proposed a modified method, demonstrating its effectiveness through simulation experiments. Levén and Segerstedt propose another modification of Croston's method following the wake of Syntetos and Boylan
	Cavalieri et al. (2008): Forecasts based on the adoption of the proper method selected in accordance to the demand pattern (i.e., TSF forecasting methods, such as a simple exponential smoothing and its derivatives or ARMA models, are suitable for the smooth and erratic demand; more customized models should be adopted for the intermittent demand and lumpy demand)
TSF methods based on bootstrap	Willemain et al. (2004): Bootstrap creates an empirical distribution of pseudo-data by sampling with replacement, e.g. using Monte Carlo random sampling from the individual observations available in the real history. Then, forecast methods are applied to this bootstrap distribution
	Bootstrap is a simple method to estimate a distribution from sample statistics when the number of observations in the sample is low. This requires, however, that the sample on hand is representative of the real population

9.2.3 Stock Management Policy Selection

Based on the result of the forecasting step, this step aims at selecting an inventory model (implementing a stock management policy or, briefly, a stocking policy), and subsequently defining the stock size in each warehouse for those items that is advisable to detain. Well known inventory models are:

- the continuous review, with fixed reorder point (r) and fixed order quantity (Q), referred to as (Q, r);
- the continuous review, with fixed reorder point (s) and order-up-to level (S), referred to as (s, S);
- the periodic review, with fixed ordering interval (T) and order-up-to level (R), referred to as (T, R);
- the continuous review and order-up-to level (S) in a one-for-one replenishment mode, referred to as (S − 1, S).

Specific models have been proposed in many case studies, mainly based on a set of rules and algorithms tailored for each single case. As an example of this kind of approach, special applications in many sectors like computers (Ashayeri et al. 1996), airline (Tedone 1989), bus fleets (Singh et al. 1980), power generation (Bailey and Helms 2007), and military (Rustenburg et al. 2001) are reported. Being industry-specific, the portability of these models to other industrial settings is generally poor.

In general, a clear application grid of the different stocking policies with an unambiguous understanding of the assumptions, starting hypothesis, area of application (in terms of classes of items) and expected performance (in terms of inventory or hidden costs) is still lacking today. According to Guide and Srivastava (1997) and Cavalieri et al. (2008), the most important and critical drivers that should be considered in selecting a proper inventory model for stock sizing decisions are: (i) the demand pattern (deterministic or stochastic), (ii) the degree of repairability of the spare items, (iii) the level of centralization/decentralization (with inventory located either in a central site, in multiple decentralized sites or with a mixed configuration between centralized and decentralized sites).

By combining the possible values of these drivers specific models can be properly selected, as reported in Table 9.5.

9.2.4 Policy Test and Validation

Last step of the decision making process involves simulation in order to ensure that the selected stocking policy is the most appropriate. Simulation is suitable to verify whether the stock sizing decision taken at the previous step is robust and consistent under different stochastic scenarios of consumption and supply/repair of the spare items. In fact, it can be used to generate and register, on a simulated time scale, the stochastic behaviors of repairable and non repairable systems. Based on a random generation mechanism—the well known Monte Carlo sampling, see (Dubi 1999)

Table 9.5 Short review of models for stock management policy selection

Types of models	Main references
Inventory models for stochastic demand with non-repairable items	Archibald and Silver (1978): Inventory policies for a continuous review system with discrete compound Poisson demand. The paper presents a recursive formula to calculate the cost for any pair (s, S) and relations among s, S, $S - s$ and the cost that leads to an efficient determination of the optimal s and S
	Dekker et al. (1998) and Jardine and Tsang (2006): Poisson model for calculating the stock size S of a ($S - 1$, S) inventory model, such that the demand may be directly fulfilled from stock on hand at a given target fill rate (i.e. probability of not running out of stock when a failure occurs) during the replenishment lead time T (which is a constant supply time of the new item)
	The Poisson model can be adopted as a good approximation for the stock sizing of spare parts when the demand rate in a period T is "not very high". In practice, this is valid for the slow moving parts
Inventory models for stochastic demand with repairable items	Jardine and Tsang (2006): The Poisson model can be also adopted for the case of repairable items, considering the replenishment lead time T as a constant time to repair (of the repairable items). This means to assume an infinite repair capacity
	Several issues related to the repair activities should be included in the model, like (i) the replacement by new items, when the repairable items are worn beyond recovery, so that they cannot be repaired anymore and have to be condemned (Muckstadt and Isaac (1981), Schaefer (1989)); (ii) the finite repair capacity of the repair shop (Balana et al. (1989), Ebeling (1991))
Inventory models for deterministic demand with non repairable items	Cobbaert and Van Oudheusden (1996): Modified EOQ models for fast moving parts undergoing the risk of unexpected obsolescence. These models are applicable mainly to fast moving consumable items with regular demand volumes; they can be thought as modifications of existent models for inventory management in manufacturing

(continued)

Table 9.5 (continued)

Types of models	Main references
Inventory models for deterministic demand with repairable items	Mabini et al. (1992): Modified EOQ model to account for multiple items that share a common and limited repair capacity. Similar considerations to those related to models for deterministic demand with non repairable items apply also to these models: they are applicable to fast moving consumable items with regular demand volumes, as modifications of existent models for inventory management in manufacturing
Inventory models for centralized logistical support/decentralized logistical support (also known as multi-echelon inventory models)	Guide and Srivastava (1997): Most of the multi-echelon models adopt an $(S - 1, S)$ inventory model at each echelon, in case of repairable items. Anyhow, different variants of other inventory models, based on continuous review or periodic reviews and order quantities, can be found
	Birolini (2004): It is opportune to compare the two extremes: the centralized versus the decentralized logistical support. In the former, the demand of the spare items issued from all the decentralized sites is supposed to be satisfied by the stock retained into the central site; in the latter, the demand of a spare part at each decentralized site is supposed to be satisfied only by the stock retained into that site

for details—which generates values according to well defined probability density functions (as for example the time to failure and time to repair), simulation results in a number of "random walks"—the simulation histories—where a system operation state is progressively registered as a sequence of up and down times. The simulation histories can then be used in order to estimate system characteristics (e.g., the point and the average system availability, the mean downtimes, etc.) for the plant/equipment supported by the maintenance logistics. This eventually may help to verify maintenance decisions based also on costs.

Table 9.6 provides a few examples of the potential applications of simulation for supporting spare parts decisions.

9.3 Motivations for Using Action Research

This study has been carried out using Action Research (AR) as the empirical methodology for applying, testing and evaluating the consistency and support to industrial decision makers of the proposed spare parts management framework.

Table 9.6 Short review of applications of the simulation method for policy test and validation

Potential applications	References (only representative examples)
Verification of the service levels guaranteed by a stocking policy	Dekker et al. (1998): Simulation is adopted to verify the service level in a system, where the stocks are already established based on a critical-level policy sized by means of analytic approximation
Verification of the availability of a system based on different settings of maintenance logistics support	De Smidt-Destombes et al. (2006): The availability of a k-out of-n system with deteriorating components and hot standby redundancy can be influenced by the combined decision on different variables concerning the maintenance logistics support: the conditions to initiate a maintenance task, the spare parts inventory levels, the repair capacity and repair job priority settings
Joint verification of the age replacement and spare parts provisioning policy	Zohrul Kabir and Farrash (1996) and Sarker and Haque (2000): a joint optimal age replacement and spare parts provisioning policy is searched for, based on the analysis of the effects resulting from factors like age based item replacement, shortage and inventory holding costs, order supply lead time

Such a qualitative methodology is important for studying complex, multivariate, real-world phenomena that cannot be reduced for study with more positivist approaches. It is significant in situations where participation and organizational change processes are necessary. It merges research and practice thus producing exceedingly relevant research findings (Baskerville and Pries-Heje 1999).

According to Coughlan and Coghlan (2002) AR has several broad characteristics, among which:

- AR focuses on research in action, rather than research about action. Action researchers are not merely observing something happen, but they actively work at making it happen.
- AR is participative. It requires co-operation between the researchers and the industrial counterpart personnel. Members of the client system are in some way co-researchers as the action researcher works tightly with them.
- AR is a sequence of events, which comprises iterative cycles of gathering data, feeding them back to those directly involved in their interpretation, planning, implementing and evaluating actions, leading to further data gathering, and so forth.

The selection of AR as the research methodology for the current study resides on several reasons, among which: (i) as the research is concerned not only with quantitative data but also with rich, subjective, qualitative data, the research philosophy cannot be considered positivist—i.e. providing universal knowledge—

but situational; (ii) the action researcher is immersed in the problem setting and not a mere observer of phenomena; (iii) the research situation has demanded responsiveness, as the research occurs in a changing environment in real-time.

On the other hand, as Conboy and Kirwan (2009) assert, there are some evident limitations a researcher should be well aware of when applying such a methodology. In particular, AR is much harder to report and implies an heavy involvement of the action researcher in the research situation, with the opportunity for good learning, but at the potential cost of sound objectivity.

9.4 The Action Research in the Avionic Sector

The action research deals with a company producing and servicing radars, whose customers are military air forces operating a relatively wide number of jet fighters. The OEM is contractually committed to execute the repairing of the radars installed on customers' jet fighters. Since the radar is a critical part of the jet fighter (a jet cannot be operated if the radar is not available) it represents an important cause of Aircraft On Ground (AOG) time when radar replacements are unavailable.

In order to reduce the AOG risk, the customer itself can stock radar modules to be substituted on field: these are the so called *Line Replaceable Units* (LRUs) and are directly replaced at military hangars without the direct intervention of the OEM. A customer requires support from the OEM on other modules which are sub-components of a LRU: these are the so called *Shop Replaceable Units* (SRUs). Figure 9.1 outlines a simplified bill of material of a generic radar.

SRU and LRU (the latter in case that the damaged or malfunctioning SRU cannot be identified at the customer's premise) must be sent to the OEM in order to be repaired. Each air force tends to accumulate a batch of failed SRUs/LRUs, in order to send them all together for being repaired, reducing the number of

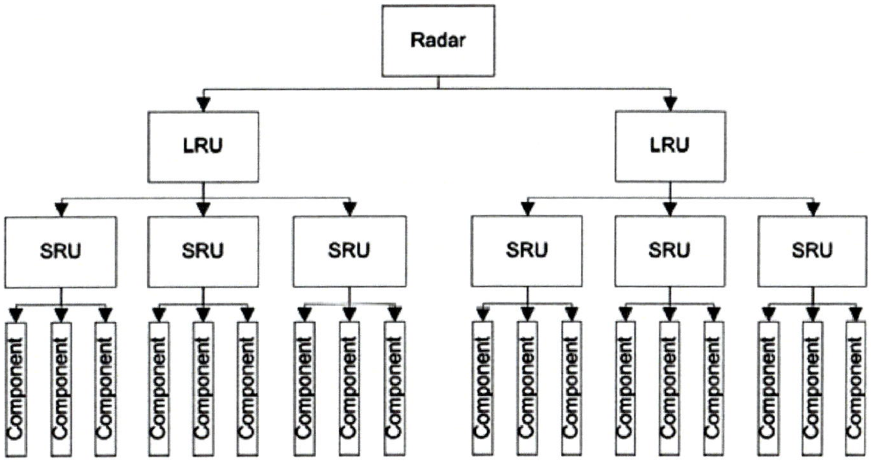

Fig. 9.1 Radar simplified BOM

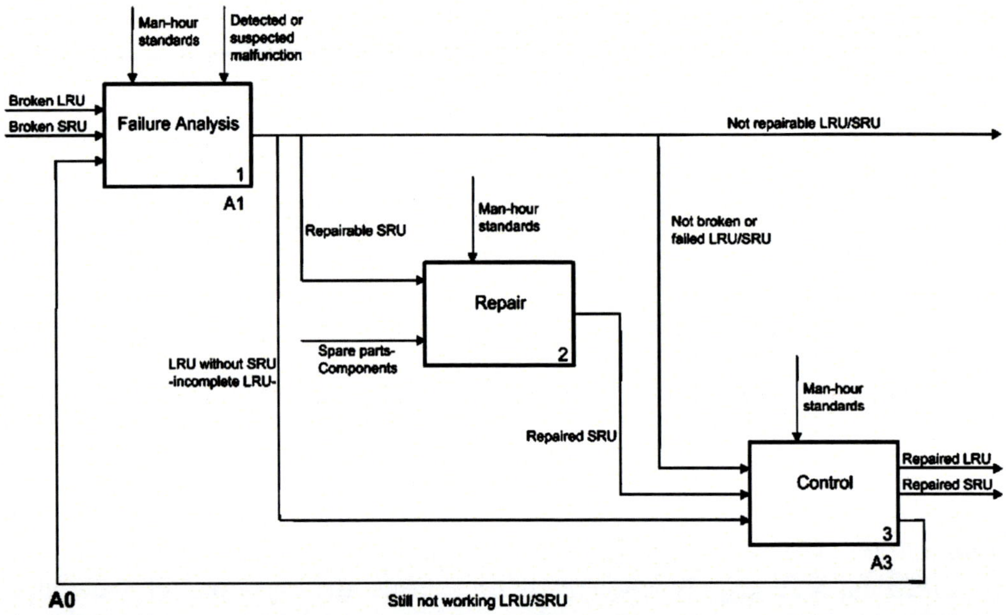

Fig. 9.2 Logistics flows

shipments needed, and to control the costs of the outbound logistics. However, due to the scarce integration of the OEM with its customers, the decision on the batch size is up to the customer. Along with the relatively low number of customers, the customers batching policy is recognized as a source of lumpy demand (Bartezzaghi et al. 1999).

At OEM's premises the broken SRUs/LRUs are first sent to the Failure Analysis unit where they are tested for defects. After detecting a malfunction, the SRU is sent to the Repair unit if it is assumed that it can be repaired. The LRU without SRU (i.e., incomplete LRU) is sent to the Control unit, where it should be reassembled together with the repaired SRU.

Another possibility is that the LRU or SRU is not repairable. Then, a complete new product has to be sent to the customer. Moreover, it is possible that the Failure Analysis unit did not detect any malfunction. Then the "not broken or failed" SRU/LRU is sent to the Control unit for the Final Inspection.

Figure 9.2 reports in a IDEF format the main logistics flows involved in the radar repair process.

The overall logistics performance is evaluated by measuring the Time to Restore (TTR). Along with the active time required for isolating and replacing the fault components (that is diagnostics and repair time in Fig. 9.3), there are specific time components which are due to the logistics support to the maintenance activities.

As an example, if the spare part is not detained, there could be a supply time needed in order to contact and negotiate with the supplier the delivery of the components to be replaced during the repair activity. Also the administrative and logistics delay counts for the outbound logistics: this mainly refers to the movement of the batch of SRUs from and to the customer's site.

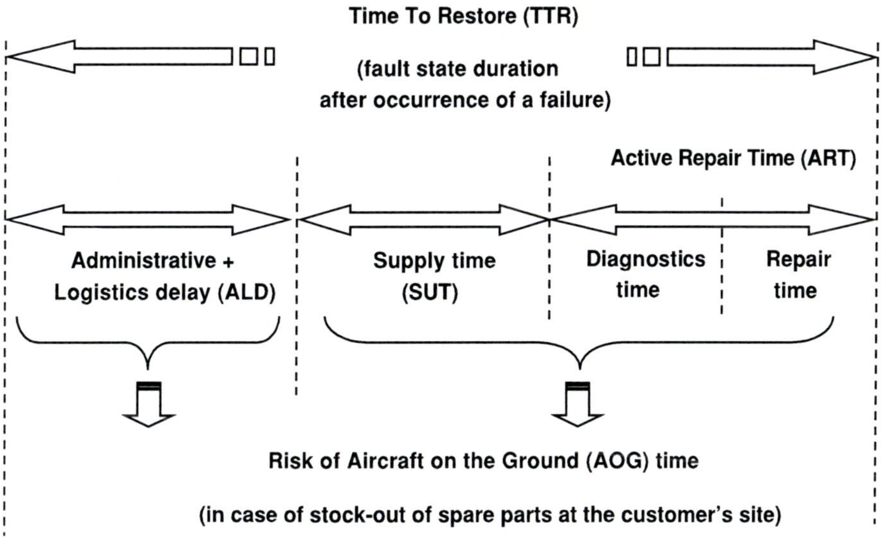

Fig. 9.3 Typical time components of the Time to Restore (TTR) of a repairable item

Clearly a long TTR negatively impacts on the performance of the OEM and the service level provided to the customer, since it increases the risk of AOG (Fig. 9.3).

The OEM is responsible for a portion of the whole TTR, that in turn is composed by three elements:

- the active repair time (ART);
- the supply time (SUT) for any spare item required for replacement;
- any administrative and logistics delay (ALD) registered within the scope of its service workshop, that is from the entrance until the exit of a batch of SRUs, before moving to the customer's site.

The SLA established in the contract is applied to this part of the TTR (hereafter referred to as TTR_{OEM}).

9.4.1 Problem Setting

A fast delivery from the OEM (that is, a low TTR_{OEM}) entails a reduction of AOG risk. Hence, in order to induce quicker repair activities, the OEM is rewarded with a *bonus* when it keeps the TTR_{OEM} lower than an established threshold (in the remainder this is referred as $TTR_{OEM}^{LowerBound}$). Conversely, the OEM has to pay a *malus*, as a penalty cost, whenever it exceeds a threshold (in the remainder referred to as $TTR_{OEM}^{UpperBound}$). If the TTR_{OEM} falls between these thresholds the customers pay the regular price defined in the contract. $TTR_{OEM}^{UpperBound}$ and $TTR_{OEM}^{LowerBound}$ are the service levels agreed by the OEM in the contract.

The following expressions provide the mathematical formulation of the *bonus/malus*. Firstly, the number of repaired items generating a *bonus* (*NB*) or a malus

(*NM*) can be evaluated starting from the repair orders delivered back by the OEM to its customers (*i* being the index of each order, see Eqs. 1 and 2).

$$NB(i) = \begin{cases} Order\ Size_i & if\ TTR_{OEM} \leq TTR_{OEM}^{LowerBound} \\ 0 & if\ TTR_{OEM} > TTR_{OEM}^{LowerBound} \end{cases} \quad (1)$$

$$NM(i) = \begin{cases} Order\ Size_i & if\ TTR_{OEM} > TTR_{OEM}^{UpperBound} \\ 0 & if\ TTR_{OEM} \leq TTR_{OEM}^{UpperBound} \end{cases} \quad (2)$$

As a result, the following Eq. 3 holds:

$$\sum_{i}^{I(T)} [NB(i) + NM(i)] \leq \sum_{i}^{I(T)} Order\ Size_i \quad (3)$$

NB(*i*) and *NM*(*i*) are indexed in accordance to the number of the repair orders delivered back to the customer, starting from i = 1 (first order delivered back to the customer) until i = *I*(*T*) (last order delivered back to the customer). *I*(*T*) is dependent on the horizon length *T* considered for decision making. Therefore, based on such a set of orders, it is possible to define the cash flow (CF_T) of *bonus/malus* (Eq. 4), which, as well, depends on the horizon length T.

$$CF_T = \sum_{i=1}^{I(T)} [Bonus \times NB(i) - Malus \times NM(i)] \quad (4)$$

In this case the *bonus/malus* are constants, not dependent on the type of items under repair, the overall number of repaired items (no discount or "economy of scale"-like effects) and the customer.

The objective of the OEM is to maximize the cash flow, either by increasing the *bonus* gained or reducing the *malus* to be paid. The leverages to this end are ART_{OEM}, SUT_{OEM} and ALD_{OEM} (Fig. 9.3). Keeping constant the capacity of the repairing shop floor, the inbound logistics and the administrative management, the only time component that might change the performances is the SUT_{OEM}. This can be reduced through a sound management of spare parts inventory.

The following objective function (*OF*) is then introduced (Eq. 5).

$$OF_T = CF_T - IHC_T \quad (5)$$

Different spare items *s* can be kept in stock (up to N types), each spare item having its own level of inventory (*S*(*t*,*s*)); hence, the Inventory Holding Cost (IHC_T) is evaluated by integrating, over the horizon *T*, the value progressively reached by the level of Inventory at each time *t* for each spare item *s* kept in stock (*S*(*t*, *s*)) (see the following Eqs. 6 and 7).

$$IHC_T = \sum_{s=1}^{N} \int_{0}^{T} IHC(s) \cdot S(t,s)dt \quad (6)$$

$$S(t,s) = IH_s - D(t,s) + R(t,s) \quad (7)$$

where IH_s inventory level for spare item *s*, representing the decision variable of the problem; *D*(*t*,*s*) total number of spare items *s* demanded by the repair orders issued over the horizon *t*; *R*(*t*,*s*) total number of spare items *s* restored over the horizon *t*; *IHC*(*s*) inventory holding cost per unit of stocked item and unit of time for spare items.

In order to guarantee the profitability of the contract, the objective for the OEM is to achieve a good balance between (i) the improvement of the total bonus and reduction of the total malus and (ii) the reduction of the inventory holding costs (Eq. 5).

D(*t*, *s*) and *R*(*t*, *s*) are the two stochastic variables which makes this balance probabilistic in nature. As better discussed later, indeed, *D*(*t*,*s*) has been considered as the most influencing variable, so a more robust study of its behavior is needed.

9.4.2 Adopting the Decision Making Framework in the Case

After having described the industrial context of the case study and the main issues and motivations for the company to change its current practice of managing its spare parts logistic system, the following subsections will provide a detailed explanation on how the decision making framework has been applied by following the logical steps illustrated in Table 9.2. During this phase of the action research, the industrial counterpart personnel has been involved by the action researcher, according to the specific competences and role required.

9.4.2.1 Part Classification

A quantitative method based on multiple drivers has been applied in order to classify the criticality of components. Three drivers representing *process* and *control criticality* have been adopted:

- The time to restore is adopted to express the process criticality. This measure is not directly bounded to the AOG process metric but to the TTR performance measure for which the OEM is responsible. This is clearly an effect due to the OEM perspective as well as the restricted information sharing with its military clients: since a quicker TTR_{OEM} is rewarding in accordance to the bonus-malus formula established in the contract (as described in Eqs. 1–4), it is worth considering this measure as a driver when assigning process criticality to spare items.
- The inventory holding cost, representing both the financial and control viewpoint, is another component of the profit function in the OEM perspective. The inventory holding cost per unit of stocked item and unit of time (*IHC*(*s*)) is the first driver used in order to evaluate the control criticality, to which attention should be carefully devoted in order to guarantee that their high unit cost is rewarded by proper grants resulting from the bonus-malus.

– The demand predictability, representing the logistics viewpoint. The demand predictability is indeed affected by the fact that the repair orders occur intermittently in time as well as they are erratic in size: the variation of the order sizes is particularly challenging and generally leads to overstock spare inventories in order to protect from the erratic demand.

These drivers are aligned with the profit function defined in the contract (see Eqs. 4–6): indeed, they can be considered as the main factors influencing the financial flows, directly as financial values (i.e., the $IHC(s)$) and indirectly as non financial values influencing the potentials grants (i.e., the time to restore and the inventory holding required for the spare items to face the erratic and intermittent demand). Their implication with respect to the financial flows is better discussed in the next sub-sections.

Process criticality—Keeping stock of SRUs guarantees shorter TTR_{OEM} and reduces AOG risk letting the repair activities start just after the failure analysis enables to identify the broken SRUs to be repaired. To this concern, an analysis has been carried out by measuring the *mean* TTR_{OEM} ($MTTR_{OEM}$) on the past orders of each type of SRU supported in the repair service: this enables to define the "as-is" situation where no SRU inventory is kept. Such an average value represents the comparison term against future improvements in the "to-be" scenario, aimed at faster deliveries of functioning SRUs to the client.

A VED (*Vital Essential Desirable*) analysis has been then applied by identifying the VED classes which eventually correspond to the different grants achievable in the cash flows of *bonus-malus* (measured as in Eq. 4) thanks to faster deliveries:

- a spare item is vital if, kept on stock, guarantees to avoid a malus and to achieve a bonus;
- a spare item is essential if, kept on stock, guarantees to achieve a bonus, while the malus is on average avoided also without stock;
- a spare item is desirable, if kept on stock, is highly likely to guarantee to achieve a bonus.

Table 9.7 shows an example presenting the results of the VED analysis built up by using the $MTTR_{OEM}$ (for sake of privacy, the $MTTR_{OEM}$ related to the part codes are fictitiously defined, even if they are representative of the different situations incurred for the SRUs in the action research).

SRU A is clearly the most critical spare item, classified as *vital*: a malus is on average expected in the "as-is" situation, without inventory holding of SRU A;

Table 9.7 Example of VED analysis based on MTTR

Part code	$TTR_{OEM}^{LowerBound}$ (months)	$TTR_{OEM}^{UpperBound}$ (months)	$MTTR_{OEM}$ (months)	VED class
SRU N	2	4	0.5	None
SRU M	2	4	1.8	Desirable
SRU C	2	4	3.6	Essential
SRU A	2	4	6.1	Vital

indeed, the TTR_{OEM} is higher than the upper bound set in the contract clause. Conversely, a bonus can be expected in the "to-be" scenario, with inventory holding of SRU A: TTR_{OEM} is reduced to the diagnostics time (which is part of the ART component) required for the first failure analysis plus some administrative and logistics delays (the ALD time component); these are far below the lower bound of the contract clause.

SRU C is ranked lower in criticality; hence it is classified as *Essential* and a malus is never incurred, on average, also without inventory holding in the "as-is" situation. Last but not least, SRU M and SRU N differs only for the confidence that they have to reach the bonus: both guarantee a bonus, even without inventory holding; however, SRU M is closer to the lower bound than SRU N, hence the decision maker is less confident that the bonus can be achieved; in this concern, SRU M is considered *Desirable*, while SRU N is not considered at all in further steps of the analysis.

Control criticality—Decisions about SRUs stocks impact on inventory holding costs differently if compared to keeping stocks of other components at lower levels of the BOM, like modules of electromechanical components of SRUs or, directly, the electromechanical components. At this step, however, it is not easy to provide a detailed assessment for balancing costs and grants, since the alternative stocking policies, possible at different levels of the product BOM, lead to a combinatorial problem. After having carried out the criticality analysis, the stocking policy is selected working only on a subset of spare items—i.e. the critical items thus reducing the size of the search space.

In a first step, an ABC classification is performed, by assuming the *IHC*(*s*) (for unit of stocked item and time of item *s*) as the main driver for the cumulated Pareto analysis: this provides a criticality ranking of most valuable SRUs (i.e. items having the highest financial impact when kept on stock). This is clearly not complete, even if it enables to identify, specifically amongst the SRUs, the class A items, critical for their high values with respect to other SRUs (the B and C class items), for which is less expensive keeping inventories.

In addition, when the demand is lumpy, the intermittent occurrences and high variation of the order sizes finally lead to criticality in logistics, since the demand predictability is challenging. According to the classification method presented in Sect. 9.2.1 and the cut off values provided in Syntetos (2001), herein just considered as a reference benchmark, the demand is lumpy if $CV^2 > 0{,}49$ and ADI $> 1{,}32$. In this concern, all the SRUs shown in Fig. 9.4 represent critical items, because they are expression of a lumpy demand.

Multi-dimensional criticality analysis—Three types of methods have been applied leading to three independent spare parts criticality rankings. The three rankings are now combined together for selecting the SRUs that deserve attention in the further steps of the decision making process.

- The VED classification is helpful in order to define a priority list to be considered in the following steps by the OEM, according to the process criticality. Keeping into account the maximization of the profitability of the contract,

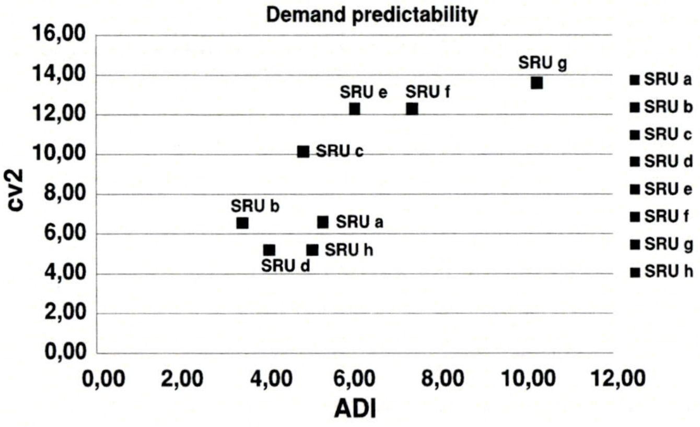

Fig. 9.4 Demand predictability of SRUs (calculated from historical data available from 2000 to 2005)

specifically the grants in the bonus-malus formula, the stock management policy is selected starting first from all vital items, which guarantee better grants incoming from achievement of bonus and avoidance of malus.

- Referring to the control criticality, the ABC classification, based on the *IHC(s)s* (for each unit of stocked item and time of spare item s), is also helpful: it would be preferable to stock first the SRUs pertaining to class B and C, which cause less financial loads.
- Finally, three different classes of control criticality can be also defined based on the demand pattern and demand predictability, in particular considering only the average demand interval (ADI): preference for selecting a stocking policy is assigned to the most frequent issues of repair orders, equivalent to lower ADIs (so a reduced intermittency of order arrivals). Three classes are defined to this concern: a class with more frequent issues (less intermittent arrivals), which, in the action research, is set to when ADI is lower than 6 months; a class with less frequent demand (more intermittent arrivals), with ADI higher than 6 months but lower than 12 months; a class with very low frequency, i.e. ADI higher than 12 months. The last class is not kept into account in the remainder, having considered more than 12 months a long time span to wait for a new order to consume the spare item.

The flow chart in Fig. 9.5 summarizes the combined adoption of the three methods. On one hand, the priorities are assigned to SRUs whose repair orders occur more frequently (i.e., ADI lower than 6 months, which guarantees a less intermittent consumption of inventories), and less valuables with respect to their unit inventory holding cost (B and C classes in the ABC classification). On the other hand, only if budget is still available, the remaining *vital* and *desirable* items are considered for stock management policy, also amongst the most valuable items (class A of *IHC*(*s*)*s*) and less frequent (ADI more than 6 months but lower than 12 months).

Only the SRUs selected from criticality analysis, according to the flow chart in Fig. 9.5, have been considered from now on: indeed, the results, achieved until the second criticality ranking lead to select eight critical SRUs.

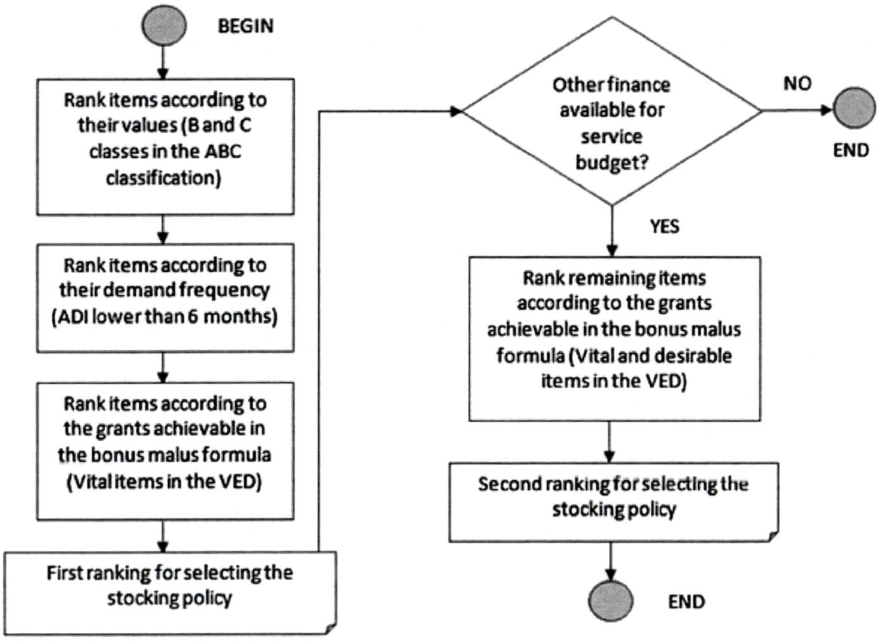

Fig. 9.5 Flow chart for combining the methods adopted for spare parts criticality analysis

9.4.2.2 Spare Parts Forecasting

The partial or null visibility on how the Radars, and so their SRUs, are actually utilized on field by the military customers on their jet fighters make forecasting activities extremely hard: number of operating hours, operating conditions and mission profiles are unknown, due to the military reluctance to reveal data. Considering the uncertainties on this operational information, a *RBF method* (see Table 9.4) *based on life data analysis* of different items on field is not applicable.

The scarce information about the real conditions of use and the duty cycle of each item operating on field obstacle, in practice, also the *RBF method based on data ban*ks. The last option is the *time series forecasting* (TSF), since data related to the time series of the spare items consumption and repair records are available in the Maintenance Information System and are duly completed during the contractual management of the service; indeed, the commitment to proper contractual management can be considered a relevant factor for guaranteeing the good quality of the data records. The main issue to solve regards demand predictability. A clear definition of which forecasting method is more suitable for the demand type of each quadrant of the classification matrix of Fig. 9.4 is not an easy task in general and this is particularly true for the lumpy demand quadrant, for which the demand predictability is the most challenging. In the scope of this action research, it has been preferred to test the stocking policies for the critical SRUs in different simulated demand scenarios, instead of making use of a forecasting method to obtain accurate spare parts forecasts. In this concern, policy test and validation (step 5 of the *decision making framework*) has become relevant. The reason for this preference is strictly subsequent to the high character of lumpiness observed for the demand, especially due to the high variation of the order sizes: CV^2 is in the range between 5.21 and 13.62 (Fig. 9.4) revealing a quite relevant variation in the demand size, combined with its high intermittency. This certainly imposes high requirements on the forecasting method, potentially asking for a complex one, with the hidden risk of only a partial capability to precisely fit, in the forecast, the repair order distribution in time and quantity.

Accordingly, it has been decided to rely on the empirical probability density functions of two primary stochastic variables, in the remainder symbolized as DI (Demand Interval) and ROS (repair order size). These functions can be easily extracted from the time series, expressed in the form of histograms of frequencies: indeed, the length of the period kept under observation (i.e., 6 years) guarantees a robust sample for building the histograms.

Looking ahead at the next steps, averages—like ADI—are the only statistics being adopted for the rough stock sizing (step 4), before passing to policy test and validation (step 5), wherein the histogram of frequencies themselves is directly being exploited. This last issue means that other summary statistics, characterizing the probability density functions (as CV^2), have been considered during policy test and validation. The empirical probability density functions of each critical SRU—hence, the histograms and their related summary statistics—are the basis to enable a Monte Carlo simulation of different demand scenarios. Further details about that are provided in the step of policy test and validation (see Sect. 9.2.4).

Figure 9.6 summarizes the links between spare parts forecasting and the other steps of the decision making framework, by showing the respective expressions of the empirical observations used by each further step.

9.4.3 Stock Management Policy Selection

Among the different models available in literature, the continuous review and order-up-to level (S) in a one-for-one replenishment mode, referred to as ($S - 1, S$),

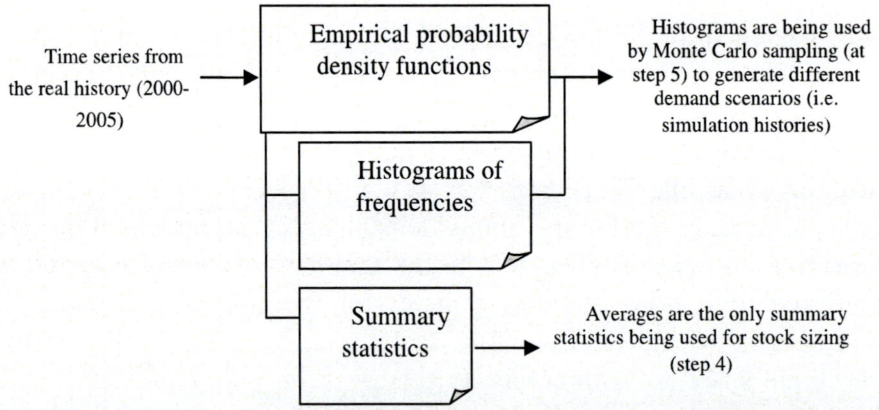

Fig. 9.6 From spare parts forecasting to next steps in the decision making framework

is more advisable for this case. It should be kept into account that the repair order size varies, depending on each customer: so the one-for-one replenishment is applied directly to the repair order that is received, meaning that one batch of SRUs is repaired and is used, without being splitted, to replenish the stock level on hand, previously reduced of the same quantity delivered to the same customer.

The selection of the $(S-1, S)$ model is in line with many references in literature of similar examples where the demand rate in a period T is assumed "not very high" (see for example Jardine and Tsang 2006). Also in our case, as the ADI shown in Fig. 9.4 demonstrates, the situation implies a demand rate in a period T which is "not very high", leading to a slow moving character of the spare items. The Poisson model can be adopted as a good approximation for the stock sizing of spare parts in such a situation (as already outlined in Table 9.5).

Applying the Poisson distribution, a stock size S to be kept can be calculated, based on a target level of fill rate R (i.e. the probability of not running out of stock when a failure occurs). Equation 8 expresses the general formula of the Poisson model.

$$\Pr\{\tau_1 + \cdots + \tau_n > T\} = \sum_{i=0}^{S-1} \frac{(d \cdot T)^i}{i!} \cdot e^{-d \cdot T} \geq R \tag{8}$$

where $\tau_1, \ldots, \tau_n$ represent the times to each failure, requiring the spare part (until the n-th request), and are assumed to be independent positive random variables; d is the demand rate per unit period and it is, in general, estimated according either to a reliability based or a time series based forecasting; T expresses the time interval taken as a reference for the target fill rate.

Applying the formula to the data of the case study:

- due to the simplification done for spare parts forecasting, at this step d is estimated only on average, simply as 1 repair order for each ADI (1/ADI), so using only part of the summary statistics available for each SRU;
- T is defined by considering the Poisson model calculated in the case of repairable items (according to the classification illustrated in Table 9.5), hence it represents the $MTTR_{OEM}$ for the repair orders;
- the stock level S is fixed at a number of items close to the Average Repair Order Size (AROS) keeping into account the fact that, at each ADI, the number of items requested from stock is AROS.

Table 9.8 shows the rough stock sizing resulting from applying the Poisson model: S has been rounded to the closer integer number (lower or higher); R is the resulting target rate for this choice.

It is worth pointing out that R represents an analytic approximation that should be further verified; besides, R means the instantaneous or point reliability, providing that spares are available on demand, at any given moment in time; while the interval reliability (i.e., spares are available at all moments in a given interval) is not estimated and would be "more demanding" (Jardine and Tsang 2006): the following step of policy test and validation may help to provide more accurate

Table 9.8 Stock sizing based on the Poisson model

Part code	AROS (# items)	S (# items, rounded)	R (%)
SRU a	1.33	1	61.8
SRU b	1.22	1	88.1
SRU c	1.63	2	99.6
SRU d	3.13	3	99.7
SRU e	1.88	2	99.9
SRU f	3	3	99.9
SRU g	2	2	99.9
SRU h	1.27	1	98.5

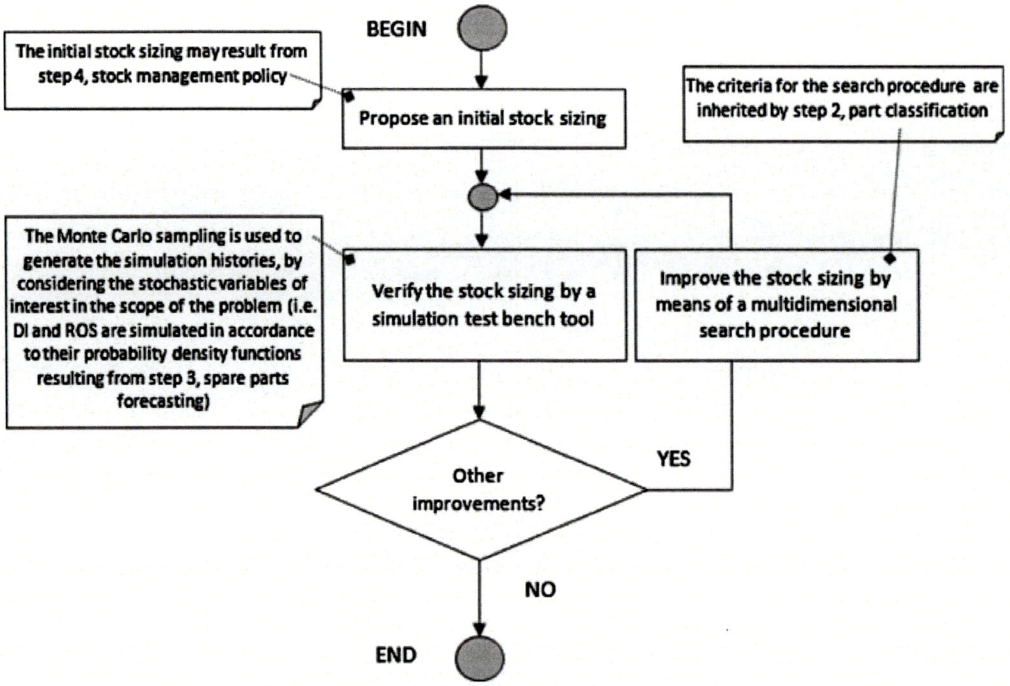

Fig. 9.7 Policy test and validation: a conceptual scheme

estimates of the performances expected by the service workshop, both for R and other additional measures.

Furthermore, S is clearly not the final solution: it is instead an initial and rough solution that will be improved again in the next step; hence, rounding to the lower or upper integer is an acceptable choice at the current phase.

9.4.4 Policy Test and Validation

Policy test and validation enables to verify the stocking policies selected for the SRUs. The initial stock sizing provided in the previous step 4 is at this phase verified and progressively improved thanks to a search procedure enacted together with the decision maker and the aid of the simulation used as a test bench tool (Fig. 9.7).

Table 9.9 Rules associated to the spare parts criticality ranking based on the IHC(s)s

Part code	Cumulated IHC(s)s (% calculated on the eight critical items) (%)	Action
SRU b	20	Decrease
SRU d	39	Decrease
SRU c	55	Decrease
SRU h	70	Decrease
SRU a	85	Keep
SRU e	93	Keep
SRU f	98	Increase
SRU g	100	Increase

Table 9.10 Rules associated to the spare parts criticality ranking based on the ADIs

Part code	ADIs (calendar months, ascendant order)	Action
SRU b	3.38	Increase
SRU d	4.00	Increase
SRU c	4.80	Increase
SRU h	5.00	Keep
SRU a	5.25	Keep
SRU e	6.00	Keep
SRU f	7.34	Decrease
SRU g	10.25	Decrease

Table 9.11 Rules associated to the spare parts criticality ranking based on the VED classes and $MTTR_{OEM}$

Part code	VED (according to $MTTR_{OEM}$ in descendent order)	Action
SRU a	Vital	Increase
SRU d	Vital	Increase
SRU f	Vital	Increase
SRU c	Essential	Keep
SRU b	Essential	Keep
SRU g	Essential	Keep
SRU e	Essential	Decrease
SRU h	Essential	Decrease

The search procedure is derived directly from the multi dimensional criticality analysis played out during the step of part classification.

In order to improve the solution, three actions are defined: *increase* or *decrease* the stock size (by one or more items) or *keep* it constant. Each action is associated with the spare parts criticality rankings of the critical items (Tables 9.9, 9.10 and 9.11) resulting from the multi dimensional criticality analysis of part classification.

The associations express the behavior of the logistics expert who generally tends to:

1. decrease stocks of items when they have a high *IHC(s)* or, vice versa, increase stocks of items having a low *IHC(s)s* (Table 9.9 shows the rules associated to the ABC classification, cumulated Pareto analysis, of *IHC(s)*);
2. increase stocks of items frequently requested, while decrease those not frequently requested (Table 9.10 reports items sorted from low ADIs—orders more frequently requested—up to high ADIs—orders less frequently requested—rules are likewise associated);
3. increase or at least keep the stocks when the items are vital or essential, to gain grants from the bonus-malus contract (the ranking in Table 9.11 is descendent, being the top ranking occupied by the item with the highest $MTTR_{OEM}$, so the highest risk, on average, to pay a malus and not be rewarded with a bonus).

The rules are then combined in order to form an integrated rule set (Table 9.12) as a guideline for deciding on how to improve an initial solution (i.e. the stock size). Using this rule set, a trade off may emerge in some cases. In order to manage in a simple way these uncertain cases, alternative criteria have been selected: (i) a random criterion is the simplest one; (ii) a criterion where either the *process criticality* or the *control criticality* are dominant in the multidimensional analysis.

Table 9.12 demonstrates an example of improvement of an initial solution by means of the search procedure, passing from an initial stock *SI* to a final stock *SF*: (i) in clear situations prevail the most voted action (as an example, when at least 2 out of 3 *increase* are present, *increase* is selected; see these situations highlighted in italic in the same table); (ii) uncertain situations are the remainder (i.e., three out of eight SRUs highlighted in bold); in this specific case, the *decrease* or *keep* action are prevailing and the control criticality is considered dominant. It is however worth noticing also that, when trying a possible improvement, it has been kept in mind the criticality of the eight items, as a binding rule: in this context, it has been decided that a "minimum stock" should anyhow be guaranteed (i.e. S equal to 1 item as minimum). This applies to one case in the example, SRU b, where, due to this binding rule, the *keep* decision prevailed.

Table 9.12 Finding out a solution during the search procedure driven by the integrated rule set

Part code	Integrated rule set			Improving from SI to SF	
	Action (due to IHC(s)s)	Action (due to ADIs)	Action (due to VED)	SI (initial)	SF (final)
SRU a	Keep	Keep	Increase	1	1
SRU b	Decrease	Increase	Keep(*, minimum)	1	1
SRU c	Decrease	Increase	Keep	2	1
SRU d	Decrease	Increase	Increase	3	4
SRU e	Keep	Keep	Keep	2	2
SRU f	Increase	Decrease	Increase	3	4
SRU g	Increase	Decrease	Keep	2	1
SRU h	Decrease	Keep	Keep	1	1

Further improvements may be obtained by progressively correcting the solutions: i.e. by deciding to change the search direction in other different trials for the uncertain cases (e.g., try to *increase* the stock of some SRUs instead of *decrease*, action done at the previous improvement step), or, conversely, continue to follow a given search direction for the certain ones (e.g., continue to *increase* or continue to *decrease/keep*).

Therefore, the whole search procedure may start from the rough stock sizing resulting from step 4 of the *decision making framework*. Then, the procedure is applied together with simulation, in accordance to the conceptual scheme (Fig. 9.7): progressively, by using a stepwise approach (increase or decrease by 1 or, in general, a small number of items) different solutions for the stock size are decided, in accordance to the guidelines of the rule set fixed in Table 9.12, and tested in simulation histories. Some trials are summarized in the Table 9.13: S1 represents the solution which guarantees the "minimum stock" in the "to-be" scenario (i.e. at least 1 item on stock is guaranteed for each critical SRU), S2 is the initial and rough stock sizing resulting from step 4 (Table 9.8); S3 is the stock sizing after one improvement step (this is the step shown in Table 9.12, where S3 is SF) and S4 is the stock sizing resulting after a further improvement step, when a different trial of the uncertain cases has been carried on.

The simulation trials of Table 9.13 have been evaluated through some selected performances: the Total Number of Bonus (*TNB*), Malus (*TNM*), Profitable (*TNP*) and Non Profitable (*TNNP*) items along the horizon length of interest *T* (indeed, at this step, *T* is the horizon of a simulation history, equal to the 6-year time span available).

$$TNB_T = \sum_{i=1}^{I(T)} NB(i) \tag{9}$$

$$TNM_T = \sum_{i=1}^{I(T)} NM(i) \tag{10}$$

$$TNP_T = \sum_{i=1}^{I(T)} NP(i) \tag{11}$$

Table 9.13 Solutions used as data inputs to be verified by simulation

Part code	S1	S2	S3	S4
SRU a	1	1	1	1
SRU b	1	1	1	1
SRU c	1	2	1	1
SRU d	1	3	4	4
SRU e	1	2	2	2
SRU f	1	3	4	4
SRU g	1	2	1	2
SRU h	1	1	1	1

$$TNNP_T = \sum_{i=1}^{I(T)} NNP(i) \quad (12)$$

These performances are clearly defined in order to be aligned with the profit function (see the previous Eqs. 4–6). Indeed:

- the Total Number of Bonus and Total Number of Malus (Eqs. 9 and 10) directly influence the grants coming from the cash flow of bonus/malus (Eq. 4): when assuming, for example, the same value for the bonus and malus of an item, profitability is guaranteed if the total number of bonus (i.e., items achieving the bonus) is higher than the total number of malus (i.e. items achieving the malus); a similar reasoning can be applied when knowing that the ratio r between the bonus and malus parameters of an item is different than 1; in this case, profitability is guaranteed if the total number of bonus is higher than 1/r times the total number of malus;
- on the other hand, inventory holding costs are also part of the profit function (Eqs. 5 and 6); hence, the Total Number of Profitable items (Eq. 11) is used in order to measure a subset of Total Number of Bonus; this subset counts only those items for which the achieved Bonus is higher than the Inventory Holding Cost spent for the time that the same items are kept in stock before being used; vice versa, in the case of Total Number of Non Profitable items (Eq. 12), the achieved Bonus is lower than the Inventory Holding Cost; hence, these performances are also aligned to the profit function, now having a specific concern to the ratio between benefits (the bonus achieved by each item) and costs (the *IHC(s)s* spent for each item on stock for some given units of time).

The results of the progressive search, as a whole, are shown in Table 9.14: the marginal improvement progressively observed for the selected performances decreases, passing from the "minimum stock" situation (S1) to the last trial (S4) for the "to-be" scenario. The last trial in the table provides a solution which, amongst all the trials, is characterized by the highest service level: it shows the highest number of bonus *TNB* (high service due to quick response) and a limited number of malus *TNM* (low service due to slow response). Besides, it is worth noticing how this last trial guarantees that, among the orders reaching the bonus, the total number of profitable items TNP is clearly higher than *TNNP*. These are possible "to-be" solutions and have to be compared, of course, with the "as-is" case, without inventory holding, so to assess how much improvement each solution could guarantee, with respect to the existent operation.

Table 9.14 Results from simulation of different trials done during the search for improved solutions

Part code	S1	S2	S3	S4
TNB_T	87	120	120	123
TNM_T	62	29	29	26
TNP_T	83	115	115	117
$TNNP_T$	4	5	5	6

As a concluding remark of this step, it is important to remind that Table 9.14 results from the same simulated demand scenario, built up in accordance to the empirical probability density functions of the SRUs. This is the first simulation history (or "random walk") allowing to prove, against the stochastic variations both of the Demand Interval as well as the Repair Order Size, the different solutions proposed for the stock sizing. It is worth noticing that the stochastic variations of this "random walk" resembles the real history, in particular its summary statistics of ADI and CV^2. Hence, the assessment of the stocking policies is certainly extended with respect to the previous step (i.e. step 4), where only the ADI and AROS have been considered for stock sizing.

A last step of financial assessment, based on the profit function (Eqs. 4–6), would help complete the analysis of performance measures (illustrated in Table 9.14) by providing the traditional financial indicators, useful to support the decision on stocking policies.

9.5 Conclusions and Managerial Implications

Most of the chapter has been devoted to a detailed description of the way the multi-step decision making procedure has been planned to be properly applied in the specific industrial context. The following main outcomes emerged from the feedbacks gotten from the management involved in the action research.

- Complex, analytical models sometimes assume stringent hypothesis and a reduced scope of application. Due to the peculiarities of the presented process in the avionic sector, an analytical model would have been too complex and difficult for industrialists to tackle with reasonable effort. On the other hand, a combination of simple models, organized within a well structured procedure, can overcome the practical issues imposed by larger models providing a robust decision support.
- The judgment of the decision-making operators can be effectively supported by quantitative measures that reduce the ambiguity of qualitative decisions and improve the effectiveness of the choice. As an example in the action research, the assignment of criticality of spare items is one relevant step in order to initially define the strategies for developing the maintenance spare part logistics, and there is little room for subjectiveness thanks to the quantitative measures adopted.
- A simulation model is a powerful and user-friendly tool suitable for the analysis and validation of the decisions taken about stock sizing. Besides, its combined use with the judgment of the operators can effectively improve the outcome deriving from first rough solutions.
- From the managerial point of view, the usage of a well defined decision making framework can improve the overall performance by aligning the decisions to the specific business scenario, giving sound evidence to the results of the strategic

managerial choices in a company. The proper definition of the stock sizes can definitely lead to a better exploitation of the financial capital, while preserving (or even improving) the required logistic service level.

Considering the main theoretical knowledge gained from this empirical research, some further observations can be drawn.

First, the presented framework can be classified as a *rule based* one. Hence it is quite easy to implement in the real practice of a company without the need of complex and costly software or hardware supports: the way the framework was implemented in the Action Research can be considered a live demonstration of its implementation in corporate practice. Moreover, the overall framework has been designed in order to let the human operator "keep the control" on the final decision, so to avoid the negative feeling that a completely automated decision-making system can sometimes generate.

References

Archibald BC, Silver EA (1978) (s, S) Policies under continuous review and discrete compound Poisson demand. Manage Sci 24:899–909

Ashayeri J, Heuts R, Jansen A, Szczerba B (1996) Inventory management of repairable service parts for personal computers. Int J Oper Prod Manag 16(12):74–97

Bailey GJ, Helms MM (2007) MRO inventory reduction—challenges and management: a case study of the Tennessee Valley Authority. Prod Plan Control 18(3):261–270

Balana AR, Gross D, Soland RM (1989) Optimal provisioning for single-echelon repairable item inventory control in a time varying environment. IIE Trans 21(3):202–212

Bartezzaghi E, Verganti R, Zotteri G (1999) A simulation framework for forecasting uncertain lumpy demand. Int J Prod Econ 59:499–510

Baskerville R, Pries-Heje J (1999) Grounded action research: a method for understanding IT in practice. Acc Manage Inf Technol 9:1–23

Birolini A (2004) Reliability engineering: theory and practice, 4th edn. Springer, Berlin

Cavalieri S, Macchi M, Garetti M, Pinto R (2008) A decision making framework for spare parts inventory planning: proposal and validation in industrial cases. Prod Plan Control 19(4):379–396

Cobbaert K, Van Oudheusden D (1996) Inventory models for fast moving spare parts subject to "sudden death" obsolescence. Int J Prod Econ 44(3):239–248

Conboy K, Kirwan O (2009) An action research case study of the facilitators and inhibitors of e-commerce adoption. Int Bus Res 2(2):48–56

Coughlan P, Coghlan D (2002) Action research for operations management. J Prod Oper Manage 22(2):220–240

Croston JD (1972) Forecasting and stock control for intermittent demands. Oper Res Q 23(3): 289–303

De Smidt-Destombes KS, van der Heijden MC, van Harten A (2006) On the interaction between maintenance, spare part inventories and repair capacity for a k-out-of-N system with wear-out. Eur J Oper Res 174(1):182–200

Dekker R, Kleijn MJ, de Rooij PJ (1998) A spare parts stocking policy based on equipment criticality. Int J Prod Econ 56–57(3):69–77

Dhillon BS (2002) Engineering maintenance: a modern approach, 1st edn. CRC Press, Boca Raton

Dubi A (1999) Monte Carlo applications in systems engineering, 1st edn. Wiley, New York
Ebeling CE (1991) Optimal stock levels and service channel allocations in a multi-item repairable asset inventory system. IIE Trans 23:115–120
Gajpal PP, Ganesh LS, Rajnedran C (1994) Criticality analysis of spare parts using the analytic hierarchy process. Int J Prod Econ 35(1–3):293–297
Ghobbar AA, Friend CH (2002) Sources of intermittent demand for aircraft spare parts within airline operations. J Air Transp Manag 8:221–231
Guide VDR Jr, Srivastava R (1997) Repairable inventory theory: models and applications. Eur J Oper Res 102(1):1–20
Haffar I (1995) "SPAM": a computer model for management of spare parts inventories in agricultural machinery dealerships. Comput Electron Agric 12:323–332
Harland C, Brenchley R, Walker H (2003) Risk in supply networks. J Purch Supply Manag 9(2):51–62
Huiskonen J (2001) Maintenance spare parts logistics: special characteristics and strategic choices. Int J Prod Econ 71(1–3):125–133
Jardine AKS, Tsang AHC (2006) Maintenance, replacement and reliability: theory and applications, 1st edn. CRC Press, Boca Raton
Kennedy WJ, Patterson JW, Fredendall LD (2002) An overview of recent literature on spare parts inventories. Int J Prod Econ 76(2):201–215
Kumar R, Markeset T, Kumar U (2004) Maintenance of machinery—negotiating service contracts in business-to-business marketing. Int J Serv Ind Manag 15(4):400–413
Lawless JF (2003) Statistical models and methods for life time data, 2nd edn. Wiley, New York
Levén E, Segerstedt A (2004) Inventory control with a modified Croston procedure and Erlang distribution. Int J Prod Econ 90:361–367
Mabini MC, Pintelon LM, Gelders LF (1992) EOQ type formulations for controlling repairable inventories. Int J Prod Econ 28(2):21–33
Muckstadt JA, Isaac MH (1981) An analysis of a single item inventory system with returns. Naval Res Logistics Q 28:237–254
Mukhopadhyay SK, Pathak K, Guddu K (2003) Development of decision support system for stock control at area level in mines [online]. http://www.ieindia.org/publish/mn/0803/aug03mn3.pdf. Accessed 9 Jan 2008
Murthy DNP, Xie M, Jiang R (2004) Weibull models, 1st edn. Wiley, New York
Petrovic D, Petrovic R (1992) SPARTA II: further development in an expert system for advising on stocks of spare parts. Int J Prod Econ 24(3):291–300
Rustenburg WD, Van Houtum GJ, Zijm WHM (2001) Spare part management at complex technology-based organizations: an agenda for research. Int J Prod Econ 71(1–3):177–193
Saaty TL (1988) Multicriteria decision making: the analytic hierarchy process, 2nd edn. RWS Publications, Pittsburgh
Saaty TL (1990) How to make a decision: the analytic hierarchic process. Eur J Oper Res 48(1):9–16
Sarker R, Haque A (2000) Optimization of maintenance and spare provisioning policy using simulation. Appl Math Model 24(10):751–760
Schaefer MK (1989) Replenishment policies for inventories of repairable items with attrition. OMEGA Int J Manage Sci 17:281–287
Schultz CR (2004) Spare parts inventory and cycle time reduction. Int J Prod Res 42(4):759–776
Singh N, Shah KS, Vrat P (1980) Inventory control of recoverable spares in a two echelon repair-inventory system: case study. Terotechnica 1:257–264
Syntetos AA (2001) Forecasting of intermittent demand. Unpublished PhD thesis, Buckinghamshire Business School, Brunel University, UK
Syntetos AA, Boylan JE (2001) On the bias of intermittent demand estimates. Int J Prod Econ 71:457–466
Taracki H, Tang K, Moskowitz H, Plante R (2006) Maintenance outsourcing of a multi-process manufacturing system with multiple contractors. IIE Trans 38(1):67–78
Tedone MJ (1989) Repairable part management. Interfaces 19(4):61–68

Tucci M, Bettini G (2006) Methods and tools for the reliability engineering: a plant maintenance perspective. In: Proceedings of the 2nd maintenance management MM2006, Sorrento, Italy, April

Willemain TR, Smart CN, Schwarz HF (2004) A new approach to forecasting intermittent demand for service parts inventories. Int J Forecast 20:375

Zohrul Kabir ABM, Farrash SHA (1996) Simulation of an integrated age replacement and spare parts provisioning policy using SLAM. Reliab Eng Syst Saf 52(2):129–138

Chapter 11
Optimal and Heuristic Solutions for the Spare Parts Inventory Control Problem

Ibrahim S. Kurtulus

Abstract One inventory problem that has not changed much with the advent of supply chains is the control of spare parts which still involves managing a very large number of parts that face erratic demand which occurs far in between. For most items it has not been feasible to establish a control system based on the individual item's demand history. Hence either the parts have beenbundled into one group as Wagner did and only one demand distribution has been used for all or they have been divided into different groups and group distributions have been used as in Razi. We picked two popular rules by Wagner and extensively tested their performance on data obtained from a Fortune 500 company. Comparisons were made with the optimal solution and optimal cost obtained from our procedure based on the Archibald and Silver's optimizing algorithm. For some problems with very small mean demand and small average number of demand occurrences, finding the optimal solution required an inordinate amount of CPU time, thus justifying the need to use heuristics.

11.1 Introduction

The problem studied is an (s, S) type of inventory control system for a single item, which is subject to sporadic demand that is also highly variable in size. There is a fixed (replenishment) lead time of L. Item replenishment costs consist of a setup cost K and a unit cost c. Units carried in excess of demand incur a holding cost of h

I. S. Kurtulus (✉)
School of Business, Virginia Commonwealth University, 301 W. Main St., Richmond VA, USA
e-mail: ikurtulu@vcu.edu

N. Altay and L. A. Litteral (eds.), *Service Parts Management*,
DOI: 10.1007/978-0-85729-039-7_11,

per unit and demand not satisfied is backordered and incurs a penalty cost of d per unit. The objective function is to minimize the expected value of the sum of carrying, backordering and (fixed) ordering costs. Under these assumptions, it has been shown (Iglehart 1963) that a policy of type (s, S) will minimize the undiscounted expected cost per period over an infinite horizon, where ordering decisions are restricted to demand occurrences (Archibald 1976 and Silver 1978; Beckmann 1961; Ehrhard 1979; Vienot and Wagner 1965). A replenishment order of size $(S - y)$ is placed when inventory on hand plus on order minus backorders y is less than or equal to s.

In testing the performance of heuristics and optimal procedures, various assumptions have been made with respect to demand. Common ones included Poisson (Schultz 1987; Gelders and Looy 1978; Hadley and Whitin 1963) and normal distributions (Croston 1972; Bartakke 1981; Porteus 1985; Vereecke and Verstraeten 1994; Sani and Kingsman 1997). Unfortunately, test of fit of certain parametric distribution to actual demand data has been difficult to show and rarely reported in the literature. So far the most convincing arguments have been made in favor of the compound Poisson distribution, which is what we have used in this research. The assumption of random demand occurrence (usually assumed to be Poisson as in queuing theory) and a nonparametric distribution of demand size given as a frequency distribution are combined to form the compound Poisson distribution (Adelson 1966; Feller 1968; Vienot and Wagner 1965; Feeney and Sherbrooke 1966; Archibald and Silver 1978; Razi 1999).

Another distribution recommended by researchers (i.e., Dunsmuir and Snyder 1989; Segerstedt 1994; Yeh 1997) has been a positively skewed gamma distribution with a large spike at zero. It has been recommended both for demand size and occurrence.

Two variations of the problem have also been investigated. One shows how the algorithm is simplified if orders are placed when the reorder point s is reached exactly (Zheng and Federgruen 1991). Since it is a model more appropriate for continuous review systems, it is not used in this paper. And yet another addresses the issue of nonstationary (variable) mean demand and develops a myopic heuristic (Bollapragada and Morton 1999).

Over the years, various heuristics have also been recommended to solve the problem. We pick two that were developed by Wagner with very realistic and convenient assumptions for managers. Both heuristics can also be considered variations of the same method. When the exact probability distribution (for demand) is used Wagner has called his heuristic the algorithmic method and when the probability distribution is approximated, the normal approximation. In past research, heuristics' performance has been tested by using simulation (i.e., Porteus 1985; Sani and Kingsman 1997). This paper deviates from this approach and compares heuristics' performance with the optimal cost obtained when the compound Poisson distribution is used.

11.2 Wagner's Heuristics and Optimization

Wagner in 1975 developed a heuristics as probabilistic extensions of the basic EOQ model. In his formulation Wagner first derives the formula for expected average inventory based on an exact distribution. If the distribution is known, Wagner calls his method the algorithmic solution since it has to go through a number of iterations to determine s and $Q(S = s + Q)$. We will call it the exact method (WEX). The solution will be optimal and identical to the optimal solution obtained by the Archibald and Silver algorithm, if compound Poisson is used for the demand distribution and the same approximations are made in both derivations: In Wagner's derivation, expected average inventory is approximated first. In Archibald–Silver's derivations, all approximations are made when limits are taken.

However, for an executive who neither has the time nor a readily accessible data base, it is very convenient to approximate the probability distribution in the heuristic by a parametric distribution such as the normal. If one distribution is to be used for all demand, then normal approximation is a reasonable choice. With normal, only two parameters of the distribution, mean and standard deviation have to be estimated, and the tables are readily available to compute the cumulative probabilities. In this research, we used the discrete version of Wagner's normal approximation heuristic (WNOR). A brief discussion of both heuristics is provided in Appendix 2.

The optimizing procedure (Kurtulus 2004) used is based on the first part of the algorithm (without the new bounds on $n = S - s - 1$) developed by Archibald and Silver (1978). It was tested on the data reported in their paper and the same optimal values were found (Kurtulus 2004). The Fortran code had to be written from scratch because: (1) The Archibald–Silver code is based on 1978 technology; (2) The treatment of the lower bound on s is different in this paper because one more lower bound was developed; (3) It would have taken a long time to understand the original 1978 code and the authors were no longer researching the problem; (4) It might not have been possible to modify their algorithm to accommodate for the non-optimal values of (s, S) found by the heuristics, which may fall outside the lower and upper bounds developed by their algorithm.

In order to evaluate the performance of a heuristic, the (s, S) values created by the heuristic were plugged into the objective function of the optimizing algorithm and the corresponding cost was determined. In most cases, the heuristic solution was not optimal. Percentage increase over the optimal cost was computed and the results are reported in the tables.

We had few concerns in using simulation to rank the performance of the heuristics: (1) the objective function is not convex, leaving the researcher with no clear indication of when to stop; (2) Compound Poisson is two distributions. First the interarrival time is generated to determine when the demand occurs and then the demand size with all of its possible convolutions. For example, if a demand of 5 was generated, all possible (i.e., $1 + 4$, $2 + 3$, $4 + 1$, $3 + 2$, 5) ways of obtaining a total demand of 5 had to be simulated. This further complicated the issue of determining the number of runs needed to reach a steady state (if one could define it). (3) Different parameters used for λ, μ, K and h affected the CPU time needed to find the optimal solution, which clearly indicated that they would also affect the number of runs needed to reach a steady state, or some version of it.

11.3 The Design of the Data Base

The actual data used in the study was obtained from a Fortune 500 Company (Razi 1999). The manager in charge of spare parts inventory kept monthly demand history for a 2 year period for each of the 22,500 parts. Those items with no demand history (i.e., 30% of the items.) in the 2 year period were excluded from the study. Of the remaining items, any item with total demand of 50 units or less (in the 2 year period) was classified as slow moving. Then these items were divided into 12 groups based on replenishment lead time (2, 3, 4, 6 weeks) and total demand (i.e., 1–10, 11–20, 21–50) for the 2 year period. Finally, a frequency distribution for demand size was generated for each group. All of these probability distributions for demand were extremely skewed to the right, with peaks occurring at one. Plots of the five distributions we picked are provided in Appendix 1.

The manager estimated ordering cost as $50 per order, carrying cost as 15% of the item cost, backordering cost (cost of delay) as $20 if the item's cost was less than $100 and 30% of the cost of the item if its cost was more. Implicit in these computations was the fact that the cost of work stoppage was not a problem and hence was not considered as part of the cost of backordering. In the worst case scenario the part could be obtained without work stoppage by Federal Express. Lambdas, the average number of demand occurrences per year, were computed as follows: First we found the average number of occurrences per month based on the 24 month demand history for each item. We picked the minimum and the maximum in each group. Averaged them and multiplied the average by 12 to find the yearly figure. Item cost used for each group was also a simple average of the minimum and maximum in the group. The data is summarized in Tables 11.1 and 11.2 in Appendix 1.

Peterson (1987) and Peterson et al. (2000) reported that Poisson distribution adequately reflects the demand distribution for spare parts inventory. He used ordering cost of ($3 or $20) and carrying cost of ($0.1, $0.2, $0.7) which he reports as being similar to those used by the US Air Force inventory policies. His two backordering cost levels (4× carrying cost, 9× carrying cost) correspond to 80 and 90% service levels and represent the percentage of time an optimally controlled system (in his simulations) incurs no backorders. Also, these service levels are around the 86% service level the Air Force targeted at the time. His lead times are (0, 2, 4 time units) and mean demands are (0.1 and 1.0) and Poisson distributed. When compared to Razi (1999), his ratios involving ordering-to-carrying cost and ordering-to-backordering cost were higher and his mean demands were much lower.

11.4 Results and Suggestions for Future Research

The results involving the data given in Table 11.1 and the two replenishment lead times, 2 weeks ($L = 0.04$) and 4 weeks ($L = 0.083$) with WNOR are summarized in Tables 11.3 and 11.4 (Appendix 1). In the two tables, the average increase in cost (from optimal) are 97.9 and 71.04%, respectively. When the ratio of ordering cost to holding cost is increased (or h is reduced) by 12-fold, the results improve to 76.9 and 66.4% as shown in Tables 11.5 and 11.6. The best results are obtained when the ratio of ordering to holding cost is high and the replenishment lead time is long (i.e., Table 11.6).

As expected, WEX does better than WNOR under both replenishment lead times, giving only an average 42.8% increase from the optimal in Table 11.7 and 27.2% in Table 11.8 when the data in Table 11.1 is used. When the ratio of ordering cost to holding cost is increased (or h is reduced) by 12-fold, the results improve to 15.0 and 20.1% in Tables 11.9 and 11.10, respectively.

We are puzzled with the improvement in performance of both rules when the re-supply lead time is doubled and the ratio of ordering to holding cost is not high. In Tables 11.4 and 11.8, in case of groups 1, 2, 6, and 14, the percent increase from optimal cost is less when $L = 0.083$, and approximately equal with group 3.

Under WEX, the best results are obtained when the ratio of ordering to holding cost is high but when the replenishment lead time is short (i.e., Table 11.9). The short lead time part is contrary to what we had observed in other cases. We believe it is due to interaction affects not yet defined. Under WEX, in all cases Group 3 provides the worst results. However, the same cannot be said for WNOR. Hence Group 3 cannot be treated as an outlier?

Given that all actual group distributions were skewed to the right with peak at one, can we use a triangular distribution that has a peak at 1 and uses the range of values for demand comparable to actual data being simulated. If successful, using a triangular distribution with easily definable parameters, will free the practitioner from the task of actually developing the exact distributions required by WEX? Future research will answer some of these questions.

Appendix 1

Table 11.1 Characteristics of the original Razi data used in the study

Groups	Ord$	Mean	Variance	Hold$	BckOrd$	Lambda
1	50	1.734	1.050	38.52	77.0	2.76
2	50	2.745	9.293	82.64	165.3	9.30
3	50	3.854	14.68	6.34	20.0	18.30
6	50	2.740	8.167	54.00	108.0	7.74
14	50	3.110	13.654	261.15	522.5	13.02

Table 11.2 Ratios of ordering and backordering to holding cost used in the study

Groups	Regular Ord$/ Hold$	Regular BckOrd$/ Hold$	Reduced Hold$	Reduced Ord$/ Hold$	Reduced BckOrd$/ Hold$
1	1.30	2	3.21	15.6	24
2	0.60	2	6.89	7.3	24
3	7.90	3.2	0.53	94.3	37.7
6	9.3	2	4.50	11.1	24
14	0.19	2	21.76	2.3	24

Table 11.3 Wagner normal approximation ($L = 0.04$)

Group	WNOR s	WNOR S	TC$	Arch–Silver s	Arch–Silver S	% Increase
1	1	2	212.59	−2	2	99.0
2	4	5	854.36	−2	3	110.7
3	2	10	360.20	−7	29	89.4
6	3	4	596.02	−2	4	107.4
14	5	6	2145.00	−1	2	83.0
Average						97.9

Data in Table 11.1 was used producing these results

Table 11.4 Wagner normal approximation ($L = 0.083$)

Group	WNOR s	WNOR S	TC$	Arch–Silver s	Arch–Silver S	% Increase
1	1	2	211.14	−2	2	85.2
2	3	5	804.94	−1	5	64.7
3	3	11	375.52	−5	33	89.8
6	3	4	601.40	−2	5	82.8
14	5	6	2224.51	0	5	32.7
Average						71.0

Data in Table 11.1 was used producing these results

Table 11.5 Wagner normal approximation (reduced holding costs, $L = 0.04$)

Group	WNOR s	WNOR S	TC$	Arch–Silver s	Arch–Silver S	% Increase
1	2	9	50.60	−1	11	29.3
2	6	12	229.73	0	20	50.5
3	6	33	129.06	−1	116	107.6
6	5	13	247.74	−1	22	133.6
14	8	10	732.67	3	19	63.5
Average						76.9

From Table 11.2, $h = 3.21, 6.89, 0.53, 4.5, 21.76$, are used. The rest of the data is from Table 11.1

Table 11.6 Wagner normal approximation (reduced holding costs, $L = 0.083$)

Group	WNOR s	WNOR S	TC$	Arch–Silver s	Arch–Silver S	% Increase
1	2	9	50.60	−1	12	23.2
2	6	12	250.36	2	24	45.8
3	6	33	134.47	4	123	110.5
6	5	13	165.63	1	25	42.4
14	9	10	1108.52	8	24	110.1
Average						66.4

From Table 11.2, $h = 3.21, 6.89, 0.53, 4.5, 21.76$, are used. The rest of the data is from Table 11.1

Table 11.7 Wagner exact method ($L = 0.04$)

Group	WEX s	WEX S	TC$	Arch–Silver s	Arch–Silver S	% Increase
1	1	4	179.39	−2	2	67.9
2	1	6	528.61	−2	3	30.4
3	1	13	284.61	−7	29	49.7
6	1	6	377.21	−2	4	31.3
14	1	7	1575.79	−1	2	34.5
Average						42.8

Data in Table 11.1 was used producing these results

Table 11.8 Wagner exact method ($L = 0.083$)

Group	WEX s	WEX S	TC$	Arch–Silver s	Arch–Silver S	% Increase
1	1	4	175.25	−2	2	53.8
2	1	6	543.53	−1	5	11.2
3	1	13	298.00	−5	33	50.7
6	1	6	380.70	−2	5	15.7
14	1	7	1751.95	0	5	4.5
Average						27.2

Table 11.9 Wagner exact method ($L = 0.04$)

Group	WEX s	WEX S	TC$	Arch–Silver s	Arch–Silver S	% Increase
1	1	12	43.80	−1	11	12.0
2	1	18	155.01	0	20	1.5
3	1	41	97.31	−1	116	56.5
6	1	18	110.32	−1	22	4.0
14	1	19	452.35	3	19	1.0
Average						15.0

From Table 11.2, $h = 3.21, 6.89, 0.53, 4.5, 21.76$, are used. The rest of the data is from Table 11.1

Table 11.10 Wagner exact method ($L = 0.083$)

Group	WEX s	WEX S	TC	Arch–Silver s	Arch–Silver S	% Increase
1	1	12	44.01	-1	12	7.0
2	1	18	179.58	2	24	4.6
3	1	41	105.41	4	123	65.0
6	1	18	121.41	1	25	4.6
14	1	19	629.70	8	4	19.4
Average						20.1

From Table 11.2, $h = 3.21, 6.89, 0.53, 4.5, 21.76$, are used. The rest of the data is from Table 11.1

Appendix 2

Wagner's Heuristics

Wagner (1975) has developed his heuristics as probabilistic extensions of the basic EOQ model. If the exact distribution is known, Wagner calls his method the algorithmic solution since it has to go through a number of iterations to determine s and $Q(S = s + Q)$. We will call it the exact method (WEX). The solution will be optimal with respect to the objective function assumed in the model. Let $p_L(x_L)$ be the probability mass function of demand (nonparametric) during lead time and μ_L its mean and $P_L(s) = \sum_{x_L=0}^{s} p_L(x_L)$. Then the exact method goes through the following steps to find the solution:

Step 1: Let initial trial value of

$$Q = \sqrt{2K\mu_L/h} \tag{1}$$

Step 2: Using the trial value of Q, compute:

$$R = 1 - \frac{hQ}{h\mu_L/2 + \mu_L d} \tag{2}$$

Find a trial value for s such that it is the smallest positive integer for which:

$$P_L(s) \geq R \tag{3}$$

From the definition of (3) we can think of the value of $P_L(s)$ as the minimum service level acceptable to the company.

Step 3: Stop if the new trial value of s is the same as before. Otherwise, calculate a new trial value for Q by using (4) below and go back to Step 2.

$$Q = \sqrt{(2K\mu_L)/h + (\mu_L + (2\mu_L d/h)) \sum_{x_L \succ s} (x_L - s) p_L(x_L)} \tag{4}$$

Since as s is incremented (or decremented) Q in (4) converges, the algorithm will end in a finite number of steps. However, for an executive who neither has the time nor a readily accessible data base, it is very convenient to approximate the probability distribution $P_L(x)$ by a parametric distribution such as the normal. Then only two parameters of the distribution, $\mu_L = L\mu$ and $\sigma_L = \sigma\sqrt{L}$ have to be estimated, and the normal tables can be used for the cumulative distribution. Dropping the subscript for the replenishment lead time (L), the discrete version of Wagner's normal approximation heuristic (WNOR) is defined by:

Step 1: Let initial trial value of

$$EOQ = \sqrt{2K\mu/h} \tag{5}$$

Step 2: Compute:

$$R_N = \frac{hEOQ}{d\sqrt{(T+1)\sigma}} \tag{6}$$

and find the f_s value of the unit loss function (or standardized normal loss integral) $I_N(f)$ such that:

$$I_N(f_s) = R_N \tag{7}$$

Step 3: If $\mu < 0.8888\ (K/h)$, then let s be determined by:

$$s = (L+1)\mu + f_s\sqrt{(L+1)\sigma} \tag{8}$$

and $S = s + Q$. Otherwise go to Step 4.

Step 4: Compute:

$$R = \frac{d}{(h+d)} \tag{9}$$

and find the value of the standard normal variable f_v such that

$$P_N(f_v) = R \tag{10}$$

Define:

$$f_m = \min(f_s, f_v) \tag{11}$$

Then (s, S) are determined by:

$$s = (L+1)\mu + f_m\sqrt{(L+1)\sigma} \tag{12}$$

and

$$S = \left((L+1)\mu + \min\left[f_s\sqrt{(L+1)\sigma} + EOQ, f_v\sqrt{(L+1)\sigma}\right]\right) \tag{13}$$

For further discussion and justification for both WEX and WNOR, please refer to Wagner (1975).

References

Adelson RM (1966) Compound Poisson distributions. Oper Res Q 17:73–75

Archibald BC (1976) Continuous review (s, S) policies for discrete compound poisson demand processes. Unpublished Ph.D. thesis, University of Waterloo, Waterloo

Archibald BC, Silver EA (1978) (s, S) policies under continuous review and discrete compound Poisson demand. Manag Sci 24:899–904

Bartakke MN (1981) A method of spare parts inventory planning. Omega 9:51–58

Beckmann M (1961) An inventory model for arbitrary interval and quantity distributions of demands. Manag Sci 8:35–37

Bollapragada S, Morton TE (1999) A simple heuristic for computing nonstationary (s,S) policies. Oper. Res. 47:576–584

Croston JD (1972) Forecasting and inventory control for intermittent demands. Oper Res Q 23:289–303

Dunsmuir WTM, Snyder RD (1989) Control of inventories with intermittent demand. Eur J Oper Res 40:16–21

Ehrhard R (1979) The power approximation for computing (s, S) inventory policies. Manag Sci 25:777–786

Feeney GJ, Sherbrooke CC (1966) The (s-1, s) inventory policy under compound Poisson demand. Manag Sci 12:391–411

Feller W (1968) An introduction to probability theory and its applications. Wiley, New York

Gelders LF, Van Looy PM (1978) An inventory policy for slow and fast movers in a petrochemical plant: a case study. J Oper Res Soc 29:867–874

Hadley G, Whitin TM (1963) Analysis of inventory control systems. Prentice-Hall, New Jersey

Iglehart D (1963) Optimality of (s, S) policies in the infinite horizon dynamic inventory problem. Manag Sci 9:259–267

Kurtulus IS (2004) Programming the (s, S) Archibald–Silver algorithm for the modern day PCs. In: National DSI proceedings, pp 291–296

Peterson DK (1987) The (s, S) inventory model under low demand. Unpublished Ph.D. thesis, University of North Carolina, Chapel Hill

Peterson DK, Wagner HM, Ehrhardt RA (2000) The (s, S) periodic review inventory model under low mean demand and the impact of constrained reorder points. In: National DSI proceedings, pp 1014–1016

Porteus EL (1985) Numerical comparison of inventory policies for periodic review systems. Oper Res 33:134–152

Razi M, Kurtulus IS (1997) Development of a spare parts inventory control model for a Fortune 500 company. In: National DSI proceedings, pp 1402–1404

Sani B, Kingsman BG (1997) Selecting the best periodic inventory control and demand forecasting methods for low demand items. J Oper Res Soc 48:700–713

Schultz CR (1987) Forecasting and inventory control for sporadic demand under periodic review. J Oper Res Soc 38:453–458

Segerstedt A (1994) Inventory control with variation in lead time, especially when demand is intermittent. Int J Prod Econ 35:365–372

Silver EA, Peterson R (1985) Decision systems for inventory management and production planning, 2nd edn. Wiley, New York

Veinot AF, Wagner HM (1965) Computing optimal (s, S) inventory policies. Manag Sci 11: 525–552

Vereecke A, Verstraeten P (1994) An inventory management model for an inventory consisting of lumpy items, slow movers, and fast movers. Int J Prod Econ 35:379–389

Wagner HM (1975) Principles of management science with applications to executive decisions. Prentice Hall, New Jersey

Yeh QJ (1997) A practical implementation of gamma distribution to the reordering decision of an inventory control problem. Prod Inventory Manag J 38:51–57

Zheng YS, Federgruen A (1991) Finding optimal (s,S) policies is about as simple as evaluating a single policy. Oper. Res. 39 no.4, pp 654–665

Chapter 12
Reliable Stopping Rules for Stocking Spare Parts with Observed Demand of No More Than One Unit

Matthew Lindsey and Robert Pavur

12.1 Introduction

Many studies describe challenges facing large manufacturers who must efficiently control an inventory of tens of thousands of finished products, maintenance and replacement or spare parts (Ward 1978; Gelders and Van Looy 1978; Dunsmuir and Snyder 1989; Hua et al. 2007). Wagner and Lindemann (2008) have urgently called for future research on strategic spare parts management. When stocking spare parts, a few parts often represent the bulk of the investment and the majority of the demand. However, it is important to be able to forecast the demand rate for the slow-moving items as well as the heavily used parts. If a product has not had a demand over a specified duration of time, its demand would be projected to be zero based on many of the popular forecasting models, such as simple exponential smoothing or moving averages. Yet, this product may still be required and be worth carrying, particularly if the inventory cost is well managed.

This study examines the demand for these types of products and develops a methodology to address related issues. In particular, we will propose a methodology to determine a one-sided prediction interval for predicting demand rates for intermittent or slow-moving spare parts that is adapted from statistical procedures developed for software reliability. The one-sided prediction interval is an upper-sided interval and the upper endpoint could be compared to a threshold value so that a decision can be made on whether to continue carrying a group of products. In essence, a stopping rule can be employed to determine whether to stop stocking certain products. A stopping rule procedure is a rule used in a decision-making

M. Lindsey (✉)
Stephen F. Austin State University, Nacogdoches, TX, USA
e-mail: lindseymd@sfasu.edu

R. Pavur
University of North Texas, Denton, TX, USA
e-mail: pavur@unt.edu

N. Altay and L. A. Litteral (eds.), *Service Parts Management*,
DOI: 10.1007/978-0-85729-039-7_12,

process in which a decision is required on whether to stop carrying a product or group of products. A stopping rule procedure can incorporate an estimate of the future demand rate of products into its analysis to determine the continuation decision of stocking inventory. A stopping rule typically provides an objective and empirically supported rule to determine when stocking items, such as spare parts, is no longer justified. Stopping rules may be based on economic policies, such as the cost effectiveness of continuing to carry a spare part that was seldom used.

For slow-moving spare parts having no demand or limited demand over an observed period of time, an effective rule may be difficult to establish. Managers may re-evaluate the future demand rate of a group of slow-moving spare parts at the end of specified time periods. An easy-to-implement stopping rule can be used to discontinue spare parts whose projected demand rate is less than a given threshold value. Typically, such a threshold is based on financial considerations. If the upper endpoint of a one-sided prediction interval for the product's future demand rate is below the threshold, for instance, then the decision is to liquidate.

12.2 Predicting Future Demand Rate for Slow-Moving Spare Parts

Companies may easily carry inappropriate quantities of slow-moving spares in which future demand rates are difficult to forecast. Because of the importance of maintaining a given service level as suggested by Miragliotta and Staudacher (2004) organizations can compensate for poor forecasts by increasing assets or working capital, but these options may prove costly. Service levels may be determined by customer satisfaction. Often, some level of support may be needed for an expensive product that sporadically required the replacement of a long-lasting high tech component. Being able to accurately predict demand for slow-moving spare parts and maintain an optimal inventory level for spare parts is one way to cut waste. Unfortunately, traditional forecasting techniques often result in stocking higher than needed levels of inventory for slow-moving spares. Porras and Dekker (2008) point out that Enterprise Resource Planning (ERP) packages such as SAP R/3, which uses cycle service levels, are not adequate for the control of spare parts that are more appropriately measured by the fill rate.

While traditional forecasting methods are applicable for spare parts with regular usage, the application of intermittent demand forecasting techniques is often needed. Ghobbar and Friend (2002) classify demand patterns into erratic, lumpy, smooth and intermittent demand and suggest appropriate techniques for each category. Willemain et al. (2004) classify techniques for forecasting items with intermittent demand into the categories of methods such as non-extrapolative approaches, variations of the Poisson process model, smoothing methods, variations of Croston's method, and bootstrapping methods. Boylan et al. (2007) provide the definitions for terms used in the study of slow-moving inventory. Products with infrequent demand occurrences are classified as intermittent. Slow-moving items are intermittent and have low average demand.

Cavalieri et al. (2008) provide a decision-making framework for managing maintenance spare parts. The framework consists of coding and classifying parts, forecasting demand for parts, determining a stock management policy and then testing and validating. The middle step, part demand forecasting, can be difficult when demand is low, including intermittent or lumpy demand. Cavalieri et al. (2008) suggest that traditional time series-based forecasting methods are suitable when the demand is smooth or erratic, but suggest that customized models are needed for intermittent demand and lumpy demand. The method proposed in the current paper fills the need identified by Cavalieri et al. (2008).

Many of the existing approaches to estimating demands for slow-moving products find their roots in research related to predicting usage rates for military spare parts, especially those onboard ships (Haber and Sitgreaves 1970). Haber and Sitgreaves (1970) survey several forecasting methods for goods with sporadic demand patterns, including a method that relies on expert opinion to pool usage figures for products with similar designs. These items would for the most part be classified as having lumpy demand and would not follow the assumption of demand following a Poisson distribution. A limitation is that the number of categories to which parts are classified needs to be determined to provide sufficient data to obtain reliable estimates of demand. An implicit assumption is made that demand for each product is independent of the demand for other products.

The use of reliability analysis has also provided a context for the study of controlling spare parts. Yamashina (1989) structured the problem of forecasting demand for spare parts in terms of the product manufacturing pattern, the product life characteristics and the part life characteristics. This view of forecasting demand for spare parts utilizes the coefficient of variation of the product life probability density function, shape of the demand curve, and the production pattern to determine when spare parts are required.

12.3 Stopping Rules for Deciding to No Longer Carry Inventory

Browne and Pitts (2004) state that a stopping rule can be used in the decision-making process to make a judgment based on the information gathered about the sufficiency of that information and the need to acquire additional information. That is, a stopping rule is some test or heuristic invoked by the decision-maker to determine the sufficiency of the information obtained. They remark that stopping rules have been investigated extensively in decision-making research. Brown and Zacks (2006) studied a stopping rule for the problem of quick detection of a change-point in the intensity of a homogeneous ordinary Poisson process. In general, finding optimal stopping rules is quite difficult.

Ross (1985b) considered a stopping rule to determine when a software package was ready to be released after testing for errors from bugs. Often, software packages are released with some minimal number of bugs to accept a reasonable risk of errors at the end of a testing phase. He proposed a procedure similar to that used in a quality control setting. Assume that $\varepsilon(t)$is the estimate of $\Lambda(t)$, the true future error rate of the software package. Let A be the minimal acceptable error rate that management is willing to allow in a software package that is released to the market. The stopping rule for testing can be the rule that calls for stopping the testing procedure at the first time value of t such that the following upper limit on the error estimate is below A: $\varepsilon(t) + 3\sqrt{E\,[\varepsilon(t) - \Lambda(t)]^2} < A$. An important question is: how reliable is this rule? This answer may depend on the accuracy of the estimate of of $\Lambda(t)$. The robustness of one-sided prediction intervals may provide insights into the feasibility of this guideline.

In the context of inventory management, a stopping rule procedure may be used in the decision-making process to determine whether to stop carrying a spare part or group of spare parts. A viable stopping rule procedure must incorporate an accurate estimate of the future demand rate of spare parts into its analysis to optimally determine the continuation of inventory. Upper one-sided prediction intervals for forecasting future demand rate for spare parts showing no demand over a specified time frame could be assessed for robustness with respect to their nominal Type I error rate and thus be validated for its usefulness in being part of a stopping rule procedure. The study in this paper considers the robustness of such one-sided prediction intervals and also extends these intervals to the case where spare parts display no more than one demand over a specified time frame. The reliability of one-sided prediction intervals is shown to depend on a combination of parameters: the observed time frame, the true demand rate, the number of products in inventory, and the appropriateness of assuming a normal approximation for the distribution of the estimator.

Decision rules for determining whether a group of spare parts should be discontinued must involve a high degree of confidence that the true future demand rate of the spare parts will be below a certain threshold. The threshold is typically based on economic considerations, which will not be explored in this study. For example, see Horodowich (1979) for a model that considers cost of capital, income taxes, selling price inflation, scrap value, and magnitude of inventory on hand, when considering when to discontinue a product.

Lindsey and Pavur (2009) proposed an extension of the Ross estimator to use in forming two-sided prediction intervals for demand rates of inventory that has displayed no demand over a specified period of time. The technique for forming reliable one-sided prediction intervals for the future demand rate of existing products with observed demand of zero was adapted from Ross (2002)'s methodology on software reliability. Lindsey and Pavur (2009) used a simulation study to examine the reliability of the two-sided prediction intervals for the future demand of products with no observed demand across experimental conditions that included product group size, mean time between demand, and Type I error levels.

Ranges of these parameters over which the two-sided prediction intervals were reliable were noted.

The current study differs from Lindsey and Pavur (2009) in that the reliability of one-sided prediction intervals is investigated. Although it may seem that the parameters under which two-sided predictions for products with no observed demand are the same as that for one-sided prediction intervals, this is not the case as is illustrated in a comparison of the two in the simulation study section of this paper. The two-sided prediction interval allows for more error in one tail of the prediction interval to be compensated by less error in the other tail. In addition, this paper investigates an extension in which the demand rate of products which display no more than one unit of demand are investigated. Although the prediction intervals are adapted from Ross (1985a, b, 2002)'s estimators, he did not assess the distribution or reliability of these estimators.

12.4 Estimating a Poisson Rate of Random Variables with Zero or One Demand Units

If a set of Poisson random variables are observed over a specified time frame and no observations occur, then an estimate of zero may be inappropriate simply because the time frame may not have been long enough. This type of problem has appeared in the software reliability literature. Ross (1985a, b, 2002) derived an estimator for the future demand of bugs in a software package that has no occurrences. However, he did not examine the distribution of this estimator or the reliability of prediction intervals constructed from this estimator. He also did not extend this estimator to the case of estimating the future demand rate of Poisson random variables with zero or one observed occurrences. Many models based on the Poisson distribution have been considered for estimating failure rates in software (Abdel-Ghaly et al. 1986; Kaufman 1996).

A description of this estimator will now be presented using the context of errors in a software package during testing. Then a description will follow in which the context will be translated into demand for a group of products with zero or one observations. Suppose that there are n bugs contained in a software package. The number of errors caused by bug i is assumed to follow a Poisson distribution with a mean of λ_i, $i = 1, 2, \ldots, n$. Ross (2002) defines $\Psi_i(t) = 1$ if bug i has not caused a detected error by time $t > 0$ and 0 otherwise, $i = 1, 2, \ldots, n$. These indicator variables allow the future error rate of the bug with no observed error to be $\Lambda(t) = \sum_{i=1}^{n} \lambda_i \Psi_i(t)$. This expression has unknown rate parameters, λ_i, that are difficult to estimate without using a time frame that is long enough to provide an accurate estimate. In the context of software applications, a high error rate by this expression would be unacceptable to the customer. Ross (1985a, b) used the following notation.

n	Number of errors
λ_i	Error rate for bug i, $i = 1, 2, \ldots, n$
t	Length of time period over which errors are observed
$\Psi_i(t)$	Equal to 1 if bug i has not caused an error and 0 otherwise
$\Lambda(t)$	Theoretical error rate of bugs
$M_j(t)$	Number of bugs that have caused j errors by time t

Assume that a specified time frame is denoted by t. To estimate $\Lambda(t)$, Ross (2002) used $M_j(t)$ to denote the number of bugs responsible for j detected errors by time t, j = 1, 2, …, n. That is, $M_1(t)$ is the number of bugs that cause exactly one error, $M_2(t)$ is the number of bugs that cause exactly two errors, and so on. Using the assumption of a Possion process the expected future error of the bug displaying no occurrences is $E[\Lambda(t)] = \sum_{i=1}^{n} \lambda_i E[\Psi_i(t)] = \sum_{i=1}^{n} \lambda_i e^{-\lambda_i t}$. Interestingly, $\frac{M_1(t)}{t}$ and $\Lambda(t)$ have the same expected value according to Ross (2002). $\frac{M_1(t)}{t}$ is an attractive estimator since it can be easily computed in practice. Thus, $E\left[\Lambda(t) - \frac{M_1(t)}{t}\right] = 0$, which is key to establishing that $\frac{M_1(t)}{t}$ is an unbiased estimate of $\Lambda(t)$. For it to be a good estimator of $\Lambda(t)$, its difference with $\Lambda(t)$ should be small. The second moment of $\Lambda(t) - \frac{M_1(t)}{t}$ is the same as the expected value of $\frac{M_1(t)+2M_2(t)}{t^2}$, which is a function of $M_1(t)$ and $M_2(t)$. Therefore, the mean squared difference between $\Lambda(t)$ and $\frac{M_1(t)}{t}$ can be estimated by $\frac{M_1(t)+2M_2(t)}{t^2}$.

The following results in Eq. 12.1 summarize that $\frac{M_1(t)}{t}$ is an unbiased estimator of $\Lambda(t)$ and that the variance of the difference, or equivalently the expected squared difference, between the estimator and the unknown population rate $\Lambda(t)$ decreases over time.

$$\begin{aligned} E[M_1(t)] &= \sum_{i=1}^{n} \lambda_i\, t e^{-\lambda_i t} \\ E\left[\frac{M_1(t)}{t}\right] &= E[\Lambda(t)] \\ E[M_2(t)] &= \frac{1}{2}\sum_{i=1}^{n} (\lambda_i t)^2 e^{-\lambda_i t} \\ E\left\{\left[\Lambda(t) - \frac{M_1(t)}{t}\right]^2\right\} &= \sum_{i=1}^{n}\left(\frac{\lambda_i^2 e^{-\lambda_i t} + \lambda_i e^{-\lambda_i t}}{t}\right) \\ &= \frac{E[M_1(t) + 2M_2(t)]}{t^2} \end{aligned} \tag{12.1}$$

The underlying assumptions for occurrences of bugs are the same assumptions often made for occurrences of demand for products. That is, the demand for products can be assumed, as is frequently stated in the literature, to follow a Poisson process. In addition, an assumption can be made that one product's demand is independent from another. This assumption makes the problem of estimating demand, in the case of inventory, or error rates, in the case of software applications, tractable. The prediction of software reliability uses some methodologies that maybe adapted for the study of slow-moving inventory.

Although there are many differences between software reliability and inventory management, the proposed estimators can be used in both applications. One could argue that bugs are not always known until they are detected whereas products are always known to exist. Ross (1985a) addresses this limitation in software applications by extending the use of estimators to a case where there is a probability p of detecting a bug. However, this extension is not needed for the case of inventory management. An implicit assumption is that sales are detected with 100% certainty.

As an extension to the above approach, an estimator of the future demand rate of a pool of products experiencing no more than one unit sold is proposed. Define $\Lambda(t) = \sum_{i=1}^{n} \lambda_i I_i(t)$ as the future unknown demand rate for products having exactly one unit of sale by time t where $I_i(t) = 1$ if product i with demand rate λ_i has exactly one unit of sale by time t and 0 otherwise, $i = 1, 2, \ldots, n$. The future demand rate for products with sales of no more than one unit is the sum of the random variables $\Lambda(t)$ and $\Delta(t)$. The proposed estimator of $\Lambda(t) + \Delta(t)$ is $\frac{M_1(t)+2M_2(t)}{t}$. This estimator is unbiased since its expected value is $\sum_{i=1}^{n} (\lambda_i e^{-\lambda_i t} + \lambda_i^2 t e^{-\lambda_i t})$, the same as the expected value of $\Lambda(t) + \Delta(t)$. An unbiased estimator of the squared difference between $\Lambda(t) + \Delta(t)$ and $\frac{M_1(t)+2M_2(t)}{t}$ is $\frac{M_1(t)+2M_2(t)+6M_3(t)}{t^2}$ since the expected value of either is $\sum_{i=1}^{n} (t^{-1}\lambda_i e^{-\lambda_i t} + t\lambda_i^2 e^{-\lambda_i t} + t^2\lambda_i^3 e^{-\lambda_i t})$. If this expected squared difference is small, then $\frac{M_1(t)+2M_2(t)}{t}$ is a reasonable estimator of $\Lambda(t) + \Delta(t)$.

12.5 One-Sided Prediction Intervals for Future Demand Rate of Products with No Demand or with No More than One Demand

Inventory managers may need a forecast for not only spare parts with no demand but also spare parts that had a few demand occurrences. That is, a one-sided prediction interval for slow-moving spares with less than some minimum number of demands over a specified time period may be desired. For this reason, a one-sided prediction interval is considered for the case in which there is zero or one demand. Although this extension could be carried further, a simulation study of many different one-sided prediction intervals would become too involved.

Two-sided prediction intervals were considered in Lindsey and Pavur (2009). Since the upper limit of the prediction interval may be of more importance in decision making on liquidating or no longer carrying a product, one-sided prediction intervals (OSPIs) are proposed in this section for the case of estimating future demand for products with no observed demand or with no more than one observed demand. The proposed OSPIs with $100(1 - \alpha)\%$ confidence will be the estimate of future demand plus the appropriate $100(1 - \alpha)\%$ normal distribution percentile multiplied by an estimate of the standard error of the estimate. However, for certain parameters this OSPI may not be reliable. For example, the estimator $\frac{M_1(t)}{t}$ for the demand rate of products with no demand may be too skewed to be assumed to be approximately normally distributed. Under certain parameter values such as a small product group size, the normal approximation may make the upper limit of an OSPI too small. That is, the nominal (stated) confidence level of the OSPI will no longer hold if the normal distribution is not a good approximation of a distribution that is skewed or heavy tailed. Thus, the reliability of such one-sided intervals will be assessed over a variety of demand rates and numbers of products. The upper endpoint of a proposed OSPI for the case of estimating the future demand of products with no observed demand is as follows.

$$\frac{M_1(t)}{t} + Z_\alpha \sqrt{\frac{M_1(t) + 2M_2(t)}{t^2}} \tag{12.2}$$

A proposed estimator for the future sales rate of products having no more than one sale over a specified time frame is $\frac{M_1(t)+2M_2(t)}{t}$ since it is an unbiased estimator of the underlying demand rate. The unbiased estimator for the expected squared difference of this estimator and the future demand rate is $\frac{M_1(t)+2M_2(t)+6M_3(t)}{t^2}$. The upper end point of the proposed OSPI for the future demand rate of products having no more than one sale by time period t are as follows.

$$\frac{M_1(t) + 2M_2(t)}{t} + Z_\alpha \sqrt{\frac{M_1(t) + 2M_2(t) + 6M_3(t)}{t^2}}. \tag{12.3}$$

Equation 12.2 will be referred to as the Zero Sales prediction intervals. That is, the Zero Sales prediction intervals determine future demand rate for products exhibiting no sales over a specified time frame. Equation 12.3 will be referred to as the Zero and One Sales prediction interval. These prediction one-sided intervals determine the future demand rate for spare parts having no more than one sale unit of demand over a specified time frame. A value of zero can be used for the lower end-point for these OSPIs.

12.6 Monte Carlo Simulation Study to Assess Reliability of Proposed OSPIs

A Monte Carlo simulation with 5,000 replications was conducted to assess the reliability of OSPIs for the demand rate of slow-moving products. One group of slow-moving products that have not exhibited any demand over the specified time frame will be referred to as the Zero Sales group. Another group of slow-moving products that have exhibited no more than one demand will be referred to as the

Zero and One Sales group. The reliability of the proposed OSPIs will be assessed by simulating their Type I error rate. The next section will provide the empirical Type I error rates for the Zero Sales OPSIs across various Product Group Sizes and MTBDs. This section is followed by the results for the Zero and One Sales OSPIs. Next, a comparison of the reliability of the two different OSPIs is presented. Finally, an illustration is provided as to how the Zero and One Sales OSPIs compare to the TSPIs with respect to their reliability across a variety of MTBDs.

Monte Carlo parameters similar to those selected in Lindsey and Pavur (2009) in studying the reliability of two-sided prediction intervals for the Zero Sales group are used in this study. A specified time frame of 100 units of time was selected for the entire simulation study as 100 units of time could be converted into hourly, daily or weekly data, but not longer periods, such as monthly or yearly. One hundred time units would be roughly equal to 3 months or about a quarter if the unit was a day or about 2 years if the unit was a week. The time frame of 100 units was arbitrary, but it is reasonable to assume it would be a common time frame to study for data either collected on a daily or weekly basis. This amount of time would be an appropriate amount of time in which a manager would need to make critical decisions if products were not moving. The total number of products in a group that are observed for their demand range from 10 to 1,000. MTBD range from 10 to 1,000 and include a case in which there is a mixture of MTBDs.

The confidence levels for the OSPIs are selected to be at the 90%, 95%, and 99% levels. That is, the nominal Type I error rates (alphas) are 10%, 5%, and 1%. A Type I error occurs if a known demand rate is not within the one-sided prediction interval constructed from the simulated demand of the products. An empirical Type I error is considered near its nominal alpha value if this error is within plus or minus two standard deviations of the nominal value (Zwick 1986; Harwell 1991). For example, at the 99% confidence level with a nominal Type I error rate of 1%, the empirical Type I error for 5,000 simulations must be between $0.01 \pm 2\sqrt{\frac{(0.01)(1-0.01)}{5{,}000}}$ or from 0.007 to 0.013 for the prediction interval to be considered reliable. For the 95 and 90% confidence levels, these intervals are 0.044 to 0.056 and 0.092 to 0.108, respectively.

The next section will provide the empirical Type I error rates for the Zero Sales OPSIs across various Product Group Sizes and MTBDs. This section is followed by the results for the Zero and One Sales OSPIs. Next, a comparison of the reliability of the two different OSPIs is presented. Finally, an illustration is provided as to how the Zero and One Sales OSPIs compare to the TSPIs with respect to their reliability across a variety of MTBDs.

12.7 Reliability of Zero Sales OSPIs across Product Group Sizes

Empirical Type I errors of OSPIs for a group of products with zero sales are assessed across a variety of conditions for the number of products and the demand rates to determine if OSPIs are reliable as a measure to compare with a threshold value for making critical decisions about a subgroup of non-selling products. Product groups are formed by managers based on the expected demand of the products. Items that are expected to have similar demand levels would be grouped together based on historical data or by expert opinion. The number of products in each group ranges from 50 to 1,000 products across a specified demand rate for three confidence levels. The effect of the group size on the performance of the one-sided prediction intervals was assessed at four demand levels in terms of MTBD: 100, 200, or 300, or mixture of 50 and 400.

Product group sizes were increased by 50 up to 500 products and then by 100 up to 1,000 products. Product group sizes were incremented in this fashion to keep the total number of simulations reasonable while still gaining insight into the performance of the models when product group size is large. Product group sizes under 500 were considered more practical to investigate than group sizes over 500. Throughout the simulation, product group sizes and MTBD will not be evenly spaced so that simulations can reveal information over wider ranges of parameters while still examining performance over parameter values where performance is thought to be changing quickly. The product mixture of 50 and 400 MTBDs consisted of 25 products having an MTBD of 400 and the remaining 25 to 975 products having an MTBD of 50.

As illustrated in Fig. 12.1, OSPIs are generally reliable for higher demand rates (shorter MTBD) with relatively larger product group sizes. The OSPIs should be used with caution for product group sizes below 300 or with very low demand rates. Reliable OSPIs can be obtained for use in a stopping rule. The conditions under which OSPIs should be considered reliable include product group sizes that are large and an observed time period that should approximate or be close to the MTBD of the product group. The OSPIs appear to be less stable when the demand rate is relatively low (i.e., MTBD = 300).

12.8 Reliability of Zero and One Sales OSPIs across Product Group Sizes

Empirical Type I errors of OSPIs for a group of products with zero and one sales are assessed in a similar fashion to the OSPIs for a group of products with zero sales. The parameters used in the simulation in the previous section were used to determine the reliability of the OSPIs for zero and one sales. Figure 12.2 illustrates that for MTBDs of 100 and the mixture of MTBDs, that the Zero and One Sales OSPIs maintain their nominal Type I error rate as the product group size increases. A product group size of at least 200 should be used with these MTBDs. For low demand rates (high MTBDs), the demand may be too slow to provide sufficient information, to provide accurate statistics for a reliable OSPI for the demand rate of products which have exhibited one demand.

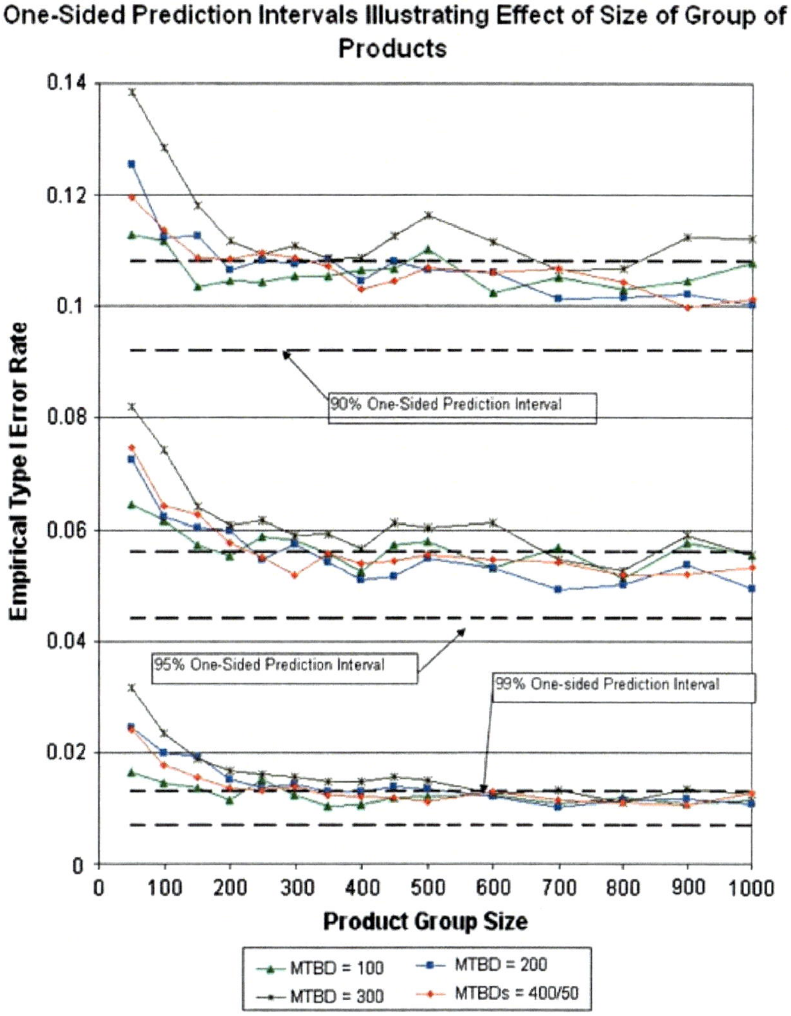

Fig. 12.1 Empirical Type I error versus product group size for OSPI of Zero Sales group

12.9 Comparison of the Reliability of Zero and One Sales OSPIs and Zero Sales OSPIs

Since the Zero and One Sales OSPIs are estimating a demand rate for potentially more products than the Zero Sales prediction interval, the Zero and One Sales OSPI is expected to perform better with products having higher demand, since more demand will occur more often even for the slowest moving products. To assess this characteristic, a comparison is made between these two prediction intervals.

Figure 12.3 reveals the empirical Type I error rates for the Zero Sales OPSIs and the Zero and One Sales OPSI across MTBDs ranging from 10 to 1,000. A group product size of 200 was selected since this group size appears to be where the OSPIs start to show robustness with respect to maintaining their nominal Type I error rates. In general, the empirical Type I error rates for the Zero and One Sales OSPIs are lower than those of the Zero Sales OSPIs for the first half of the graph

Fig. 12.2 Empirical Type I error for Zero and One Sales OSPIs across product group size

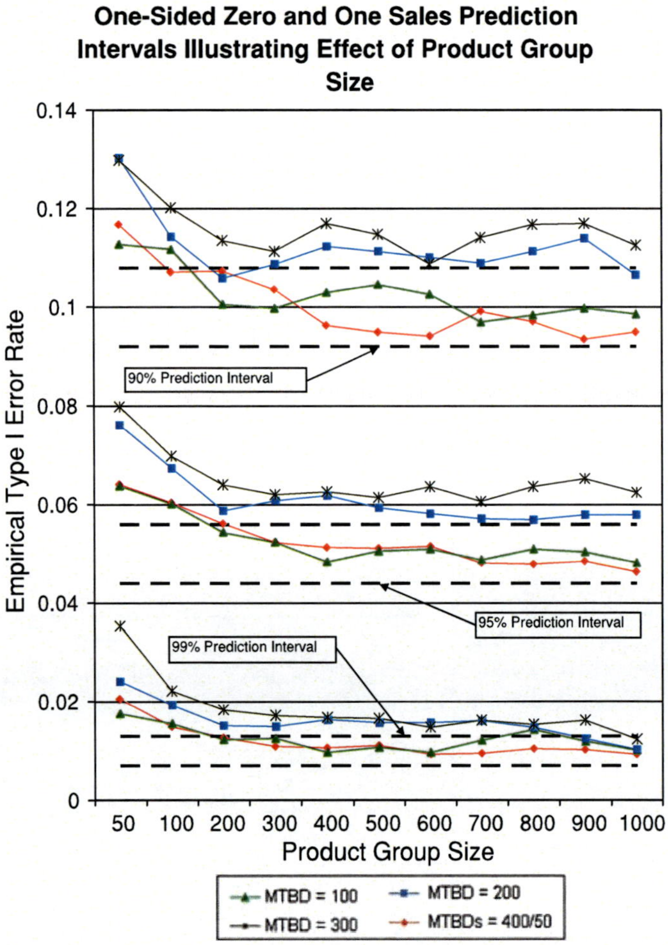

and then reverse. These results indeed reveal that for high demand rates, such as MTBD = 15 and MTBD = 20, that the Zero and One Sales prediction intervals have empirical Type I error rates closer to their nominal Type I error rates. This occurs because the variability of the demand rate estimates for the Zero and One Sales OPSIs tend to be lower at relatively higher demand rates. However, as the MTBD increases, the Zero and One Sales prediction interval does appear to have higher empirical Type I error rates than the Zero Sales prediction interval. Although these empirical Type I errors tend to improve with an increase in the number of products, the general pattern showing the relationship between these two OSPIs appears to remain.

12.10 Comparison of the Reliability of Zero and One Sales OSPIs and Two-Sided Prediction Intervals

Two-sided prediction intervals for the demand rate of products exhibiting zero demand over a fixed time frame were the focus of the study by Lindsey and Pavur (2009). To illustrate that a separate simulation study investigating OSPIs is needed

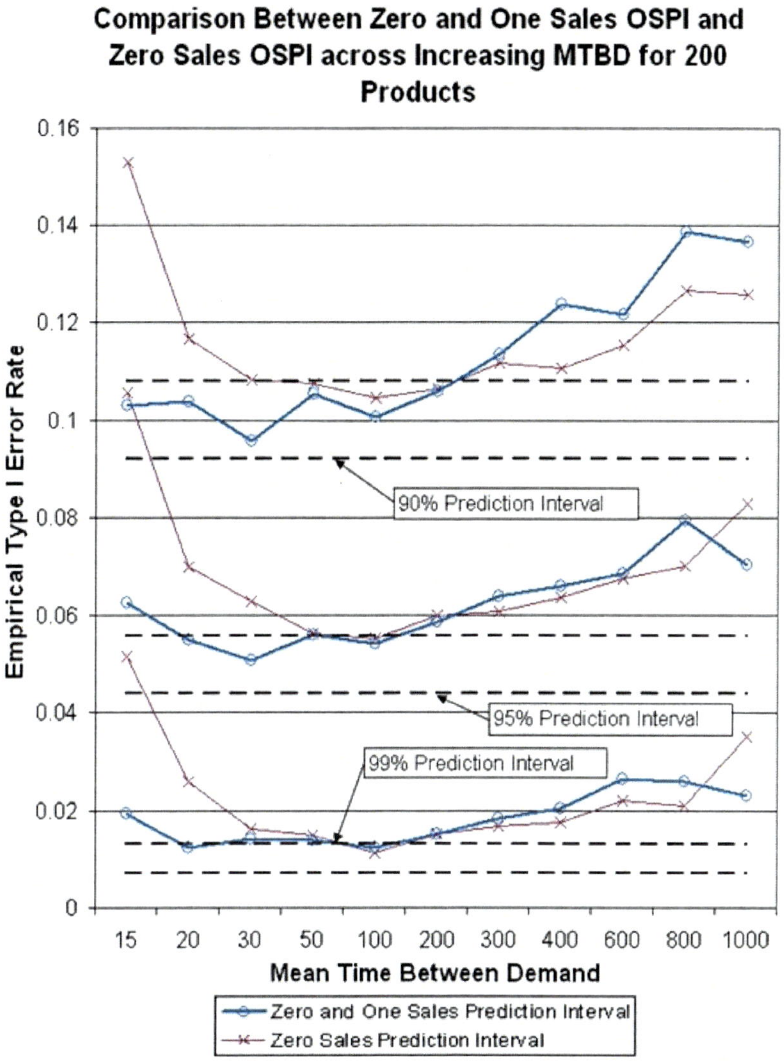

Fig. 12.3 Empirical Type I error for Zero and One Sales and Zero Sales OSPIs

and that OSPIs should be used more cautiously then the two-sided prediction intervals, a comparison of the empirical Type I errors of the two types of prediction intervals for Zero and One Sales is illustrated in Fig. 12.4. The parameters selected are the same as that used in Fig. 12.3 to compare the performance of the Zero and One Sales OSPIs and the Zero Sales OSPIs.

Generally, the empirical Type I error rates for the two-sided prediction intervals are lower than those for the Zero and One Sales OSPIs. Clearly, at the 95% and 99% confidence levels, an MTBD of 10 yields a greatly inflated Type I error. The 90% two-sided prediction interval is generally robust across the MTBDs in maintaining its nominal Type I error rate. For the 95% and 99% two-sided and one-sided prediction intervals, very large MTBDs can easily affect the reliability of these intervals.

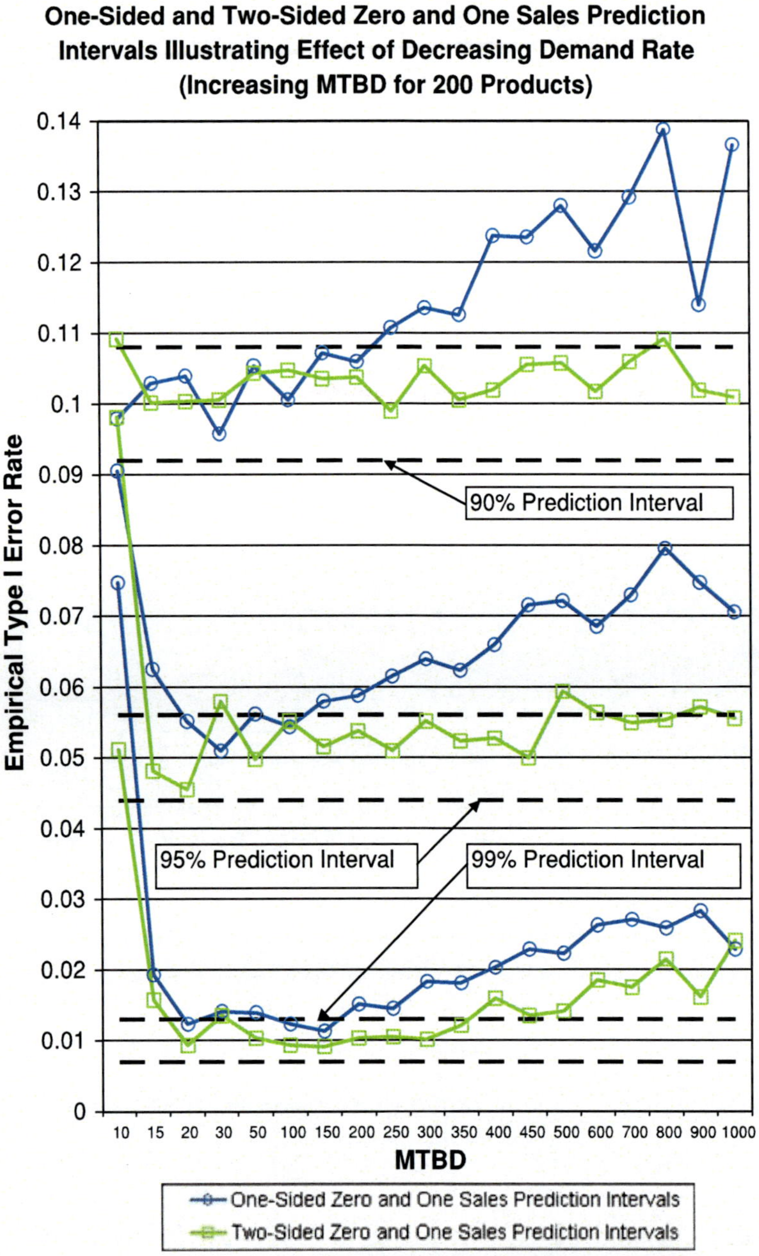

Fig. 12.4 Empirical Type I error for Zero and One Sales OSPIs and two-sided prediction intervals

12.11 Discussion of Results and Conclusion

Estimation of the future demand rate of a group of products without sales or with no more than one sale over a specified time period is difficult due to lack of data. There are limited demand rate estimation procedures for this type of slow-moving inventory. The proposed prediction intervals for the future demand rate of these slow-moving products are unique in that one-sided prediction intervals addressing this problem have not been presented. The Monte Carlo simulation results presented in this paper provide insight into conditions under which these prediction intervals are reliable. Knowledge of reliable upper endpoints of a one-sided prediction interval allows managers to compare this value to some threshold value for decision-making purposes.

The simulation study suggests that the OSPIs are not as robust with respect to maintaining their nominal Type I error as the two-sided prediction intervals. Generally, the OSPIs are reliable for higher demand rates (shorter MTBD) with relatively larger product group sizes. The OSPIs should be used with caution for product group sizes below 300 or with very low demand rates. Reliable OSPIs can be obtained for use in a stopping rule. The conditions under which OSPIs should be considered reliable include product group sizes that are large and an observed time period that should approximate or be close to the MTBD of the product group. The Zero and One Sales OSPI provides an interval for a demand rate for potentially more products than the Zero Sales OSPI. This proposed prediction interval performs better with products having higher demand.

One-sided prediction intervals are applicable to stopping rules to help determine when product demand rates are below threshold limits set by managers for carrying the merchandise. Knowledge of the upper endpoint of an OSPI allows managers to compare this value to some threshold value for decision-making purposes. As long as estimated future demand rates are above an acceptable minimum determined by management, products will likely be kept in stock. Once the minimum demand rate (threshold value) is reached, products may be considered for liquidation. There are few options for managers to use in estimating the future demand rate of products with no demand or with little demand. This study provides an additional tool that can be used in the decision-making process of deciding whether to continue carrying spare parts with little demand.

12.12 Unique Contribution of Research and Future Research Ideas

Only a finite number of experimental conditions were investigated for the proposed methodology. Additional simulations should be completed to extend this research for values outside the ranges tested and even between the parameter values selected. For example, the proposed prediction intervals were studied over a specified range of product group sizes. General trends were identified, but running additional simulations with group sizes in between the points selected would support the general trend or possibly identify potential anomalies resulting from some particular group size.

Modified prediction intervals for future demand should be investigated to determine approaches to making them reliable to a wider range of demand rates and product group sizes. There may be a correction factor that may be developed to enhance the performance of the prediction intervals. In addition, prediction intervals for products having no more than two sales, or some given number of sales, should be developed and investigated.

Two limitations to the current study were the assumption of independence of the demand of products and the assumption that sales and demand were equivalent. Assuming a correlation structure for the demand of inventory products would require newly proposed methodology that might be difficult to implement. Future research should address the issue in which independence of product demand does not hold and address the issue of estimating demand that may not result in a sale. These issues would require formulations that were more involved than those presented in this research. Most importantly, future research should provide extensive guidelines to inventory managers to select reliable models to optimize inventory levels over a wide variety of product types. Furthermore, the performance of any proposed model must be interpreted so that inventory managers may use them appropriately. Future research should assess methodology that performs well under assumptions that mimic real world conditions.

References

Abdel-Ghaly AA, Chan PY, Littlewood B (1986) Evaluation of competing software reliability predictions. IEEE Trans Softw Eng 12(9):950–967

Boylan JE, Syntetos AA, Karakostas GC (2008) Classification for forecasting and stock control: a case study. J Oper Res Soc. 59(4):473–481

Brown M, Zacks S (2006) A note on optimal stopping for possible change in the intensity of an ordinary Poisson process. Stat Probab Lett 76:1417–1425

Browne GJ, Pitts MG (2004) Stopping rule use during information search in design problems. Organ Behav Hum Decis Process 95:208–224

Cavalieri S, Garetti M, Macchi M, Pinto R (2008) A decision-making framework for managing maintenance spare parts. Prod Plann Control 19(4):379–396

Dunsmuir WTM, Snyder RD (1989) Control of inventories with intermittent demand. Eur J Oper Res 40(1):16–21

Gelders LF, Van Looy PM (1978) An inventory policy for slow and fast movers in a petrochemical plant: a case study. J Oper Res Soc 29(9):867–874

Ghobbar AA, Friend CH (2002) Sources of intermittent demand for aircraft spare parts within airline operations. J Air Trans Manag 8:221–231

Haber SE, Sitgreaves R (1970) A methodology for estimating expected usage of repair parts with application to parts with no usage history. Nav Res Logist Q 17(4):535–546

Harwell MR (1991) Using randomization tests when errors are unequally correlated, Comput. Comput Stat Data Anal 11(1):75–85

Horodowich P (1979) Evaluating the write-off of slow-moving finished goods inventory. Manag Account 60(9):35–39

Hua ZS, Zhang B, Yang J, Tan DS (2007) A new approach of forecasting intermittent demand for spare parts inventories in the process industries. J Oper Res Soc 58(1):52–61

Kaufman GM (1996) Successive sampling and software reliability. J Stat Plan Inference 49(3):343–369

Lindsey MD, Pavur R (2009) Prediction intervals for future demand of existing products with an observed demand of zero. Int J Prod Econ. 119(1):75–89

Miragliotta G, Staudacher AP (2004) Exploiting information sharing, stock management and capacity over sizing in the management of lumpy demand. Int J Prod Res 42(13):2533–2554
Porras E, Dekker R (2008) An inventory control system for spare parts at a refinery: An empirical comparison of different re-order point methods. Eur J Oper Res 184:101–132
Ross SM (1985a) Statistical estimation of software reliability. IEEE Trans Softw Eng SE11(5):479–483
Ross SM (1985b) Software reliability: the stopping rule problem. IEEE Trans Softw Eng SE11(12):1472–1476
Ross SM (2002) Introduction to probability models, 6th edn. Academic Press, New York
Wagner SM, Lindemann E (2008) A case study-based analysis of spare parts management in the engineering industry. Prod Plan Control 19(4):397–407
Ward JB (1978) Determining reorder points when demand is lumpy. Manag Sci 24(6):623–632
Willemain TR, Smart CN, Schwarz HF (2004) A new approach to forecasting intermittent demand for service parts inventories. Int J Forecast 20(3):375–387
Yamashina H (1989) The service parts control problem. Eng Costs Prod Econ 16:195–208
Zwick R (1986) Rank and normal scores alternatives to Hotelling's T2. Multivar Behav Res 21(2):169–186

Chapter 13
Reactive Tabu Search for Large Scale Service Parts Logistics Network Design and Inventory Problems

Yi Sui, Erhan Kutanoglu and J. Wesley Barnes

Abstract This chapter documents a study of a reactive tabu search (RTS) approach to the integrated service part logistics (SPL) network design and inventory stocking problem. The integrated problem of designing and stocking an SPL network has attracted more attention recently. The two sets of decisions (network design and inventory stocking) usually have been considered separately and sequentially in practice as well as in the research literature, although interdependency between them exists and integration is necessary for overall system performance optimization. However, the integrated mathematical programming model and solution development for this problem are often intractable due to the time-based service constraints which confine the lower bound of the demand percentage satisfied within the specified time windows.

We use a RTS method to efficiently find very good solutions to this problem. Tabu search combines a hill climbing strategy with a memory structure which guides the search. The reactive mechanism dynamically adjusts the tabu tenure during the search. An escape mechanism is activated when the search is trapped in a local attractor basin. We also apply heuristic techniques to construct the initial solution and use rule based comparisons to determine the best non-tabu solution in a neighborhood about the current incumbent solution.

By applying this metaheuristic method to problem sets of different sizes, we obtain high-quality solutions with remarkably small amounts of

Y. Sui (✉)
MicroStrategy, Inc, McLean, VA, USA
e-mail: sui11yi3@yahoo.com

E. Kutanoglu · J. W. Barnes
The University of Texas at Austin, Austin, TX, USA
e-mail: erhank@me.utexas.edu

J. W. Barnes
e-mail: wbarnes@mail.utexas.edu

N. Altay and L. A. Litteral (eds.), *Service Parts Management*,
DOI: 10.1007/978-0-85729-039-7_13,

computational effort. For the smaller problems, the tabu search solution is identical or very close to the optimal solution provided by classical optimization-based methods. For the larger problems, RTS obtains solutions superior to those obtained by classical approaches.

13.1 Introduction

Pre-sale profit margins of manufacturing companies continue to shrink due to intense global competition. Fortunately, in the United States, manufacturing companies now earn up to 40–50% of their profit after the original product sale and make nearly 25% of their revenues by providing parts, maintenance and other services to customers (Poole 2003). Hence, post-sale activities such as Service Part Logistics (SPL) have become increasingly important. SPL refers to the set of operations and infrastructure that responds to existing customer problems by providing replacements for failed parts in the customers' products and by dispatching a technician if required.

Providing fast, high quality post-sale service to existing customers as promised in high-dollar service contracts greatly enhances customer retention and increases customer loyalty, significantly benefitting both before and after sales supply chains. SPL plays a critical role in a wide domain of industries including computer hardware (IBM, HP, Dell), medical equipment (GE, Siemens), heavy machinery (Caterpillar), manufacturing equipment (Applied Materials), aerospace (Boeing) and defense and military logistics. Post sales service is an important product differentiator which enhances overall brand value within each industry. Because of their business requirements, many SPL customers require service recovery in hours, if not minutes. This extensively modifies the corresponding strategic, tactical and operational SPL problems, rendering traditional supply chain methodologies ineffective.

The typical strategic SPL problem includes locating and staffing the part stocking facilities, allocating customer demands to these facilities, and selecting the proper stock levels to satisfy the stringent and time-sensitive service requirements. Traditional approaches to such network design problems fail to consider inventory issues while locating facilities and allocating customers to those facilities. Moreover, time-based service level requirements in SPL (requiring very flexible network responsiveness) strongly tie the network design and inventory decisions. Earlier optimization-based methods that have been introduced to solve such integrated SPL problems (Jeet et al. 2009; Candas and Kutanoglu 2007) are promising in obtaining comprehensive (and improved) solutions to small and medium scale problems. However, these classical optimization-based methods developed through new formulation, decomposition and lower bounding discoveries, even when carefully customized and specifically designed for these problems, fail to effectively scale up to solve industry-scale SPL problems. This study introduces a more scalable, alternative approach, a Reactive Tabu Search (RTS) algorithm, to solve larger SPL network and inventory design problems.

13.2 Literature Review

Some classical network design and facility location problems such as the set covering location problem (Toregas et al. 1971), the p-median problem (Hakimi 1964), and the uncapacitated facility location (UFL) problem (Kuehn and Hamburger 1963) have been widely studied. Magnanti and Wong (1984) summarized early work on the location problem and Drezner (1995) gave a survey of applications and methods for solving the facility location problem. Melkote and Daskin (2001a) provided a comprehensive overview of network location models. Some research papers in this area also study problems with service constraints or reliability issues. Simchi-Levi (1991) studied the traveling salesman location problem with a capacitated server at the service location. Melkote and Daskin (2001b) examined a combined facility location/network design problem in which the facilities have limited capacities on the amount of demand they can serve.

There is a significant amount of inventory management literature. In this study, we focus only on literature closely related to the research documented here which includes papers for specialized SPL inventory models. Cohen et al. (1988) presented a model of an (s, S) inventory system with two priority classes of customers, Chen and Krass (2001) investigated inventory models with minimal service level constraints, and Agrawal and Seshadri (2000) derived the bounds to the fill-rate constrained (Q, r) inventory problem. Early work in the area of multi-facility inventory models includes Sherbrooke (1968, 1986) and Muckstadt (1973). SPL systems have been successfully implemented in different industries (Cohen et al. 1990, 1999, 2000) and in the military domain (Rustenburg et al 2001). Cohen et al. (1997) summarized a benchmark study of after-sales service logistics systems.

In recent years, the integrated problem of facility location and inventory stocking has attracted some attention. Barahona and Jensen (1998) studied the integrated model considering a low level of inventory for computer spare parts. Nozick and Turnquist (1998) investigated a distribution system model with inventory costs as part of the facility fixed cost while maximizing the service level. A variation of this model minimizes the cost subject to the service constraints (Nozick 2001). Daskin et al. (2002) introduced a facility location model which incorporates working inventory and safety stock inventory costs at the facilities. They also proposed a Lagrangian relaxation solution algorithm for this model. Shen et al. (2003) considered a joint location inventory problem involving a single supplier and multiple retailers and developed a column generation method to solve this model. Jeet (2006) solved the integrated facility location and inventory stocking problem while considering part commonality by using a fill rate outer approximation scheme. Candas and Kutanoglu (2007) showed that determining the network design and inventory stocking levels simultaneously is superior to treating them separately for the same model.

For most large scale problems, traditional integer programming or mixed integer programming (MIP) methods require excessive amounts of computational effort to achieve optimality. Tabu Search (TS), first introduced by Glover (1989, 1990), can provide good solutions to these large problems in a more time-efficient way. TS has been shown to be an effective and efficient method for combinatorial optimization problems that combines a hill climbing strategy with memory structures to guide the search (Glover 1989, 1990). Battiti and Tecchiolli (1994) developed the RTS by adding a hashing scheme to detect cyclic behavior and an intelligent way to dynamically adapt the tabu tenure based on the current status and past history of the search. Barnes and Carlton (1995) introduced the RTS to attack the vehicle routing problem with time windows and the results they obtained were superior to the results from a classical optimization algorithm or a genetic algorithm. In addition, the pickup and delivery problem with time windows can be solved by RTS very efficiently to achieve high solution quality (Nanry and Barnes 2000). Ciarleglio (2007) and Ciarleglio et al. (2008) extended traditional TS techniques with the creation of Modular Abstract Self-Learning Tabu Search which includes rule based objectives and dynamic neighborhood selection.

TS has been widely used in the network design area to achieve near optimal solutions. Crainic et al. (1995, 1996) applied TS to the capacitated multicommodity network design problem. Cooperative parallel TS (Crainic and Gendreau 2002) was developed for the capacitated network design problem; their method obtained better solutions than sequential approaches and a well designed cooperative search outperformed independent search strategies. Xu et al. (1996) introduced probabilistic TS for telecommunications network design by using movement estimations, error correction and probabilistic move selection.

13.3 The SPL Mathematical Model

This section introduces the MIP model for the integrated SPL problem. Stochastic demands are included in terms of fill rate to represent the part availability. The limitations of the mathematical programming model are listed at the end of this section as well.

13.3.1 The Integrated SPL Model with Stochastic Demands

The service level requirements imposed on the SPL system motivate the simultaneous modeling of the network design and inventory stocking decisions. Our integrated model development is based on Jeet et al. (2009). Similar to their study, we make the following assumptions:

1. We assume that the SPL network has only one echelon and all the stocking facilities serve the customers directly. We also assume these facilities can be replenished by a central warehouse with unlimited-capacity (i.e., infinite supply) and no time delay.
2. We use a one-for-one replenishment policy for all the stocking facilities. Since the demand is assumed to be very low and lead time is relatively short, we do not need to consider the batch ordering where replenishment quantity is more than one. This is a very common assumption in low demand inventory systems such as SPL.
3. We assume that all the customer service level requirements are aggregated to obtain the system target service level.
4. The demands from different customers are governed by statistically independent Poisson distributions. The Poisson distribution is a good approximation of the low-demand distribution (Muckstadt 2005).
5. Customer demand can be only assigned to one facility with at least one unit of stock. In case of a facility "stock-out," the customer demand is passed to the central warehouse, which satisfies the customer demand with a direct shipment. For the "stocked-out" facility, the demand is considered lost, hence we use the lost-sales fill rate. (Long run average fill rates are computed using the lost sales formula derived from the steady state behavior of the M/G/s/s queuing model (Zipkin 2000).
6. There is only one time-window over which the target aggregate service level is to be satisfied. When the facility can provide the part to the customer in need within the time window (depending on the distance between the facility and the customer), the satisfied unit of demand is counts towards its target service level.

13.3.2 The Model Notation

The following is used in the model:

- Sets and Indices

I Set of candidate facility locations, indexed by i.
J Set of customers/demand points, indexed by j.

- Parameters

f_i Annual cost of operating facility i (assumed constant once the facility is open).
c_{ij} Cost of shipping one unit of the part from facility i to customer j.
d_j Mean annual demand for customer j.
h_i Annual inventory holding cost at facility i.
t Stock replenishment (resupply) lead time (assumed identical for all facilities).
α Time based service level (defined as the percentage of demand that is to be satisfied within the time window).

δ_{ij} Binary parameter that is 1 if customer j is within the time window of facility i.

$S_{\max}$ Common maximum possible stock for any one facility.

- Variables

β_i Fill rate at facility i.

λ_i Mean lead time demand at facility i.

S_i Stock level at facility i.

X_{ij} Binary decision variable; 1 if demand from customer j is assigned to facility i.

Y_i Binary decision variable that is 1 if facility i is open.

13.3.3 The SPL Integer Programming Model

The model objective is to minimize the total annual cost of open facilities, transportation, and inventory stocking. The constraints of our model draw mainly from the UFL model and integrate the inventory part of the problem (fill rates and service levels):

$$\text{minimize} \quad \sum\nolimits_{i\in I} f_i Y_i + \sum\nolimits_{i\in I} \sum\nolimits_{j\in J} c_{ij} d_j X_{ij} + \sum\nolimits_{i\in I} h_i S_i \tag{13.1}$$

$$\text{subject to} \quad \sum\nolimits_{i\in I} X_{ij} = 1, \quad \forall j \in J \tag{13.2}$$

$$X_{ij} \leq Y_i, \quad \forall i \in I \text{ and } j \in J \tag{13.3}$$

$$\sum\nolimits_{i\in I} \sum\nolimits_{j\in J} \delta_{ij} d_j \beta_i X_{ij} \geq \alpha \sum\nolimits_{j\in J} d_j \tag{13.4}$$

$$\lambda_i = t \sum\nolimits_{j\in J} d_i X_{ij}, \quad \forall i \in I \tag{13.5}$$

$$\beta_i = 1 - \frac{\lambda_i^{s_i} / s_i{}^4}{\sum_{n=0}^{s_i} \lambda_i^n / n!}, \quad \forall i \in I \tag{13.6}$$

$$X_{ij} \leq S_i, \quad \forall i \in I \text{ and } j \in J \tag{13.7}$$

$$Y_i \in \{0, 1\}, \; X_{ij} \in \{0, 1\}, \quad \forall i \in I \text{ and } j \in J \tag{13.8}$$

$$0 \leq S_i \leq S_{\max} \quad \text{and integer}, \quad \forall i \in I \tag{13.9}$$

The objective function (1) is the total annual cost of operating facilities, transportation, and inventory stocking. Constraints (2) and (3) are the UFL constraints which force each customer to be assigned to an open facility. Constraint (4) requires that α fraction ($0 \leq \alpha \leq 1$) of the total annual customer demands be satisfied within the time window. Constraints (5) define the mean lead time demand for each facility which is used to calculate the facility's fill rate in constraints (6). Constraints (6) compute facility fill rates using the lost sales formula (Zipkin 2000). Constraints (7) state that every facility that has been allocated some demand must have at least one unit of stock. Constraints (8) state that the variables X_{ij} and Y_i are 0 or 1, and constraints (9) confine the range and enforce the integrality of the stock levels for all the facilities. For more details of the mathematical model, see Jeet et al. (2009).

13.3.4 The Limitations of the SPL Model

As we can see from the mathematical model above, variables λ_i are dependent on integer demand assignment variables X_{ij} in constraints (5) and fill rate variables β_i are calculated from mean lead time demand variables λ_i in constraints (6). However, on the left hand side of service level constraint (4), the variables β_i, which are already originally computed from X_{ij}, are multiplied by X_{ij}. These interactions make the problem non-linear and strongly coupled on variables X_{ij} and make it impossible to solve practically sized problems with classical MIP methods. Jeet (2006) tested the above model for different smaller problem sizes, and created benchmark results for these problem sets. These problems were modeled using CPLEX 9.0 on a microcomputer with dual Xeon 1.8 GHz processors with 1 GB RAM running the Suse Linux operating system. The largest data set had 24 potential facilities and 158 customers. Some of the problems were not solved to optimality within 30 min of computational effort. A custom method specifically developed for the integrated model using an outer approximation and a strong lower bound had to solve a still time consuming lower bounding problem multiple times. In real applications, SPL networks may have hundreds of potential facilities and thousands of customers requiring far more scalable approaches. The inability of classical approaches required the development of an efficient heuristic methodology that would provide high quality solutions within acceptable amounts of computational effort.

13.4 An RTS Approach to the SPL Problem (RTS–SPL)

TS, a metaheuristic search algorithm (Glover 1989, 1990), controls a hill climbing heuristic using a tabu memory structure to manage the search and allows non-improving solutions during the search to permit escape from local optima. Battiti and Tecchiolli (1994) first introduced Reactive Tabu Search (RTS) to dynamically change the tabu tenure by monitoring the repetitions of visited incumbent solutions and allowed to search to escape from chaotic attractor basins. The solutions visited during the search and the corresponding iteration numbers are preserved in memory. When a candidate solution is considered, the repetitions of that solution and the number of iterations between the two adjacent visits are checked. The RTS increases the tabu tenure when solutions are often repeated to diversify the search and decreases the tabu tenure to intensify the search when there are few repeated solutions.

13.4.1 Solution Representation and Neighborhood Structure

The RTS–SPL solution representation is an integer list of facility inventories, bounded between 0 and $S_{\max}$, with cardinality equal to the total number of the potential facilities. Closed facilities have a list value of zero. This list provides the values of the decision variables Y_i and S_i of the MIP model. Given a particular solution, a heuristic method assigns the customers to open facilities (stipulates the X_{ij}'s) based on the values of λ_i and β_{i} to achieve the maximal service level (given the previous assignments made in the algorithm). This demand assignment method is detailed in the next section.

The RTS–SPL neighborhood consists of alternate solutions created by three different moves that transform the current solution into a neighboring solution:

- increasing inventory in a facility by one unit,
- decreasing inventory in a facility by one unit, and
- swapping all inventory between two facilities.

For illustration, suppose there are four facilities and the current solution is $\{1, 0, 2, 1\}$ (where the second facility is closed). The RTS–SPL neighborhood of this solution is $\{2, 0, 2, 1\}$, $\{1, 1, 2, 1\}$, $\{1, 0, 3, 1\}$, $\{1, 0, 2, 2\}$, $\{0, 0, 2, 1\}$, $\{1, 0, 1, 1\}$, $\{1, 0, 2, 0\}$, $\{0, 1, 2, 1\}$, $\{2, 0, 1, 1\}$, $\{1, 2, 0, 1\}$, $\{1, 1, 2, 0\}$, and $\{1, 0, 1, 2\}$ where move types (i) and (ii) generate the first 7 neighboring solutions and move type (iii) generates the remaining 5 neighboring solutions.

13.4.2 The Solution Evaluation

The total solution cost is the sum of the fixed facility opening costs, the inventory costs, and the transportation costs. The first two of these are easily calculated using the solution representation. For transportation costs, customer demands must be assigned to facilities. The assignment of customers to facilities also directly affects the service level achieved. To this end, assignment decisions (represented by X_{ij}'s in the MIP model) are made carefully by paying attention to both transportation costs and service level contribution of an assignment. We define $\varphi_i = \beta_i \sum_{j \in J} \delta_{ij} d_j \beta_i X_{ij}$ to be the service level contribution of facility i. Summing the φ_i over all i yields the left hand side of the service level constraint (4). φ_i is the expected demand satisfied within the time window (over a year period) at facility i. If a customer within a facility i's time window is assigned to that facility, φ_i will increase depending on the total demand assigned to that facility. If a customer j outside of facility i's time window is assigned to facility i, β_i will decrease in accordance with Eqs. 13.5 and 13.6. For this case, δ_{ij}, and the service level would decrease due to a decrease in fill rate.

Consider the following simple heuristic method for assigning each customer to a facility:

1. Sort the customers in descending order according to their demands, breaking ties arbitrarily.
2. Traverse the sorted customer list, starting at the top. If there is an *open* facility with a unique smallest transportation cost for the current customer, assign the customer to that facility.
3. Traverse any remaining unassigned customers (which have two or more open facilities with the same minimal transportation cost) in sorted order and, if only one of the customer's open facilities satisfies the customer's time window, assign the customer to that facility.
4. Remaining unassigned customers have two or more open facilities with the same or "very close" minimal transportation costs (within ε of each other) that are either (a) within the customer's time window or (b) outside the customer's time window.

    a. Type (a) customers are assigned to the facility that maximally increases the total service level contribution. For example, suppose the total demand is $\sum_{j\in J} d_j = 80$ with two facilities under consideration. Facility 1 has stock level $S_1 = 1$, and with the assignments done in the first 3 steps of the heuristic, we have: $\beta_1 = 0.3946$ and $\varphi_1 = 31.5676$. Facility 2 has $S_2 = 1$, $\beta_2 = 0.5105$ and $\varphi_2 = 25.5245$ with the assignments performed so far. Suppose $d_j = 2$ for a customer to be assigned and customer j is within both facilities' time windows with the same transportation cost. Assigning j to facility 1, yields $\varphi_1' = 31.8744$ with $\Delta\varphi_1 = \varphi_1' - \varphi_1 = 0.3068$. Assigning j to facility 2 yields $\varphi_2' = 26.0357$ with $\Delta\varphi_2 = \varphi_2' - \varphi_2 = 0.5112$. Since $\Delta\varphi_2 > \Delta\varphi_1$, we assign j to facility 2.
    b. Type (b) customers are assigned to the facility that minimally decreases the total service level contribution. Using the same parameters as in the above example, assigning j to facility 1, yields $\varphi_1' = 31.097$ with $\Delta\varphi_1 = \varphi_1 - \varphi_1' = 0.4706$. Assigning j to facility 2 yields $\varphi_2' = 25.0343$ with $\Delta\varphi_2 = \varphi_2 - \varphi_2' = 0.4902$. Since $\Delta\varphi_1 < \Delta\varphi_2$, we assign j to facility 1.

5. Assigning all customers according to the first 4 steps may not achieve the required service level yielding an infeasible solution. If the difference between the required service level and the achieved service level is below an empirically determined amount, a "fine tuning" procedure is performed in an attempt to increase the service level with an acceptable cost increase. First, consider moving customers from open low fill rate facilities to an open high fill rate facilities. If moving a *single* customer achieves feasibility, choose the one with the least cost increment. Otherwise, compare the ratio of the cost increment to

```
function: Choose_Best_Neighboring_Solution
begin
old_incumbent_solution = incumbent_solution
initialize incumbent_solution.cost to a sufficiently large value
for all solutions over the neighborhood do
  begin
  evaluate the solution
  if solution.cost <= incumbent_solution.cost then
    begin
    if solution.cost < incumbent_solution.cost then
      begin
      if is_tabu = false or aspiration = true then
        update incumbent_solution
      end
    else if solution.alpha > incumbent_solution.alpha then
      begin
      if is_tabu = false or aspiration = true then
        update incumbent_solution
      end
    end
  end
if incumbent_solution = old_incumbent_solution then
  begin
  choose the best solution regardless of its own tabu status
  tabu_tenure = tabu_tenure × DECREASE
  end
end
```

Fig. 13.1 *Choose_Best_Neighboring_Solution* function

the service level increment and pick the customer with the least ratio. Continue moving customers to open higher fill rate facilities until the solution either reaches feasibility or a maximum number of customers are moved.

13.4.3 Neighborhood Solution Selection

Figure 13.1 presents the pseudo-code for the *Choose_Best_Neighboring_Solution* function. In each search iteration, the new incumbent solution is determined by a rule based comparison operator (Ciarleglio 2007) that uses the total solution cost and service level. All neighboring solutions are evaluated. Tabu neighbors are considered only when either they are superior to the previous best found solution or when all neighboring solutions are tabu. Identical total costs cause the achieved service levels to be compared with the incumbent solution's service level.

13.4.4 The Tabu Memory Structure and Aspiration Criterion

The tabu memory structure consists of a matrix of size $|I|XS_{max}$ with the rows corresponding to the facility indices (*fac_index*) and the columns to the facility stock levels (*fac_stock*). A neighbor solution is tabu if the stock level for that facility has been changed from the neighbor solution's proposed stock level in the last tabu tenure iterations. For a swap move, both facilities stock levels must satisfy this tabu criterion for the neighbor solution to be tabu.

13.4.5 Reactive Memory Structure

The memory structure is indexed by *fac_index* and *fac_stock*. Each visited (incumbent) solution is stored with the iteration most recently visited (*last_time*) and the number of times visited (*repetition*). The constants *INCREASE* and *DECREASE* determine the speed of the multiplicative tabu tenure adjustment. The exponential modifications of the tabu tenure cause the search to react faster and the *moving_average* and *tabu_tenure_change* are used for tabu tenure reduction. As exemplified by Battiti and Tecchiolli (1994) and Carlton and Barnes (1996), the variable *chaotic* stores the number of often-visited solutions and a diversifying *Escape* strategy is executed when *chaotic* is greater than a predefined constant *CHAOS*.

Occasionally, the search can stagnate and needs to be restarted with a "good" initial restart solution. For this reason, all visited solutions which improve the current best solution found are saved in the *best_solution_stack*. When the current best solution found has not been improved for *non_update_limit* iterations or the incumbent solution fails to possess a cost that is less than 1.5 times the best cost previously found for consecutive *outside_range_limit* iterations, the top solution from the *best_solution_stack* becomes the incumbent solution for restarting the search. The search restarts after reinitializing all tabu and hashing information and resetting the parameters such as *non_update_limit*, and *outside_range_limit*. The *Check_For_Restart* function, whose pseudo-code is presented in Fig. 13.2, monitors for the occurrence of these conditions required for restarting the search. Another function determines whether a termination criterion has been satisfied. The two termination criteria are: (1) the time limit, TIME_MAX, is reached, or (2) the *best_solution_stack* is empty when the restart condition is satisfied.

13.4.6 Escape Mechanisms

The *Check_For_Repetitions* function, presented in Fig. 13.3, detects the repetition of previously visited solutions. Increasing the tabu tenure discourages additional repetitions. The tabu tenure's exponential increase quickly breaks any simple

```
function: Check_For_Restart
set outside_range_counter =0; non_update_counter = 0; restart = false;
begin
if incumbent_solution.cost > best_solution_found.cost × 1.5 then
  outside_range_counter = outside_range_counter + 1
else
  outside_range_counter = 0
if outside_range_counter > outside_range_limit then
  begin
  check_for_restart = true
  return from function
  end
if incumbent_solution.cost > best_solution_found.cost then
  non_update_counter = non_update_counter + 1
else
  non_update_counter = 0
if non_update_counter > non_update_limit then
  begin
  check_for_restart = true
  return from function
  end
end
```

Fig. 13.2 *Check_For_Restart* function

cycling behavior. A more powerful reaction mechanism counts the number of often-visited solutions. When a solution has been visited $REP + 1$ times, the solution is added to an often-visited set. When the number of often-visited solutions reaches the threshold *CHAOS*, the *Escape* function, presented in Fig. 13.4, is activated. Exponentially reducing the tabu tenure keeps the tabu tenure from growing too large and quickly returns the tabu tenure to smaller values.

The escape mechanism executes a series of swap moves while preserving the tabu memory structure. The rule based comparison is also used here to select the swap move. First the swap move that achieves a lowest total cost becomes the new incumbent solution. If two different swap moves performed from the current incumbent solution obtain the same cost, then the higher service level is preferred. Swap moves assure an escape from a local attractor basin since an inventory swap between two open facilities usually will not change the total cost but will change the service level. The solution with higher service level has a greater chance of reducing inventory while satisfying the service level constraint. Inventory swaps between open and closed facilities can lead to a new search region since increasing or decreasing inventory at a single open facility does not change the solution dramatically in such a swap move. If no swap move yields a feasible neighboring solution for the current incumbent solution's neighborhood, one unit of stock level will be added to increase the total service level, increasing the likelihood of obtaining a feasible swap move in the following steps.

```
function: Check_For_Repetitions
begin
last_tabu_tenure_change = last_tabu_tenure_change + 1
search current solution in the hashing structure
point = the location of the solution if it is found
if the solution is found then
  begin
  length = iter_counter – point.last_time
  point.last_time = iter_counter
  point.repetitions = point.repetitions + 1
  if point.repetitions > REP then
    begin
    add the current solution to the often-visited set
    chaotic = chaotic + 1
    if chaotic > CHAOS then
      begin
      chaotic = 0
      check_for_repetitions = true
      return from function
      end
    end
    if length < CYCLE_MAX then
      begin
      moving_average = 0.2 × length + 0.8 × moving_average
      tabu_tenure = tabu_tenure × INCREASE
      last_tabu_tenure_change = 0
      end
  end
else
  install it into the hashing structure
if last_tabu_tenure_change > moving_average then
  begin
  tabu_tenure = max (tabu_tenure × DECREASE, 1)
  last_tabu_tenure_change = 0
  end
check_for_repetitions = false
end
```

Fig. 13.3 *Check_For_Repetitions* function

13.4.7 Overall RTS–SPL Logic

The *Initialization* function initializes the data structures for the tabu memory structure and the initial solution is generated by a heuristic method. Viewing the construction of the initial solution as a weighted set covering problem, we desire to open the facility, i, that can cover the largest amount of unassigned demand. To determine the appropriate inventory level for that facility, we compute the ratio of

```
function: Escape

begin
clean hashing memory structure
for i = 1 to 3 do
  begin
  perform swap move
  make_tabu (fac_index1, fac_stock1)
  make_tabu (fac_index2, fac_stock2)
  update incumbent solution
  end
end


function: Restart

begin
clean hashing memory structure
clean tabu memory structure
pop up the solution from the best_solution_stack
make that solution as the incumbent solution for the restarting search
initialize all the counter and adjust all the limits
end
```

Fig. 13.4 *Escape* and *Restart* functions

the opening cost, f_i, and the cost of providing one unit of inventory, h_i. If $f_i/h_i \geq S_{\max}$, opening facility i is more expensive than stocking more units in the currently open facilities which indicates the stock level of the newly opened facility i should be "high," hence is set to its maximum level. Otherwise, we set the inventory level of facility i to one unit. This facility opening procedure is repeated until all the customers are assigned. Function *Reactive_Tabu_Search*, presented in Fig. 13.5, summarizes the RTS–SPL algorithm.

13.5 Computational Results

In this section, we apply RTS–SPL to a benchmark set of problems of different sizes. For small and medium sized data sets (Jeet 2006), we compare the RTS-SPL results with the MIP solutions and with other TS methods. Since no classical MIP methodology can successfully attack practical sized problems, we compare the RTS–SPL solutions only to solutions obtained by other TS methods for the large and very large problems.

13.5.1 Test Problems

Two of our test problem sets are from Jeet (2006) and the other two larger sets are randomly generated for this study using the same approach in Jeet (2006).

```
function: Reactive_Tabu_Search
begin
while stop = false do
  begin
  Choose_Best_Neighboring_Solution
  escape = Check_For_Repetitions
  if escape = false then
    begin
    execute the best admissible move
    Make_Tabu (fac_index, fac_stock)
    iter_counter = iter_counter + 1
    restart = Check_For_Restart
    if incumbent_solution.cost < best_solution_found.cost then
      begin
      best_solution_found = incumbent_solution
      push the best_solution_found onto best_solution_stack
      end
    if restart = true then
      Restart
    end
  else
    Escape
  stop = Check_For_Stop
  end
end
```

Fig. 13.5 *Reactive_Tabu_Search* functions

RTS–SPL was executed on the same computer used to generate Jeet (2006) results: Dual Xeon 1.8 GHz processors with 1 GB RAM. During all problem runs, all RTS-SPL parameters are held constant at the following values: $INCREASE = 1.2$, $DECREASE = 0.8$, $REP = 2$, $CHAOS = 5$, $CYCLE\_MAX = 10$, $W1 = 0.1$, $W2 = 0.9$.

13.5.2 Results for the Small Problem Sets

The small problems have 15 facilities and 50 customers (15 × 50) with customer and facility locations generated randomly on a 150 × 150 grid. The transportation costs, c_{ij}, or equal to one tenth of the Euclidean distance between facility i and customer j, rounded to a positive integer. The annual mean demand values, d_j, are uniformly distributed over the range from 1 to 3. Replenishment lead times for all facilities are 7 (days). Time window indicators, δ_{ij}, are set to one if the Euclidean distance between facility i and customer j is less than 40, and zero, otherwise. $S_{\max}$

is set to 5, which is more than enough to obtain fill rates very close to 100% at all facilities regardless of the demand assigned to any facility.

Five levels of holding cost, same for all facilities, hi (h = 1, 10, 20, 50, and 100) and three service levels (40, 60, or 80% of the maximum possible service) are investigated. Fixed costs, fi, of opening all facilities are set to either 0 or 1,000. Generating one problem for every possible combination of holding costs, service levels, and facility fixed costs yields 30 problems. Three such problem sets were generated.

Tables 13.1, 13.2, and 13.3 present the comparative results including the RTS–SPL solution values, MIP solution values (Jeet 2006), the percentage difference, $\Delta = 100(\text{RTS} - \text{MIP})/\text{MIP}\%$, the number of open facilities, and the total inventory across all facilities. A bracketed value after the facility or inventory numbers indicates the difference between the RTS–SPL and MIP solutions. For example, 15(−1) means the RTS-SPL solution has 15 units of stock, 1 less than the MIP solution's total inventory. Similarly, 17(+1) means that the RTS-SPL's total stock level is 17, 1 more than the MIP solution's.

For Tables 13.1, 13.2, and 13.3, when the fixed cost is 0 (top halves of the tables), RTS–SPL obtained the same or better solutions than the MIP for all problems (Note that MIP solutions may not be optimal due to time limit used for both methods, 900 s in this case). When the fixed cost is 1000 (bottom halves of the tables), RTS–SPL found slightly inferior solution values in 8 of the 45 problems when compared to the MIP solutions with a maximal difference of 1.06% and superior solutions in 10 out of the 45 instances with a maximal difference of 2.81% improvement.

The computational time for RTS–SPL for any problem was <1 s while the average time for MIP to run for one instance was 34 s with maximum time of 806 s. The RTS–SPL obtains near optimal solutions in markedly less computational effort.

13.5.3 Results for the Medium Size Data Sets

The medium sized problem sets were based on real data provided by a service parts logistics group at a large computer hardware manufacturer. We use six representative networks (customer and facility locations), representing different regions in the United States. The sizes of the networks are different depending on the region. We use the actual transportation costs and δ_{ij} values provided in the real data. The problems were solved for service time-windows of 2 hours (TW1) and 12 hours (TW2). Longer time-windows yield more $\delta_{ij} = 1$ indicating that the demands are more easily satisfied towards time-based service since more candidate facilities can provide service to those customers. The maximum stock level for each facility is 5.

Three levels of holding cost h_i (50, 100, or 200) and three service levels (40, 60, or 80%) were investigated. Facility opening costs were all set to 0, focusing on the

Table 13.1 RTS and MIP solutions for small data set A

Instance				Solution				
n	h	α (%)	f	RTS	MIP	Δ (%)	Open fac	Total inventory
1a	1	40	0	232	232	0.00	12	12
2a	1	60	0	232	232	0.00	12	12
3a	1	80	0	232	232	0.00	12	12
4a	10	40	0	314	314	0.00	7	7
5a	10	60	0	314	314	0.00	7	7
6a	10	80	0	321	321	0.00	9	9
7a	20	40	0	369	369	0.00	5	5
8a	20	60	0	369	369	0.00	5	5
9a	20	80	0	411	411	0.00	9	9
10a	50	40	0	491	491	0.00	4	4
11a	50	60	0	519	519	0.00	5	5
12a	50	80	0	652	652	0.00	7	8
13a	100	40	0	665	665	0.00	3	3
14a	100	60	0	769	769	0.00	5	5
15a	100	80	0	1052	1052	0.00	7	8
16a	1	40	1000	2444	2444	0.00	2	5
17a	1	60	1000	3366	3366	0.00	3	7
18a	1	80	1000	4300	4300	0.00	4	9
19a	10	40	1000	2489	2489	0.00	2	5
20a	10	60	1000	3429	3435	−0.17	3	7
21a	10	80	1000	4381	4381	0.00	4	9
22a	20	40	1000	2539	2539	0.00	2	5
23a	20	60	1000	3499	3505	−0.17	3	7
24a	20	80	1000	4471	4471	0.00	4	9
25a	50	40	1000	2689	2689	0.00	2	5
26a	50	60	1000	3709	3715	−0.16	3	7
27a	50	80	1000	4741	4741	0.00	4	9
28a	100	40	1000	2939	3024	−2.81	2	5
29a	100	60	1000	4059	4065	−0.15	3	7
30a	100	80	1000	5191	5191	0.00	4	9

inventory and transportation tradeoffs. Replicating each combination of the holding costs and target service levels for each of the six data sets, we have nine instances for each network at each level of the time window. The sizes of the six different data sets are (a) 12 × 90, (b) 13 × 96, (c) 13 × 106, (d) 19 × 128, (e) 16 × 134, and (f) 24 × 158.

The results of RTS-SPL are compared with results of the MIP method in Tables 13.4, 13.5, 13.6, 13.7, 13.8, and 13.9. For the medium size data sets with time windows of 2 h, RTS–SPL is superior on 9 of the problems with a maximum improvement of 1.20% and inferior on only two instances with a negligible difference of 0.01%. With these exceptions, the better RTS–SPL results are obtained at the highest service level. RTS–SPL and MIP solutions for the much easier

Table 13.2 RTS and MIP solutions for small data set B

Instance				Solution				
n	h	α (%)	f	RTS	MIP	Δ (%)	Open fac	Total inventory
1b	1	40	0	279	279	0.00	15	15
2b	1	60	0	279	279	0.00	15	15
3b	1	80	0	280	280	0.00	15	16
4b	10	40	0	368	368	0.00	9	9
5b	10	60	0	368	368	0.00	9	9
6b	10	80	0	388	388	0.00	9	11
7b	20	40	0	443	443	0.00	7	7
8b	20	60	0	443	443	0.00	7	7
9b	20	80	0	490	490	0.00	8	10
10b	50	40	0	583	583	0.00	4	4
11b	50	60	0	628	629	−0.16	6	6
12b	50	80	0	790	790	0.00	8	10
13b	100	40	0	783	783	0.00	4	4
14b	100	60	0	928	929	−0.11	6	6
15b	100	80	0	1290	1290	0.00	8	10
16b	1	40	1000	2535	2535	0.00	2	5
17b	1	60	1000	3449	3449	0.00	3	9
18b	1	80	1000	6341	6342	−0.02	6	15(−1)
19b	10	40	1000	2580	2580	0.00	2	5
20b	10	60	1000	3530	3530	0.00	3	9
21b	10	80	1000	6464	6464	0.00	6	13
22b	20	40	1000	2630	2630	0.00	2	5
23b	20	60	1000	3620	3620	0.00	3	9
24b	20	80	1000	6594	6594	0.00	6	13(+1)
25b	50	40	1000	2780	2780	0.00	2	5
26b	50	60	1000	3890	3890	0.00	3	9
27b	50	80	1000	6984	6954	0.43	6	13(+1)
28b	100	40	1000	3030	3030	0.00	2	5
29b	100	60	1000	4340	4340	0.00	3	9
30b	100	80	1000	7634	7554	1.06	6	13(+1)

medium size data sets with 12 h time windows are virtually identical, hence are not presented here.

The run time limits of RTS-SPL are 5 s for the first three data sets and 10 s for the last three data sets. The actual average total run time over 18 instances for each of the six networks was 1.814, 2.844, 3.957, 8.361, 6.453, and 7.809 seconds and the time when the best solution was first encountered was shorter. The average run times of MIP for the six sets are 0.28, 1.67, 69.78, 4.11, 19.83, and 3.78 s. There is no apparent advantage for RTS–SPL for some data sets, but the MIP run times are highly variable while the RTS is much more consistent. The longest MIP run times are 1, 18, 386, 36, 183, and 17 s for each set. RTS–SPL found very good or optimal solutions in acceptable times for all the instances.

Table 13.3 RTS and MIP solutions for small data set C

Instance				Solution				
n	h	α (%)	f	RTS	MIP	Δ (%)	Open fac	Total inventory
1c	1	40	0	290	290	0.00	11	11
2c	1	60	0	290	290	0.00	11	11
3c	1	80	0	297	298	−0.34	11	18
4c	10	40	0	366	366	0.00	6	6
5c	10	60	0	366	366	0.00	6	6
6c	10	80	0	438	438	0.00	7	14
7c	20	40	0	416	416	0.00	5	5
8c	20	60	0	426	426	0.00	6	6
9c	20	80	0	578	579	−0.17	7	14
10c	50	40	0	533	533	0.00	3	3
11c	50	60	0	606	606	0.00	6	6
12c	50	80	0	998	999	−0.10	7	14
13c	100	40	0	683	683	0.00	3	3
14c	100	60	0	906	906	0.00	6	6
15c	100	80	0	1698	1699	−0.06	7	14
16c	1	40	1000	2462	2462	0.00	2	6
17c	1	60	1000	3390	3385	0.15	3	7(-4)
18c	1	80	1000	6336	6326	0.16	6	17(+1)
19c	10	40	1000	2516	2530	−0.55	2	6(+1)
20c	10	60	1000	3453	3453	0.00	3	7
21c	10	80	1000	6485	6470	0.23	6	16
22c	20	40	1000	2576	2580	−0.16	2	6(+1)
23c	20	60	1000	3523	3523	0.00	3	7
24c	20	80	1000	6645	6630	0.23	6	16
25c	50	40	1000	2729	2730	−0.04	2	5
26c	50	60	1000	3733	3733	0.00	3	7
27c	50	80	1000	7125	7110	0.21	6	16
28c	100	40	1000	2979	2980	−0.03	2	5
29c	100	60	1000	4083	4083	0.00	3	7
30c	100	80	1000	7925	7910	0.19	6	16

Table 13.4 RTS and MIP solutions for medium data set A

Instance				Solution				
n	h	α (%)	TW	RTS	MIP	Δ (%)	Open fac	Total inventory
1a	50	0.4	TW1	8404	8404	0.00	9	9
2a	100	0.4	TW1	8854	8854	0.00	9	9
3a	200	0.4	TW1	9738	9738	0.00	8	8
4a	50	0.6	TW1	8404	8404	0.00	9	9
5a	100	0.6	TW1	8854	8854	0.00	9	9
6a	200	0.6	TW1	9738	9738	0.00	8	8
7a	50	0.8	TW1	8454	8454	0.00	9	10
8a	100	0.8	TW1	8954	8954	0.00	9	10
9a	200	0.8	TW1	9954	9954	0.00	9	10

Table 13.5 RTS and MIP solutions for medium data set B

Instance				Solution				
n	h	α (%)	TW	RTS	MIP	Δ (%)	Open fac	Total inventory
1b	50	0.4	TW1	8437	8437	0.00	12	12
2b	100	0.4	TW1	9012	9012	0.00	11	11
3b	200	0.4	TW1	10010	10010	0.00	9	9
4b	50	0.6	TW1	8437	8437	0.00	12	12
5b	100	0.6	TW1	9012	9012	0.00	11	11
6b	200	0.6	TW1	10010	10010	0.00	9	9
7b	50	0.8	TW1	8487	8487	0.00	12	13
8b	100	0.8	TW1	9112	9212	−1.09	11	12(−1)
9b	200	0.8	TW1	10312	10437	−1.20	11(−1)	12(−1)

Table 13.6 RTS and MIP solutions for medium data set C

Instance				Solution				
n	h	α (%)	TW	RTS	MIP	Δ (%)	Open fac	Total inventory
1c	50	0.4	TW1	7397	7396	0.01	7	7
2c	100	0.4	TW1	7651	7651	0.00	5	5
3c	200	0.4	TW1	8151	8151	0.00	5	5
4c	50	0.6	TW1	7397	7446	−0.66	7(−1)	7(−1)
5c	100	0.6	TW1	7747	7751	−0.05	7(+2)	7(+1)
6c	200	0.6	TW1	8351	8351	0.00	5	6
7c	50	0.8	TW1	7547	7546	0.01	7	10
8c	100	0.8	TW1	7999	7999	0.00	6	9
9c	200	0.8	TW1	8899	8899	0.00	6	9

Table 13.7 RTS and MIP solutions for medium data set D

Instance				Solution				
n	h	α (%)	TW	RTS	MIP	Δ (%)	Open fac	Total inventory
1d	50	0.4	TW1	13468	13468	0.00	17	17
2d	100	0.4	TW1	14246	14246	0.00	15	15
3d	200	0.4	TW1	15656	15656	0.00	12	12
4d	50	0.6	TW1	13468	13468	0.00	17	17
5d	100	0.6	TW1	14246	14246	0.00	15	15
6d	200	0.6	TW1	15656	15656	0.00	12	12
7d	50	0.8	TW1	13518	13518	0.00	17	18
8d	100	0.8	TW1	14346	14377	−0.22	15(−1)	16(−1)
9d	200	0.8	TW1	15946	16077	−0.81	15(−1)	16(−1)

Table 13.8 RTS and MIP solutions for medium data set E

Instance				Solution				
n	h	α (%)	TW	RTS	MIP	Δ (%)	Open fac	Total inventory
1e	50	0.4	TW1	10055	10055	0.00	14	14
2e	100	0.4	TW1	10708	10708	0.00	13	13
3e	200	0.4	TW1	11863	11863	0.00	11	11
4e	50	0.6	TW1	10055	10055	0.00	14	14
5e	100	0.6	TW1	10708	10708	0.00	13	13
6e	200	0.6	TW1	11863	11863	0.00	11	11
7e	50	0.8	TW1	10205	10205	0.00	14	17
8e	100	0.8	TW1	11008	11008	0.00	13	16
9e	200	0.8	TW1	12511	12511	0.00	12	15

Table 13.9 RTS and MIP solutions for medium data set F

Instance				Solution				
n	h	α (%)	TW	RTS	MIP	Δ (%)	Open fac	Total inventory
1f	50	0.4	TW1	17040	17040	0.00	19	19
2f	100	0.4	TW1	17891	17891	0.00	16	16
3f	200	0.4	TW1	19480	19480	0.00	14	14
4f	50	0.6	TW1	17040	17040	0.00	19	19
5f	100	0.6	TW1	17891	17891	0.00	16	16
6f	200	0.6	TW1	19480	19480	0.00	14	14
7f	50	0.8	TW1	17040	17090	−0.29	19	19(−1)
8f	100	0.8	TW1	17990	18022	−0.18	19(+2)	19(+1)
9f	200	0.8	TW1	19691	19822	−0.66	16(−1)	17(−1)

The restart and escape features do not play very important role in the medium size data sets. The instances are "easy" for the RTS-SPL since there are certain patterns embedded in the real data. Some limitations are incorporated in the data itself to prevent some customers from being assigned to certain facilities and these limitations make the search easier. The iteration when the best solution was first found was small implying that the search finds the best solution very quickly and leaves very little room of improvement for the restart and escape mechanisms.

13.5.4 Results for the Large and Extra Large Problem Sets

The two data sets discussed in this section were generated the same way as the small data sets. The three large data sets have 50 facilities and 200 customers randomly generated on a 150 × 150 grid and the three *extra* large data sets have

100 facilities and 500 customers randomly generated on a 200 × 200 grid. Different random number seeds were used for the problem generations. The service radius is set to 40 as in the small data set. The maximum stock level for the large problem set is 7 and for the extra large problem set 10.

Three values for holding cost (1, 20, 100) and three target service levels (40, 60, 80%) were investigated. Two values for the facility fixed cost (0, 1000) were also considered. Combining all possible interactions of these factors produces 18 instances for each problem set. Thus, there are 54 instances associated with either the large problem sets or extra large problem sets.

The results of the three large data sets presented in Tables 13.10, 13.11, and 13.12 show that the average time for obtaining the best solution is 3.175 s with a 300-s limit on run time. For the extra large data sets (Tables 13.13, 13.14, and 13.15) the average run time is 117 s with a 600-s limit.

The problem sizes considered here made it *impossible* to obtain an optimal or near optimal solution for *any* of these problems using classical MIP approaches. Hence, we cannot compare RTS–SPL with a classical MIP approach. Nevertheless, we compare the RTS–SPL to its variations without escape and without restart procedures. Tables 13.10, 13.11, 13.12, 13.13, 13.14, and 13.15 show the results of (1) the standard RTS, (2) the RTS without escape mechanism (NoEsc), and (3) the RTS without restart mechanism (NoRes). The Δ after NoEsc or NoRes means the percentage differences between their solutions and RTS solutions. All the versions of TS have the same parameter settings and the same run time limit. For the large data sets, there are 29 and 6

Table 13.10 RTS and TS solutions for large data set A

Instance				RTS solution			Compare			
n	h	α (%)	f	RTS	Best iter	Best time(ms)	NoEsc	Δ (%)	NoRes	Δ (%)
1a	1	40	0	686	31	280	686	0.00	686	0.00
2a	1	60	0	686	28	170	686	0.00	686	0.00
3a	1	80	0	686	35	230	686	0.00	686	0.00
4a	20	40	0	1071	56	620	1072	0.09	1071	0.00
5a	20	60	0	1071	53	600	1072	0.09	1071	0.00
6a	20	80	0	1180	109	1780	1195	1.27	1180	0.00
7a	100	40	0	1722	153	1470	1755	1.92	1722	0.00
8a	100	60	0	2012	2447	29240	2029	0.84	2015	0.15
9a	100	80	0	2618	222	3140	2698	3.06	2618	0.00
10a	1	40	1000	3914	1427	16530	3977	1.61	3914	0.00
11a	1	60	1000	4581	1059	11660	4592	0.24	4592	0.24
12a	1	80	1000	6262	30	370	6304	0.67	6262	0.00
13a	20	40	1000	4123	1427	16460	4167	1.07	4123	0.00
14a	20	60	1000	4841	9	100	4841	0.00	4841	0.00
15a	20	80	1000	6592	163	1920	6646	0.82	6592	0.00
16a	100	40	1000	4967	5	60	4967	0.00	4967	0.00
17a	100	60	1000	5881	9	100	5881	0.00	5881	0.00
18a	100	80	1000	7952	163	1880	8086	1.69	7952	0.00

Table 13.11 RTS and TS solutions for large data set B

Instance				RTS solution			Compare			
n	h	α (%)	f	RTS	Best iter	Best time(ms)	NoEsc	Δ (%)	NoRes	Δ (%)
1b	1	40	0	639	42	420	639	0.00	639	0.00
2b	1	60	0	639	43	440	639	0.00	639	0.00
3b	1	80	0	639	49	530	639	0.00	639	0.00
4b	20	40	0	1052	67	730	1052	0.00	1052	0.00
5b	20	60	0	1052	110	1420	1062	0.95	1052	0.00
6b	20	80	0	1155	1244	18420	1189	2.94	1159	0.35
7b	100	40	0	1671	131	1350	1679	0.48	1671	0.00
8b	100	60	0	1976	1477	16330	1987	0.56	1978	0.10
9b	100	80	0	2605	1043	13130	2640	1.34	2623	0.69
10b	1	40	1000	3879	20	290	3914	0.90	3879	0.00
11b	1	60	1000	4480	568	4270	4547	1.50	4480	0.00
12b	1	80	1000	5268	42	380	5308	0.76	5268	0.00
13b	20	40	1000	4104	5	40	4104	0.00	4104	0.00
14b	20	60	1000	4750	1499	11300	4794	0.93	4759	0.19
15b	20	80	1000	5629	42	380	5669	0.71	5629	0.00
16b	100	40	1000	4904	5	40	4904	0.00	4904	0.00
17b	100	60	1000	5813	21	190	5834	0.36	5813	0.00
18b	100	80	1000	7072	23	190	7189	1.65	7072	0.00

Table 13.12 RTS and TS solutions for large data set C

Instance				RTS solution			Compare			
n	h	α (%)	f	RTS	Best iter	Best time(ms)	NoEsc	Δ (%)	NoRes	Δ (%)
1c	1	40	0	641	38	380	641	0.00	641	0.00
2c	1	60	0	641	38	390	641	0.00	641	0.00
3c	1	80	0	641	45	510	641	0.00	641	0.00
4c	20	40	0	1016	34	250	1016	0.00	1016	0.00
5c	20	60	0	1016	38	350	1016	0.00	1016	0.00
6c	20	80	0	1139	216	3330	1176	3.25	1139	0.00
7c	100	40	0	1685	784	7900	1685	0.00	1685	0.00
8c	100	60	0	1940	46	470	1967	1.39	1940	0.00
9c	100	80	0	2624	2	30	2603	−0.80	2624	0.00
10c	1	40	1000	4022	5	90	4022	0.00	4022	0.00
11c	1	60	1000	4565	9	110	4565	0.00	4565	0.00
12c	1	80	1000	5311	20	300	5365	1.02	5311	0.00
13c	20	40	1000	4212	5	60	4212	0.00	4212	0.00
14c	20	60	1000	4812	9	110	4812	0.00	4812	0.00
15c	20	80	1000	5672	20	290	5745	1.29	5672	0.00
16c	100	40	1000	5012	5	60	5012	0.00	5012	0.00
17c	100	60	1000	5852	9	110	5852	0.00	5852	0.00
18c	100	80	1000	7192	20	290	7345	2.13	7192	0.00

Table 13.13 RTS and TS solutions for extra large data set A

Instance				RTS solution			Compare			
n	h	α (%)	f	RTS	Best iter	Best time(ms)	NoEsc	Δ (%)	NoRes	Δ (%)
1a	1	40	0	1572	75	3850	1572	0.00	1572	0.00
2a	1	60	0	1572	83	4650	1572	0.00	1572	0.00
3a	1	80	0	1574	91	7750	1574	0.00	1574	0.00
4a	20	40	0	2455	211	18800	2457	0.08	2455	0.00
5a	20	60	0	2455	209	18590	2457	0.08	2455	0.00
6a	20	80	0	2771	1767	205600	2858	3.14	2775	0.14
7a	100	40	0	3919	2603	177460	3933	0.36	3925	0.15
8a	100	60	0	4700	100	8960	4762	1.32	4700	0.00
9a	100	80	0	6095	507	45400	6151	0.92	6095	0.00
10a	1	40	1000	8332	86	9400	8643	3.73	8332	0.00
11a	1	60	1000	8963	39	3720	9075	1.25	8963	0.00
12a	1	80	1000	10491	130	14330	10532	0.39	10491	0.00
13a	20	40	1000	8714	31	3150	9023	3.55	8714	0.00
14a	20	60	1000	9533	39	4080	9626	0.98	9533	0.00
15a	20	80	1000	11358	458	48070	11406	0.42	11358	0.00
16a	100	40	1000	10314	31	3600	10623	3.00	10314	0.00
17a	100	60	1000	11933	39	4080	11946	0.11	11933	0.00
18a	100	80	1000	14958	458	50480	15086	0.86	14958	0.00

Table 13.14 RTS and TS solutions for extra large data set B

Instance				RTS solution			Compare			
n	h	α (%)	f	RTS	Best iter	Best time(ms)	NoEsc	Δ (%)	NoRes	Δ (%)
1b	1	40	0	1618	62	3360	1618	0.00	1618	0.00
2b	1	60	0	1618	66	3510	1618	0.00	1618	0.00
3b	1	80	0	1621	110	10850	1622	0.06	1621	0.00
4b	20	40	0	2440	146	12500	2442	0.08	2440	0.00
5b	20	60	0	2440	156	15120	2442	0.08	2440	0.00
6b	20	80	0	2758	4717	516530	2799	1.49	2756	−0.07
7b	100	40	0	3853	4259	355130	3866	0.34	3854	0.03
8b	100	60	0	4568	7223	598410	4613	0.99	4613	0.99
9b	100	80	0	5877	6016	575340	6030	2.60	5925	0.82
10b	1	40	1000	7911	46	5460	8045	1.69	7911	0.00
11b	1	60	1000	8838	5836	557400	8997	1.80	8839	0.01
12b	1	80	1000	10292	43	3960	10408	1.13	10292	0.00
13b	20	40	1000	8386	46	5470	8501	1.37	8386	0.00
14b	20	60	1000	9409	99	10880	9586	1.88	9409	0.00
15b	20	80	1000	11052	43	3950	11320	2.42	11052	0.00
16b	100	40	1000	10037	174	19610	10214	1.76	10037	0.00
17b	100	60	1000	11809	91	10830	12066	2.18	11809	0.00
18b	100	80	1000	14252	43	4200	15160	6.37	14252	0.00

Table 13.15 RTS and TS solutions for extra large data set C

Instance				RTS solution			Compare			
n	h	α (%)	f	RTS	Best iter	Best time(ms)	NoEsc	Δ (%)	NoRes	Δ (%)
1c	1	40	0	1644	67	3460	1644	0.00	1644	0.00
2c	1	60	0	1644	72	4320	1644	0.00	1644	0.00
3c	1	80	0	1647	133	16250	1648	0.06	1647	0.00
4c	20	40	0	2467	146	11830	2478	0.45	2467	0.00
5c	20	60	0	2461	475	49650	2481	0.81	2461	0.00
6c	20	80	0	2768	3928	462650	2853	3.07	2769	0.04
7c	100	40	0	3772	1486	103140	3798	0.69	3772	0.00
8c	100	60	0	4595	5933	560910	4621	0.57	4608	0.28
9c	100	80	0	5921	5009	477140	6084	2.75	6025	1.76
10c	1	40	1000	8049	1042	109980	8122	0.91	8054	0.06
11c	1	60	1000	8929	4205	439850	8971	0.47	8959	0.34
12c	1	80	1000	11086	105	10230	11330	2.20	11086	0.00
13c	20	40	1000	8567	28	3930	8616	0.57	8567	0.00
14c	20	60	1000	9518	2976	311770	9579	0.64	9548	0.32
15c	20	80	1000	11799	80	8190	12147	2.95	11799	0.00
16c	100	40	1000	10157	62	7560	10189	0.32	10157	0.00
17c	100	60	1000	11998	3788	405300	12139	1.18	12028	0.25
18c	100	80	1000	14759	80	8030	15587	5.61	14759	0.00

out of 54 instances with better results when compared to the RTS–SPL without escape and without restart separately, respectively. Only 1 instance saw improvement for the RTS–SPL without escape or restart (red highlighting). For the extra large data sets, only 7 instances for the RTS–SPL without escape obtained the same result as the full version of the RTS. The other 47 instances obtained inferior results when the escape feature was disabled and 2 of them were inferior by more than 5%. Disabling the restart mechanism affects the solution quality for total 13 instances and only 1 instance improves. Hence, it appears that the larger the data set size, the more important the restart and escape mechanisms become.

13.5.5 RTS–SPL Solutions and Randomly Generated Solutions

RTS–SPL provides very good solutions for various size problems as tested computationally. However, during the search procedure, RTS–SPL visits a large number of solutions. To "validate" RTS–SPL, we randomly generated the same number of solutions that were visited by RTS–SPL and compared their best solution value found with the RTS–SPL final solution. We performed this exercise using the small problems, conjecturing that the random solutions would have the highest chance to "hit" a near optimal solution for this data set and hence would be more competitive compared to RTS–SPL solutions. For the small data sets, there are around 2,000 iterations processed for each instance and the number of solutions visited at each iteration is about 30. Therefore, we randomly generate 60,000 solutions for each instance and compare the best of these with RTS solutions. As presented in Sui (2008), the RTS solutions were always better than the best of random ones and the improvement was greater than 100% for some instances. The average improvement was about 28% for the zero fixed cost and 56% for the fixed cost of 1000.

13.6 Conclusions and Future Research

In this study, we investigated the application of a reactive tabu seach approach, RTS–SPL, to a service part logistics problem which possesses strong interactions between the network design and inventory decisions. The traditional MIP or even the customized method based on the lower and upper bounding schemes does not scale up well to solve the large SPL problems. RTS–SPL developed to overcome this computational difficulty produces high quality solutions. We show that the full featured RTS–SPL competes very favorably with MIP-based methods for small and medium size problems and finds high quality solutions for large and extra large problems. The escape and restart features for the *full* RTS greatly enhance the methodology.

As a future step, we believe combining RTS with the exact optimization-based methods for the integrated SPL problem could lead to improved results, especially in providing a solution quality guarantee. Similar hybrid methods have been developed in the literature. Dumitrescu and Stutzle (2002) studied several local search approaches that were strengthened by the use of exact algorithms. The direct flight network design problem was solved efficiently by a method that combines TS and branch and bound (Budenbender 2000). Chelouah and Siarry (2005) designed a hybrid method combining continuous TS and Nelder–Mead simplex algorithms for global optimization for multi-minima functions. One way to combine RTS and optimization would be using optimization based methods to explore the TS neighborhood. We could enlarge the TS neighborhood to make it more efficient and effective by solving the facility inventory and customer assignment at one time. Large neighborhoods can reach much better solutions in one search step than simple and small neighborhoods. However, large neighborhoods have the associated disadvantage that a large amount of time may be spent finding the best neighboring solution. It may be possible to model the problem of searching a large neighborhood as an optimization problem which can be solved by an optimization based method. Another way of combining the two methods is solving the problem by an optimization based method while considering the structure of good solutions. First, TS is conducted to collect the attributes of good solutions by defining a set of good solutions. Then, the problem is solved using the optimization method whose solution space is confined within the attributes of the good solutions.

References

Agrawal V, Seshadri S (2000) Distribution free bounds for service constrained (Q, r) inventory systems. Naval Res Logist 47:635–656

Barahona F, Jensen D (1998) Plant location with minimum inventory. Math Program 83:101–111

Barnes JW, Carlton WB (1995) Solving the vehicle routing problem with time windows using reactive tabu search. INFORMS Conf, New Orleans

Battiti R, Tecchiolli G (1994) The Reactive Tabu Search. ORSA J Comput 6:126–140

Budenbender K (2000) A hybrid tabu search/branch-and-bound algorithm for the direct flight network design problem. Transp Sci 34:364–380

Carlton WB, Barnes JW (1996) Solving the traveling salesman problem with time windows using tabu search. IEE Transactions 28, 617–629.

Candas MF, Kutanoglu E (2007) Benefits of considering inventory in service parts logistics network design problems with time-based service constraints. IIE Trans 39:159–176

Chelouah R, Siarry P (2005) A hybrid method combining continuous tabu search and Nelder-Mead simples algorithms for the global optimization of multiminima functions. Eur J Oper Res 161:636–654

Chen FY, Krass D (2001) Inventory models with minimal service level constraints. Eur J Oper Res 134:120–140

Ciarleglio MI (2007) Modular abstract self-learning tabu search (MASTS) metaheuristic search theory and practice. PhD Dissertation, the University of Texas at Austin

Ciarleglio MI, Barnes JW, Sarkar S (2008) ConsNet—a tabu search approach to the spatially coherent conservation area network design problem. J Heuristics. http://dx.doi.org/10.1007/s10732-008-9098-7 (to appear)

Cohen MA, Kleindorfer PR, Lee HL (1988) Service constrained (s, S) inventory systems with priority demand classes and lost sales. Manage Sci 34:482–499

Cohen MA, Kamesam PV, Kleindorfer PR, Lee HL, Tekerian A (1990) Optimizer: IBM's multi-echelon inventory system for managing service logistics. Interfaces 20:65–82

Cohen MA, Zheng Y-S, Agrawal V (1997) Service parts logistics: a benchmark analysis. IIE Trans 29:627–639

Cohen MA, Zheng Y-S, Wang Y (1999) Identifying opportunities for improving Teradyne's service-parts logistics system. Interfaces 29:1–18

Cohen MA, Cull C, Lee HL, Willen D (2000) Saturn's supply-chain innovation: high value in after-sales service. MIT Sloan Manag Rev 41:93–101

Crainic TG, Gendreau M (2002) Cooperative parallel tabu search for capacitated network design. J Heuristics 8:601–627

Crainic TG, Toulouse M, Gendreau M (1995) Synchronous tabu search parallelization strategies for multicommodity location-allocation with balancing requirements. OR Spect 17:113–123

Crainic TG, Toulouse M, Gendreau M (1996) Parallel asynchronous tabu search for multicommodity location-allocation with balancing requirements. Ann Oper Res 63:277–299

Daskin MS, Coullard CR, Shen ZM (2002) An inventory-location model: formulation, solution algorithm and computational results. Ann Oper Res 110:83–106

Drezner Z (1995) Facility location: A survey of applications and methods. Springer, New York

Dumitrescu I, Stutzle T (2002) Combinations of local search and exact algorithms. In: Proceedings of applications of evolutionary computing EvoWorkshop. Lecture Notes in Computer Science, vol 2611. Springer, Berlin, pp 211–223

Glover F (1989) Tabu search—part I. ORSA J Comput 1:190–206

Glover F (1990) Tabu search—part II. ORSA J Comput 2:4–32

Hakimi SL (1964) Optimum locations of switching centers and the absolute centers and medians of a graph. Oper Res 12:450–459

Jeet V (2006) Logistics network design with inventory stocking, time-based service and part commonality. PhD Dissertation, The University of Texas at Austin

Jeet V, Kutanoglu E, Partani A (2009) Logistics network design with inventory stocking for low-demand parts: modeling and optimization. IIE Trans 41(5):389–407

Kuehn AA, Hamburger MJ (1963) A heuristic program for locating warehouses. Manage Sci 9:643–666

Magnanti TL, Wong RT (1984) Network design and transportation planning: model and algorithms. Transp Sci 18:1–55

Melkote S, Daskin MS (2001a) An integrated model of facility location and transportation network design. Transp Res A 35:515–538

Melkote S, Daskin MS (2001b) Capacitated facility location/network design problems. Eur J Oper Res 129:481–495

Muckstadt JA (1973) A model for a multi-item, multi-echelon, multi-indenture inventory system. Manage Sci 20:472–481

Muckstadt JA (2005) Analysis and algorithms for service parts supply chains. Springer, New York

Nanry WP, Barnes JW (2000) Solving the pickup and delivery problem with time windows using reactive tabu search. Transp Res B 34:107–121

Nozick LK (2001) The fixed charge facility location problem with coverage restrictions. Transp Res E 37:281–296

Nozick LK, Turnquist MA (1998) Integrating inventory impacts into a fixed-charge model for locating distribution centers. Transp Res E 34:173–186

Poole K (2003) Seizing the potential of the service supply chain. Supply Chain Management Review 7(4):54–61

Rustenburg WD, van Houtum GJ, Zijm WHM (2001) Spare parts management at complex technology-based organizations: an agenda for research. Int J Prod Econ 71:177–193

Shen ZM, Coullard CR, Daskin MS (2003) A joint location-inventory model. Transp Sci 37:40–55

Sherbrooke CC (1968) METRIC: a Multi-Echelon Technique for recoverable item control. Oper Res 16:122–141

Sherbrooke CC (1986) VARI-METRIC: improved approximations for multi-indenture, multi-echelon availability models. Oper Res 34:311–319

Simchi-Levi D (1991) The capacitated traveling salesmen location problem. Transp Sci 25:9–18

Sui Y (2008) Solving service part logistics network design and inventory stocking problems with reactive tabu search. Master Report, The University of Texas at Austin

Toregas C, Swain R, ReVelle C, Bergman L (1971) The location of emergency service facilities. Oper Res 19:1363–1373

Xu J, Chiu SY, Glover F (1996) Probabilistic tabu search for telecommunications network design. Combin Optim Theory Practice 1:69–94

Zipkin P (2000) Fundamentals of inventory management. McGraw Hill-Irwin, Boston

Chapter 14
Common Mistakes and Guidelines for Change in Service Parts Management

Andrew J. Huber

Abstract Throughout this book several characteristics and approaches are presented for dealing with the unique challenges of service parts management. While there are some organizations who apply rigorous and innovative methods, service parts management is an area where practice often lags theory. This chapter presents some of the common misperceptions and failures that often occur in the management of service parts supply chains in the high technology, aviation, automotive and telecom industries.

The good news for managers of service parts supply chains is that opportunities for improvement abound. Readers of this book can evaluate the performance of any parts supply network and compare the methods actually used to each method described in this book. This can be done by asking a few simple questions relative to each technique:

- Does the problem this technique addresses exist in this service environment? Often the answer will be yes!
- How does this organization currently deal with this problem?
- Is the method described in this book appropriate for this environment?
- Are there other methods that would lead to an improvement in this area?

Through an analysis beginning with the simple questions listed above, consultants and managers can discover guidelines for improvement that can serve as the foundation for a short, medium and long term plan for transformation of the service supply chain. Due to the common mistakes fueled by misperception, the opportunity for improvement is often dramatic.

A. J. Huber (✉)
Xerox Services, Xerox Corporation, 100 Clinton Avenue South, Rochester, 14644 NY, USA
e-mail: Andrew.Huber@xerox.com

N. Altay and L. A. Litteral (eds.), *Service Parts Management*,
DOI: 10.1007/978-0-85729-039-7_14,

Some common mistakes discussed in this chapter are:

- Failure to recognize the strategic importance of service to the profitability of the company.
- Failure to apply systems thinking in the structure and operation of the parts supply chain.
- Failure to forecast demand at the point of consumption.
- Failure to incorporate causal factors into the forecasting process.
- Failure to apply advanced inventory optimization and automatic replenishment.
- Inability to effectively deal with short supply situations.

For each of the common mistakes listed above, some guidelines for improvement are discussed. The objective is the leave the reader with a model for evaluating current performance and crafting a plan for action from which service organizations will realize significant value.

14.1 Service Parts Management Misperceptions

In recent decades the level of investment in service supply chains has increased due to the recognition of their importance. Service parts management is ripe with opportunity for academicians, practitioners and managers due to several emerging trends:

1. Product commoditization and price pressure
2. Cost reduction opportunities through technology upgrades
3. Increased demand for differentiated service offerings
4. The need for improvement in underperforming service supply chains

This chapter will focus primarily on the need for improvement. There is a wide disparity between organizations in the level of sophistication brought to the practice of service parts management. On the sophisticated side of the spectrum, the United States military has long employed the use of sophisticated techniques and systems to manage over $100 billion worth of service parts that are necessary to maintain the readiness of its weapon systems. In the commercial world there are large organizations that have very sophisticated processes and systems for the management of parts, however there are many more who do not. Before we list some of the more common mistakes and methods available for correcting them, it is helpful to understand some misperceptions that cause the common mistakes and often lead to poor performing service networks.

14.1.1 Misperception #1: Service is not Viewed as Highly Profitable

When a company sells a product, it must convince its customers to purchase their product rather than its competitors. The company's effectiveness is largely a function of its ability to effectively execute the "4 Ps" of marketing: *product, price, place and promotion.* As product markets mature, competition leads to product *commoditization*, where the products are relatively indistinguishable from one another. This leads to competition based on *price* that benefits consumers but reduces profits to the supplier.

For long life cycle maintenance intensive products, the profitability associated with the service of a product is typically much larger than that achieved on the sale. For one thing, the sale of the product occurs only once, but the sale of parts, service, supplies, and sometimes the operation of the product represent a stream of payments where profits are made year after year. The manager of a medical equipment manufacturer once demonstrated to me that his company realizes *eight times* as much profit from post sale services and supplies than from the sale of the product itself. Similar parallels exist in many industries. Where maintenance is involved and the parts required are uniquely engineered to the specific product, service parts revenues are often a significant source of profitability.

14.1.2 Misperception #2: Parts Management is not that Hard

For the uninitiated, service parts management is viewed as a simple series of tasks: people place orders on suppliers, shipments are put into inventory, and orders are filled to customers. All that is needed is to monitor supplies on hand and order more when needed. As a result of this perception of simplicity, rudimentary systems are used and people with insufficient skills are employed to perform the repetitive tasks of ordering. This is where the trouble starts!

Suppliers de-prioritize orders for small quantities and shipments arrive late. End consumers incur backorders while their machines are down. Planners are blamed for poor performance and stock up on items that they've been "burned" on before. Inventory grows yet shortages persist. Millions of dollars off write-offs occur on obsolete inventories. Customers incur costs of not having their equipment available and have to resort to alternatives like redundant machines, demanding their own inventory often on consignment, sending jobs out while waiting for equipment to be restored. Jobs don't get done on time, planes don't take off on time, medical tests are delayed, and penalties are paid. "Planning" becomes just a word on an organization chart as most of the people spend most of their time *expediting* rather than planning.

14.1.3 Misperception #3: Good Planning is All that is Required

This misperception is closely related to the misperception of simplicity. Upon recognizing that service parts management is not as simple as previously thought, a typical reaction might be to get some "smart" planners or a "good" planning system and the problems will be "fixed".

In reality, effective management of the service supply chain requires effective decision making and execution at the strategic, tactical and operational levels. While planning is critical, it is primarily a tactical activity with operational elements like buying parts. Traditionally most supply chain improvement efforts have focused on the tactical methods to the detriment of strategic and operational considerations. Strategic decisions are not given the attention they deserve, resulting in high costs than necessary in the latter stages of the product life cycle. Operational decisions are dominated by simplistic approaches such as first come, first served or priority service to the most demanding customers where "the squeaky wheel gets the grease".

Much of the missed opportunity is strategic. As a rule of thumb, I contend that 75% of the effectiveness of the service parts supply chain is predetermined before the product is introduced. The predetermination comes from the several important strategic decisions including what tactical and operational systems and processes will be employed to manage the supply chain. The misperception that supply chain effectiveness is largely determined by the planning organizations sitting behind their computer screens processing transactions is not unlike the perception that the performance of one's vehicle is largely determined by the mechanic who performs the maintenance. The mechanic's performance is critical, but the design of the vehicle, components used, their availability and manner in which the vehicle is driven are also important.

14.1.3.1 Strategic Decisions

1. Design for serviceability and choice of components

   Will the product be repaired or replaced? Will parts be replaced at the component or subassembly level? Will failures be diagnosed remotely or on site? What sensors and diagnostic software will be required? Will commonly available or uniquely engineered parts be used?

2. Choice of suppliers and supplier agreements

   Who will supply which components? What manufacturing lead times will apply? How long after manufacturing discontinuance will parts continue to be manufactured for service? What warranties will apply?

3. Where service will be performed and by whom

   Will service be performed organically, contracted to others, or performed by channel partners? Will service be performed at customer sites or local or central repair and maintenance facilities? What training, diagnostic tools, documentation and knowledge management systems will be required? What skill levels or certifications are required by the technicians?

4. Number and positioning of part stocking locations

Can there be a few centrally-located facilities or must they be located close to customer sites to for rapid access? Will replenishments occur direct from the supplier or will a multi-echelon resupply network be used? Who will run the facilities? Who will own the inventories? How will they be replenished?

5. Reparability of parts

Which parts will be repaired, by whom, and where? What reverse logistics network will be used? How quickly can failed assemblies be returned, repaired, and placed back into service?

6. Capacity investments

What manufacturing and part repair capacity will be required? Who will make the investment and incur the risk? Who will own the inventory, the manufacturer, the repair provider or the end customer?

14.1.3.2 Tactical Decisions

1. Metrics

What performance goals will be applied and how will they be measured?

2. Forecasting

How will demand forecasts be generated and from what data? Different methods are generally appropriate at the pre-launch, post-launch, mid-life and end of life cycles.

3. Replenishment methods

What replenishment rules will be applied and how will they be calculated? Examples include use-one-replenish-one; economic order quantities, or periodic replenishment.

4. Replenishment order prioritization

Which parts will be repaired first? How will manufacturing prioritize allocations between production requirements versus service needs? How will critical shortage situations be managed?

5. Stock allocation

How will arriving shipments be distributed throughout the stocking network? Will transshipments between locations be possible and under what circumstances will they occur? When shortages exist, will inventory be reserved to meet critical customer needs?

14.1.3.3 Operational Decisions

Traditionally, advanced analytical methods have not been typically viable for transactional systems at the operational level. Today however, several classes of resources have become more prevalent making the potential for operational decision making in real time an expanding area of opportunity. These resources include:

- inexpensive storage
- data from transactional systems
- communications bandwidth
- inexpensive and rapid processing power has created an environment where the application of techniques that can respond in real time can improve performance. Real time analytical methods are fertile ground for research and application. In time, real time analytical modules could supplement or replace rules and thresholds developed by tactical solutions. Imagine a world where a diagnostic system indicates an item has failed, and instantaneously a work order to replace the unit is generated, the return ship instructions are provided, the part repair facility schedules the repair of the part, and components are immediately replenished from the manufacturer. The technology for this type of world exist today, but to a large degree has not been exploited.

Here are some examples of operational decisions for service parts.

1. Whether to retain, repair or scrap a failed item.
2. To which repair facility a returned item should be sent.
3. When items should be transshipped between facilities.
4. Immediate updates to forecasts triggered from transactional activity. Examples include:

    a. Orders for parts
    b. Installation of equipment
    c. Relocation of equipment
    d. Removals of equipment
    e. Changes in the usage or mission of equipment

14.1.4 Misperception #4: Forecasting is Purely Statistical

On the day of this writing, a Google search on the words "improve demand forecasting" yielded the following results on the first search results page:

- 10 articles on forecasting and planning software packages
- 2 articles containing the essential proposition was that judgmental forecasting is "bad" and statistical forecasting is "good"
- 1 web site of a panel of experts listing their findings on the challenge of forecasting global demand for health care services

Given the prevalence of attention to statistical techniques, it is no wonder that many perceive that improving forecasts is largely an exercise of finding the best statistical forecast technique. I am reminded of a discussion I once had with a software marketer to whom I had made the point that good forecasting is both an art and a science. Her response to me was that at least 90% of the people in our industry would disagree with me, and she is probably right. Forecasting experts understand that forecasts can be improved by the choice of the model that best reflects demand patterns of the environment for which forecasts are needed. This can lead to the simple yet often wrong conclusion that the best forecasts are obtained by fitting several candidate models to historical demand data and choosing the one with the best fit. Best fit is typically determined based on some metric like mean square error, mean absolute deviation or mean absolute percentage error.

In practice however, I contend that improving forecasting in a service parts environment *may* involve the use of improved statistical forecasting techniques, it *may* involve the incorporation of additional or more accurate causal factor data, and/or it may involve the use of *knowledge that is contained outside historical demand data*. It is knowledge discerned outside of the historical data where the art comes in. Several examples come to mind.

Early in my career I was provided with data on a central parts warehouse where I obtained the economic order quantities and safety stock targets of every item. In a steady state environment with a r, Q order policy where quantity Q is reordered whenever inventory drops below r, one would expect the average inventory level to be ½ Q + Safety Stock where r = (Safety Stock + Lead Time Demand). In fact I discovered that actual inventory value was *three times higher* than this calculated average. I then discovered that the "overstock" conditions were largely due to three causes: ½ of the excess inventory was for a single product. That product had expected to be a market-changing strategic offering for the company. To get ready for its introduction, the parts had been ordered in advance. However, the product introduction had been delayed and sales were well below initial projections. In addition to the excess parts inventory for that single product, 1/3 of the excess was due to end of life buys of parts for discontinued products that would require years of continued service. The remainder of the excess was due to another product that had not met sales projections. Clearly, the overstock could only have been prevented by better forecasting of the *product* sales prior to introduction and continued sources of supply for products no longer being manufactured. Ironically, management at the time did not recognize these 3 big opportunities for inventory reduction because their root causes were not evident in the summary reports they had been receiving.

Another ability that could be characterized as the "art" of forecasting comes is selecting the best forecast technique for the situation at hand. Consider this real example. A manufacturer of consumer products had done an adequate job of forecasting demand for parts for its traditional analog products but was plagued with part shortages on its newer digital products. The analog products had traditionally long product life cycles where an initial supply of parts could be purchased prior to launch then the statistical forecasting would take over and perform adequately until an all time buy had to be purchased. With the digital products however, the life cycles were much shorter. Typically by the time the product was on retail shelves manufacturing was already planning the last production runs and the parts planner was being asked to place all time buy orders. Here the statistical methods that simply projected past history for the analog products could no longer perform for the shorter life-cycle digital products. A statistical model was in fact constructed for this situation, and it incorporated channel sales volumes, sell-through time lags, and time to failure distributions derived from previous generation but similar products. This example illustrates that forecasting requires the use of both statistical methods as well as judgment derived from knowledge of the market matched to knowledge of forecasting methods.

14.2 Common Mistakes in Practice

Now that we understand some common misperceptions about service parts management, let us turn to some of the mistakes that lead to poor performance.

Mistake #1: Failure to Recognize the Strategic Importance of Service to the Profitability of the Company

I have introduced the misperception by many that equipment service is not highly profitable. This misperception results in organizations to forgo profit opportunities in either of two ways:

1. Failure to exploit service revenue potential by the development and marketing of differentiated services offerings
2. Failure to apply technologies and an information infrastructure to achieve efficiency and effectiveness.

14.2.1 Failure to Fully Exploit Revenue Opportunities

Companies who do fully exploit revenue opportunities associated with service have active service marketing departments that study the market segments for their product offerings in order to tailor services offerings that meet the unique needs of each market segment and often tailored to each specific customer. Companies who do not fully exploit the revenue potential of service tend to provide "one size fits all" service plans or they may leave the service portion of their business to channel partners or independent service providers.

At its most fundamental level, the market for differentiated offerings stems from the fact that different customers may use the same product in different ways and incur different costs if equipment is down and not available for use. For one customer a machine may be used for work that is not critical and a service offering that provides next business day response and repair within 48 h may meet their needs. For another customer, the same type of machine may be used in the operation of their business and if it cannot be repaired the same day that customer may lose business. In these environments it is common to offer three levels of service with significantly uplifted prices for higher levels of service.

When companies tailor service offerings to unique customer needs, they create win-win business for both them and their customer: they realize the greater revenue and margins from customers with the greatest need for equipment availability, and they retain the business from customers who would otherwise turn to lower cost competitors. Differentiated offerings and flexibility in terms along with consistency in delivery to those terms enable long term business relationships based on customer loyalty that enable consistent and predictable levels of profitability.

Many companies traditionally thought of as product manufacturers have been transforming themselves to become total solutions providers comprised largely of three components:

1. The products
2. The maintenance of those products
3. The operation of the product and delivery of the services produced

Thus a manufacturer of printers may sell the printers, or the management of a fleet of printers including maintenance and supplies replenishment, or it may be a provider of printing services where the product, supplies and maintenance are only the means to produce prints.

For repair and maintenance contracts, service offerings may be differentiated along three dimensions:

1. The days and hours of coverage when service is available
2. The services that are included and excluded from the offering
3. The Service Level Agreement (SLA) performance guarantees

Differentiation in service offerings represent an opportunity for profits yet a challenge for service parts managers. Consider a simple example of three tiered offerings:

- mission critical with 2 h guaranteed onsite response
- same day service
- next day service

It is clear that a very high level of parts availability must be maintained for customers who have purchased the "mission critical" service offering. Service level targets would be less stringent for the same-day customers whereas overnight shipment for the next day customers sufficient if problems can be diagnosed remotely. In this example, answers to important strategic, tactical and operational questions become more complicated compared to a "one size fits all" service environment where every customer is entitled to the same level of service. Where will what parts be stocked? For whom will they be stocked? How will orders be filled? How can we ensure that parts for mission critical customers are always readily available without having to provide the same high levels of service to same-day and next-day customers? Service marketing consultant Al Hahn emphasizes that to effectively market service, "It is important to deliver the level of service you have sold the customer, and no more." If you don't deliver to the SLAs the customer paid for, you will pay penalties and risk losing that customer. However, if you over-deliver on the 2nd tier contracts there will be no incentive to buy the higher priced (and more profitable) mission-critical contract.

One well-known equipment provider in the information technology industry solves this challenge by maintaining two separate sets of inventories with different service targets: one for its business customers, and one for its consumers. Most readers of this book however will quickly recognize that maintaining two separate inventories of the same parts is inherently inefficient. Savings could be realized if both customer sets can be supplied from the same pool of parts inventory.

Another executive I spoke with from a large service provider also in the information technology industry told me that his company manages to 82 separate sets of SLAs depending on what was negotiated with each customer. His company's problem has become even more complex. How many parts must be stocked to meet the needs of all customers? At what point do you put an order from a lower priority customer on backorder to reserve inventory for the higher priority customers? When must a replenishment order be shipped by air versus ground? To which locations should parts in short supply be replenished first if each location supports a different combination of high, medium and lower priority customers?

As companies traditionally thought of as product manufacturers expand their scope towards providers of total solutions, new challenges (and hence opportunities) have arisen for service parts managers and researchers. Aviation and defense industries are an example. Aircraft operators like Air France, United Airlines and Delta Airlines have large maintenance organizations that they leverage for other airlines as well as their own. Aircraft manufacturers Boeing and Lockheed offer Performance-Based Logistics (PBL) contracts to commercial and government customers. SLA targets associated with these contracts are typically based on targets for uptime or operational availability of aircraft systems. The challenge for the service parts manager is to provide a supply of parts that provide high probabilities of achieving the SLA targets, but at lowest possible cost. The "right" level of cost in these complex environments include all costs incurred in support of the PBL contract, such as the number and capacity of stock locations and part repair facilities, the operating costs of those facilities, the information technology infrastructure, and the cost of transportation. As new methods are developed and new systems that apply them are deployed, costs can continue to decrease with no degradation or even higher levels of performance. As a result, the "right" levels of cost are continuously moving targets.

14.2.2 Failure to Apply Technologies to Achieve Efficiency and Effectiveness

Companies who are best in class at *marketing* their service can are not necessarily the most efficient in *delivering* those services. Conversely, companies who are best in class in *delivery* of service do not necessarily fully exploit the market value of their capabilities.

One large global high technology company is often cited as best in class for its service portfolio of offerings. This company appropriately places extremely high priority on ensuring that each customer consistently receives the contracted levels of service. Everyone in the supply chain organization knows the importance of having a full compliment of parts readily available locally for its most important customers. While it does a credible job in achieving high levels of service, it does so at tremendous levels of inefficiency. While it makes very good profits from service, it clearly could do much better.

Oftentimes companies fail to achieve efficiencies that are possible because they are focused on relatively small areas of opportunity while overlooking larger opportunities. The high technology company mentioned above has been in a multi-year battle for cost efficiency due to intense price competition with its competitors. The price competition has dramatically impaired the profitability of both companies. One of the opportunities it has pursued is the off-shoring of business process services. The service parts supply chain planning groups were one of the business processes identified and the company will undoubtedly reduce its labor costs by training offshore providers to perform planning functions that had previously been done primarily in the United States and Europe. While the company is clearly benefiting from the labor cost reduction, it did so before pursuing even higher cost reductions that may have been achieved by reengineering the manner in which it manages its service parts supply chain. First, there existed opportunities to reduce global parts inventories by reducing its number of stocking locations and pooling of inventories, the savings from which could have *surpassed the entire cost of its planning organization.* Secondly, the planning processes were highly manual. New systems had been put in place to consolidate multiple sets of legacy applications it had accumulated through acquisitions. The new systems largely supported the old manual processes, whereas much of the manual decision making could have been automated reducing the need for the planning labor. In my assessment, increasing the level of automation had the potential to reduce the total labor required by *more than half.* Had the company consolidated its stock locations and improved its inventory management practices *before* it moved the work off-shore, much greater savings could have been realized.

Why didn't the company reengineer its inventory management practices? It certainly had the opportunity when it was consolidating its legacy applications and developing new systems. To me the answer was simple: the people on the project team *did not know how to transform the supply chain.* This leads me to the next mistake.

14.2.2.1 Mistake #2: Failure to Apply Systems Thinking in the Structure and Operation of the Parts Supply Chain

In large organizations people have well-defined scopes of responsibility and as a result focus primarily on their own challenges at hand. At the management levels, managers worry about their own scope of responsibility. This can result in behaviors that I have characterized as "not knowing or caring about what goes on outside your own cubical". While this helps drive *individual* accountability and performance, it often does so at the expense of *overall performance.* Optimizing individual sub-processes does not lead to overall optimization. Overall performance suffers due to disconnected silos of functions. As my friend Jack Muckstadt of Cornell University likes to say "Local optimization leads to global disharmony."

As an example, there is often one organization responsible for service parts inventories in the field, another responsible for central parts inventories. The central warehouses are the suppliers to the field locations. Field planners and central planners work in different locations and may never have even met each other. Central planners focus on fill rates and backorders, perhaps giving priority to external customers over field stock replenishments. Field planners expect but do not receive perfect delivery from the central warehouse, so they may engage in hoarding of inventories due to shortages that have occurred in the past. Similarly, central planners order parts from manufacturing and do not receive perfect delivery and adopt similar behaviors. Manufacturers may be months late on some items yet may want to "unload" its excess inventories to the service supply chain where they will eventually be used.

Compare the scenario described above to the benefits of multi-echelon multi-item inventory optimization. Multi-echelon optimization is a *system-wide* approach to setting inventory levels. Readers familiar with these techniques understand (and can prove mathematically) that this technique can provide better *overall* levels of service than managing parts on the basis of single-part single-location decisions. Despite the benefits of this approach, the organizational structure and lack of collaborative processes and the inability to make decisions based on what's best for *all* the customers rather than specific sets of internal or external customers can prevent these benefits. In order to achieve breakthrough levels of performance, opportunities must be discovered using a *systems approach* where the operation of the entire supply chain is synchronized rather than the optimization of its individual parts. (I refer to the supply chain system, not just the information technology systems.) Successful application of a systems approach may require that organizational responsibilities be redefined to support the process redesign.

If there is a simple answer to enabling systems thinking when driving transformational change, it is to *involve people who understand methods to improve the entire supply chain.* This is not to say that current employees should not be involved. John Kotter, Harvard's expert on driving transformational change in organizations, emphasizes the need to empower broad based action (Kotter 1996). While current employees must participate, it is critical that the change effort involve people with the right *mix of perspectives and skills.* To make sure that teams do not perpetuate suboptimal past practices, thought leaders with deep knowledge of available research and current technologies should be key contributors to strategies for change. The power of knowledge and the value of outside perspective should not be underestimated. It is critical that people with an expanded perspective be participants in the decision process. Sometimes the views of experts can be overruled by project leaders unable to grasp the potential of the ideas of the expert or they reject them because they don't fit within their mental model of the current process. This can be a warning sign: either the expert is really no expert at all, or the expert's ideas have been pushed aside and the value of those ideas will not be realized.

While the presence of experts is important, operational people are best able to evaluate the practicality of new approaches. Operational people must have an open mind and be able to see possibilities beyond current practices. To ensure ideas are practical, operational people must identify what additional considerations are required, based on their knowledge of customers, suppliers and products.

The need for systems thinkers and process and technology experts in driving transformational change to achieve dramatic levels of improvement cannot be overemphasized. If change agents do not drive efforts, new systems will typically be enablers of past business processes and practices. Because most operational people have the current process as their point of reference, they are likely to ask for functionality that duplicates what they already have or support their current practices with some minor improvements that are unlikely to achieve dramatic levels of improvement.

Supply chain and information technology transformation projects must be coordinated if their potential is to be realized. Substituting one set of systems for another under the assumption of "like for like" functionality substitution generally represents a missed opportunity for improvement. (Upgrading from version x to Version x.1 of a commercial off the shelf application may be an exception.) Major information technology implementations are projects where it is critical to involve experts, otherwise the end result will be to continue current processes and (at best) continue current levels of performance.

One symptom of failure to apply systems thinking to the structure and operation of the supply chain is the inflexibility imposed by not planning for change. Major information technology projects can take several years from initial planning to full deployment. Cross-functional teams of people with business subject matter knowledge are teamed with information technologists. The business people make it a point to demand the capabilities they use today, plus issues that have challenged them in the past, plus challenges they are currently facing (a new product launch for example). By the time the project is fully deployed, the technology supports business needs that were in place several years ago, but do not meet the unanticipated challenges of today. There may be new technology in place, yet the business continues to be paralyzed by inflexibility. That is why it is important to think strategically from a systems perspective. What types of products might the organization need to support, what are the range of associated services that may be offered, how will service levels be defined, and what are might be the implications for service parts management? By answering those fundamental strategic questions first, the supply chain structure, business processes and technology solutions can be designed for dynamic responses to changing support requirements.

14.2.2.2 Mistake #3: Overreliance on Internal Metrics like Fill Rates and Inventory Turns

A common mantra of the service parts manager is to have the right part in the right place at the right time at the right cost. Another common business mantra is "what gets measured gets done". Metrics like fill rates and inventory turnover ratios are convenient and easy to produce. But are they accurate reflections of the "right" parts, places, times and costs?

At its most fundamental level, the role of the service parts manager is to provide parts availability when needed. The end consumer, the user of the equipment, or the consumer of the service that the equipment produces, desires that the service be available when needed. Thus the traveler wants no flight delays, the pilot wants to fly the plane when needed, the mechanic wants to perform maintenance when scheduled, and he needs the tools, knowledge, time and service parts to perform the maintenance. Similarly, the military wants the weapon system to be available when a mission must be performed, the printer wants to complete print jobs on time, and the homeowner wants the furnace or air conditioner to be operating when its needed most. In each of these scenarios, the customer desires the *product* to be operational when needed. Thus it is the *uptime of the product* during the times the user is likely to need it that is the only relative metric to the end consumer. The customer only cares about service parts management if the parts are not available, they cost too much, or they do not perform as intended. In other words, they don't care about fill rates or inventory turns.

This is not to say that fill rates and inventory turns are not useful. Fill rates are a convenient way to gauge changes in what it represents: the percentage of demands that can be filled immediately from inventory. If the objective is to improve uptime by placing more critical parts closer to the customer, then it may be useful to measure fill rates of critical components at field locations.

Military supply chains have long targeted service parts management around *operational availability* metrics (see Sherbrooke 2004). PBL contracts also commonly use operational availability to define service levels required. In the commercial world outside the airlines, uptime-based service level agreement targets are routinely specified in contracts, yet many service supply chains continue to use metrics like fill rates and turns as their primary performance targets. Having a high fill rate is of no value to the end consumer if they are unable to use their equipment when needed. A better approach is to have a mix of service-level metrics that are more relevant to what the customer wants. Examples include backorders, backorder duration, outstanding backorders with down machines, aircrafts on ground, equipment uptime, "hard" machine down occurrences (where the equipment is not functional) or "soft" machine down occurrences (where the equipment is operating at reduced performance or without all features).

Inventory turns is another useful metric but is not the ultimate objective. The owners of the business desire to meet the needs of the customer at the least possible cost. The lower the inventory value, the higher the turnover ratio, the better. Too often however targets are set based on some arbitrary number (increase turnover by X%) rather than thorough consideration to the important strategic questions of where will parts be stocked, what replenishment rules will be applied, what are lead times for manufacture, repair, and transportation; and what levels of service must be met. Answers to these strategic questions and the resulting structure and operation of the supply chain will determine the "right" inventory turn targets. *Inventory targets should be the result of policy, not the basis for policy.*

14.2.2.3 Mistake #4: Failure to Forecast Demand at the Point of Consumption

Much has been written about the "bullwhip effect" in supply chains (Lee et al. 1997). When a systems approach is not applied to the structure of the entire supply chain and each parts facility is operated as if it were an independent entity, it is common to track demand and generate forecasts based on orders from downstream locations as if they were independent demands from external customers, when in reality they are not. In fact, the bullwhip effect will result in more variability as you move up the echelon structure, making the orders less predictable and harder to forecast, resulting in excess inventory, variable workloads, and shortages.

At a large global diesel powered vehicle manufacturer I worked with, independent repair shops purchase parts from local dealers. Local dealers purchase parts for from the manufacturer for their own service facilities as well as for sales to the local repair shops. The manufacturer supplies the dealers from a global network of large distribution facilities which in turn are resupplied from a central warehouse located within the grounds of their expansive manufacturing plant. The distribution network for truck and bus parts is *multi-enterprise:* dealers had no visibility to consumption of the repair shops, and the manufacturer had no visibility to the consumption of the dealers, only their replenishment orders. As a result, replenishment orders from the dealers to the manufacturer warehouses upstream were erratic, making them difficult to forecast and resulted in the need for expensive safety stock.

In order to improve levels of service to the dealers, the company transformed the way they replenished dealers using a more collaborative systems approach. Each dealer provided from their own inventory system daily demands for each part it used or sold. The dealers and the manufacturers agreed on the levels of service that would be provided to the end customers. Consistent levels of service and "minimum truck on lift" time for long-haul truckers is a competitive advantage for both the manufacturer and its dealer network. The logistics arm of the manufacturer developed and provided to its dealers a system to set target stock levels of every part based on its high standards for service. Critical repair parts are given priority over deferrable maintenance items with different service objectives. Advanced inventory optimization methods are used to enable the dealers to provide high levels of service at minimal inventory investment.

The primary objective of the collaborative system with its dealers was to ensure consistently high levels of parts availability throughout the dealer network so that its customers could deliver its goods and passengers on time with minimal cost of downtime. While it certainly achieved its goal of consistently high service, it was able to do so at significantly reduced inventory, both for the dealer and for the manufacturer. Because it now forecasts based primarily on *consumption* demand data rather than replenishment orders, forecasts are more accurate, safety stock is reduced, and the work levels in the warehouses are less variable. Dealer inventory was reduced due to the use of advanced inventory optimization. Warehouse work levels reflected scheduled replenishments to the dealers rather than reacting to orders from disparate dealer systems and practices. As a result, overtime cost at the warehouse was also reduced.

At a company producing high technology equipment, the global parts supply chain was largely a set of independently operated operations. Local service providers ordered from facilities in each country. In Europe country stocks were replenished from a central facility. The European facility was replenished from the US central warehouse using largely manual ordering processes. The US facility had highly erratic order patterns that were difficult to forecast accurately. When demand is not forecast at the point of consumption and the inventory replenishments are not coordinated throughout the entire network, the result is high levels of inventory and imbalances with overages in some locations while others incur backorders of the same items.

14.2.2.4 Mistake #5: Failure to Incorporate Causal Factors into the Forecasting Process

Statistical forecast methods that apply historical demand data to predict future demands are perfectly acceptable in some but not all situations. When demand is relatively steady or in some way predictable and changes come gradually over time and the lead time to replenish inventory is short relative to the time in which demand levels change, simple methods like moving averages or exponential smoothing may be all that is required. If no reliable causal data is available and the value of increases in forecast accuracy is high, it may be worthwhile to apply more advanced and sophisticated methods that can detect seasonal and cyclical patterns.

Oftentimes however, demand can be highly dynamic and causal data is readily available. Forecasts and inventories can be highly responsive to changes in causal data in dynamic environments yet many companies continue to apply simple time series-based statistical methods. Consider a field stock location that supports a team of technicians servicing equipment under service contracts. New machines are sold, older machines are retired. If a large local customer replaces all its machines with a newer model, it may take a year or more before the statistical forecasts at the field stock location "learn" of this event by reducing the forecasts of parts for the older model and increasing the forecasts of the newer model. A better method is to calculate the replacement rate for each part on each model of equipment, then apply those failure rates to the currently installed equipment population size (the causal factor). Similarly, the inventory levels at upstream locations can immediately react to known changes in the install base. Other causal factors may be applied, such as the volume of jobs run through a fleet of machines. Since the forecasts can be immediately updated to changes in the known causal factors, events such as the retirement of a fleet of equipment at one location can immediately free up that inventory for other locations that still need the parts. Without such a responsive forecast environment, forecasts take time to reflect events such as reductions in the install base of equipment, often resulting in wasted inventory that may never be sold.

Other examples of valuable causal factors include the following:

- flight schedules by airlines
- fleet deployments at military facilities
- product sales forecasts of consumer products
- historical failure rates of similar previous generation products to forecast part demands for new products
- currently contracted equipment as a predictor of geographic dispersion of future sales placements

Both forecasting at the point of consumption and incorporation of relevant causal factors are important ingredients in building lean and effective service parts supply chains.

14.2.2.5 Mistake #6: Failure to Apply Advanced Inventory Optimization and Automatic Replenishment

Failure to utilize advanced multi-item and multi-echelon inventory optimization methods that are readily available (and have been for decades) is another example of treating decisions as single-part single-location decisions rather than applying a systems approach. Take the simple example of maximizing the fill rate of a location at minimum cost of inventory. When there is a high degree of variability in demand rates and a high degree of variability in part costs as is typically the case, it can easily be shown that multi-item optimization can achieve the same fill rate for roughly half the inventory value. Similar results can be easily shown for inventory levels optimized to an uptime rather than a fill rate objective. Despite this fact many companies continue to apply simplistic sets of stocking and replenishment rules.

Significant value from a systems approach can often be obtained when applying the following tactical and operational practices:

- Forecast demand at the point of consumption.
- Apply relevant causal factors in the forecasting process.
- Optimize inventory targets using multi-item multi-echelon inventory optimization
- Automate routine replenishment transactions.

A manufacturer of printing systems wanted to improve its local availability of parts for one of its products to reduce customer downtime. A team of engineers and a parts planning specialist gathered data on historical demand patterns, equipment placements, parts reliability, and engineering changes. The team came up with 3 sets of field stocking recommendations; one each for large, medium and small cities. The process took about a month and was reasonably successful in improving the local supplies for that one product. Later however an automated solution was put in place that had a more dramatic effect. A system was developed that tracked part usage and the supported contracted equipment at the technician level. Demand rates were calculated and inventory targets set automatically for the technicians van stock, with some allowable discretion based on their local knowledge. The system computed demand rates for the local field stock locations after consideration of the inventory carried by each technician. Field stock locations were reduced to fewer, larger facilities in strategic locations with delivery services available when needed. As a result, field inventory levels were reduced by 2/3 while local fill rates improved from roughly 85 to 95%. Replenishment of both technician van stock and field stock locations was fully automated. All that was needed was to put away the parts that were shipped and return items on the return list (parts no longer required due to changes in the install base).

Levels of performance improvement possible at companies that make these fundamental mistakes (often without knowing it) can be so dramatic as to be unbelievable. An engine manufacturer sells parts to a network of independent distributors who in tern sell parts to dealers and repair shops. Like the large vehicle manufacturer already discussed, this engine producer set out to establish a collaborative system with its customers. In this companies case, its customers were the independent distributors. High levels of service were deemed to be the most critical requirement of the product. The manufacturer adopted software that utilized advanced inventory optimization methods to set inventory levels whereas their previous system applied a simplistic set of single-item safety stock calculations with targets set by traditional ABC classifications. When the new system was being configured, the company used its historically high fill rate values as its performance objective. When the resulting recommended value of the inventory was only *one third* the current inventory, the managers could not believe it was possible. In fact, they assumed it was an error in the software.

Similarly, a high technology provider of computer networking equipment was so uneasy about taking a risk of impairing service that it would run an older system concurrently with a newer optimization solution. It then wrote a program of its own to choose the higher of the two systems stock level targets for each part at each location.

Small carefully controlled field pilots are recommended for organizations implementing advanced inventory optimization for the first time. When choosing pilot metrics, make sure they are from a systems perspective and not by comparing each part at each location to previous decisions, for they will be different, and for good reason. Use customer-focused metrics like the ones mentioned: backorder occurrences, backorder delay times, first time fix rates, and equipment downtimes due to parts. Starting with small pilots and expanding to larger scale pilots are a good way to identify any technical or process bugs and glitches as well as a way to prove the concept to become more comfortable with the approach.

14.2.2.6 Mistake #7: Inability to Effectively Deal with Short Supply Situations

Systems approaches are needed to make the best of a bad situation when it comes to short supply situations. A clear hierarchy of needs should be defined in advance as part of a strategic design exercise. Many organizations fall into the trap of simply reacting and expediting when shortages occur.

An example of a more effective approach is to first supply the most important customer when their machine is down and when they need it the most. If all machine-down situations can be fulfilled, then position remaining inventory in locations where demands from machine-down customers are most likely to occur. As more parts become available, distribute them throughout the network using optimization that maximizes the expected number of critical orders filled until the next replenishment arrives. When replenishments arrive late and inventory decreases, reverse the process.

When automated controls are not put in place, human behavior can cause part shortages to go from bad to worse. Spread a rumor among the field technician workforce that a part is in short supply and hoarding behavior will ensue. Fearing that they need to protect inventory for "my own customers", each inventory manager refuses to transship parts to other locations in dire need. Inventory increases, service gets worse.

When advanced parts management techniques are applied, when systems are put in place to automate processes and when audits are put in place to ensure accurate data, there is little need for manual overriding of individual transactions. In fact environments where individuals examine and override optimized inventory targets, order quantities and order timing more typically make performance *worse* rather than better. When automated methods are effective, it reduces the need for labor and frees people to work the root causes of supply problems while the system makes best use of the inventory available at the time.

14.2.2.7 Mistake #8: Overreliance on Benchmarking Best in Class Performance

Calling the practice of benchmarking as a "mistake" may come as a surprise to the reader. The practice of benchmarking is not a mistake in itself, in fact it can be very helpful to discover which organizations achieve the best performance in certain aspects of the supply chain, then to evaluate and prioritize the implementation of similar practices within their own organization. A simple framework for evaluation is presented later in this chapter may serve as a helpful means of evaluating best practices.

The *mistake* I refer to is the *overreliance* on benchmarking that leads to "copycat" duplication of past practices from other organization that may either not be appropriate or it may not be the best technique available in light of *current* technology. Copycat duplication of *all* methods that happen to be used by leading supply chain practitioners can only lead to performance *as good as* or *almost as good as* the organization being benchmarked. In fact, if the benchmark performer is truly a leading edge practitioner, they may in fact be working on changing and improving the practice you intend to copy.

A better practice would be to utilize a combination of benchmarking, keeping up with industry innovations, attendance at conferences, awareness of new technology vendor offerings (with a healthy sense of skepticism) and industry-academia collaboration in the development and deployment of new techniques. The authors of this book are a good place to start especially if they have identified new approaches to solving problems that are directly applicable to challenges that exist in your parts supply chain. Societies such as INFORMS and POMS maintain service and supply chain management interest and practice groups. Several universities have service and supply chain centers of excellence. Companies such as IBM and UPS have been actively engaged in parts supply chain research. In my experience some of the best performing service supply chains understand the complexity of service environments and have one or more thought leaders on staff employed full time in pursuit of strategic opportunities to drive towards improved performance. These resident experts can help guide managers in charting their course for the service parts supply chain.

14.3 A Simple Framework for Improvement

If Service Parts Management is not at all simple, if an organization has suffered from misperceptions and fallen victim to some of the strategic "mistakes" we have discussed, where should one begin in charting the course toward improvement? Fortunately, the more mistakes an organization has made, the more "low hanging fruit" opportunities exist to dramatically improve performance and reduce cost.

The objective of this section is to describe a simple approach to identify strategic issues around service parts management, then derive a Pareto ranking of the opportunities listed in order of importance. Before we dive in to the investigation, it is helpful to keep in mind two principles:

1. What is the role of service parts in meeting the larger organizational strategic objectives? What levels of service are required and what is the current performance?

The truck manufacturer described earlier clearly defined their objective of maintaining consistently high levels of availability of repair parts at each dealer to enable their customers to minimize vehicle on lift time and achieve on time deliveries.

2. If service levels fall below requirements, fix them first. It makes no sense to reduce costs first if the supply chain is not meeting its fundamental responsibility of enabling operational maintenance of equipment at the uptime levels required.
3. Identify cost saving opportunities, and rank them by value.

A simple ranking approach based on the principles above will help prioritize what should be done first, and avoid the trap of haphazard yet good ideas that seem reasonable to do, have great potential but may not be the right project at the right time in light of more urgent priorities. Fix the customer first, then chase the dollars, then go after the dimes.

14.3.1 Step 1: Define the Problems to be Solved and Gather Ideas

If service levels are not meeting requirements, clear priority must be placed on finding the root causes of delays in providing parts and causes of shortages and methods to avoid them. Often times, opportunities to improve service through process and technology improvement will *simultaneously* bring about opportunities to reduce cost. If processes are not changed, service levels can only be improved at increased cost, largely by increasing inventory to mask the inefficiencies that are occurring. This may be necessary in the short term, but it is not an effective strategy for the long term.

If cost is the primary objective, it is helpful to have team members define the overall cost levels and the components of cost, including any hidden components. It is critical that the focus is on *all relevant costs,* not just those included in the current set of performance metrics and not just those in the accounting entries. Obvious costs include operational costs of planning, repair facilities, transportation, warehousing, order processing. Inventory levels represent opportunity cost of capital. Not so obvious are the costs associated with shortages and delays. For example, the average mission capable availability of aircraft in the US Air Force fleet was 72% in 2001. Fully 14% of the fleet on average was not mission capable due to supply problems

(GAO 2001). Given the size of the Air Force fleet and the cost of each aircraft, this represents and opportunity cost of billions of dollars.

Although it is important to define the problem and understand the causes of delays and shortages as well as the operational, inventory and opportunity costs; do not limit the idea gathering to only those that directly address the stated problems at hand. While this may seem counterintuitive, I emphasize this because there may be opportunities for new processes and technologies that may not be apparent. For example, the techniques described throughout this book represent ideas worthy of evaluation that will be new to most readers. Other new ideas may be gleaned from conferences, recent publications, consultants and staff.

Ideas can come from anywhere. When I managed a parts supply chain, the warehouse manager expressed that my planning organization must be doing a bad job of planning because there were so many expensive parts in the warehouse that were not moving at all. That led us to adopt a practice of what we called the "dusty parts" test. When he found high cost parts with a "thick layer of dust" on them, he would bring them to my attention and I would have the planning group find out when were they purchased and why were they purchased at that time. On several occasions we uncovered process improvement opportunities that we could go after to prevent wasted expense in the future.

14.3.2 Step 2: Evaluate Each Idea

Four fundamental questions may serve to evaluate improvement ideas:

1. Does the problem this technique addresses exist in this service environment?
2. How does this organization currently deal with this problem?
3. Is the idea appropriate for this environment?
4. Are there other methods that would lead to an improvement in this area?

Let us consider each of these simple questions using a hypothetical case study:

A manufacturer of medical diagnostic equipment provides a line of 30 machines that perform a variety of medical tests at hospitals, clinics and some doctors' offices. The company's service organization maintains its own team of approximately 250 service technicians in North America, uses a combination of direct employees and service partners in Europe, and utilizes a dealer network in other regions of the world. Approximately 70% of its business resides in the North America.

The typical machine requires service about six times per year, of which about ½ are scheduled maintenance calls and ½ are unscheduled failures. Some of the customers have contracted for same day service; others receive the standard next-day service.

About 1/3 of the field replaceable parts are high value repairable assemblies, most of which the company refurbishes in a central repair facility located on the grounds of the manufacturing plant. Some of the assemblies require special equipment and skills and are sent to component manufacturers for repair.

When assemblies are removed and replaced in the field, technicians return them in the box the new assembly was shipped in using a preprinted return label from the shipping provider. The damaged assemblies are then sent to the receiving area of the central parts warehouse, also on the grounds of the manufacturing plant.

The service parts planning organization determines the disposition of each returned item. During the early and mid life cycle stages, almost all assemblies are sent to the appropriate repair facility for refurbishing. During the later stages of the product life cycle, the planner may choose to scrap some units or store them in their damaged condition until more are needed.

In the field, technicians are expected to return each item within a week. To save trips, technicians typically keep the return parts in their vans and stop by UPS to drop off them off when they happen to be near a UPS facility. Currently, about 65% of the parts arrive at the central warehouse within 2 weeks, 85% within a month, some take much longer, and each year about 4% are never accounted for.

The receiving workload at the central warehouse is highly variable. The wait time for a returned part to be received can vary from same day to as long as two weeks, with three days being typical. When items are received, the warehouse system instructs the receiving clerk whether the item is to be scrapped (in which case it is sent to the recycler), put away in its unrepaired state, or sent to the repair facility for refurbishment.

The repair facility typically runs with 30 days of work in process. About 15 percent of the time, items wait longer because not all the required component parts are available. Some units can wait as long as 5 months before they are repaired. Repaired parts are returned to the central warehouse where they are returned to stock.

The manager of the service parts operation saw the reverse logistics process as a big opportunity for improvement. She was frustrated with the large amount of unrepaired inventory in both the receiving area, the storage area for unrepaired items, and within the repair facility. Although the used parts did not show on the books, she knew that her planners were buying newly manufactured parts when good used ones were tied up in technician vans, the warehouse, and the repair facility. A team was commissioned to study the problem and make recommendations. Two university professors of Operations Management were brought in to work with the team and make recommendations. They interviewed managers and specialists from the planning department, warehouse and repair center. They were provided with data from the planning and transaction processing systems. After a few months of part time effort, the professors made a preliminary recommendation.

The professors proposed that a real-time decision support model be developed by them over the summer that could be incorporated into the company's existing warehouse, repair center and field parts management systems. Data from the planning application would feed the decision making model that would disposition "optimal" courses of action for each failed repairable part as the failures occurred. The model would be developed by the professors and some of their students who would deliver a working prototype that the company's IT organization could transform to a production application. The new decision model would be integrated with the transaction processing systems.

Here's how the application would work. At the time a part failed, a transaction would be sent to the real time decision model along with data on that part and each of its substitutes from the planning system. The model would determine in real time where the part should be returned to and what method of transportation to use. The decision alternatives would be to send it to a reclamation facility, to the repair vendor, or to the central warehouse for long term storage. Once the model recommended scrap, store or repair; it would determine the level of urgency and recommend air or ground transportation. Technicians would be expected to deliver emergency air shipments before taking their next service call, ground shipments would be processed every Friday afternoon.

For parts in all time buy status, a complimentary forecast model would estimate the remaining lifetime supply required incorporating current the current install base and its trend rate of decline as well as actual repair yields. These lifetime supply requirements forecasts would be used to feed the real-time model in deciding whether to scrap or return parts in lifetime buy status.

Since the needed parts would go directly to the repair facility, they would no longer be tied up at the central warehouse. For the company's repair facility, the professors made two recommendations. First, there were bottlenecks that occurred frequently for several classes of expensive critical parts. For those, additional capacity would be added by purchasing additional test equipment and cross training technicians from other stations. Second, the professors would develop a 2nd set of algorithms that would optimize the order that each waiting part would be repaired in. Since it would receive real-time notification of every failed part as it happened, the repair center could prioritize the work with an optimization objective of minimizing priority-weighted customer delay times.

The professors quantified the projected reductions in cycle times, work in process, and inventory of both repaired and unrepaired parts, yielding a substantial savings. They also pointed out how quickly critical parts in short supply could flow through the entire reverse logistics supply chain. This would significantly reduce critical customer backorders and reduced customer downtime.

After the professors presented to the service parts management team, the idea could be evaluated using the four questions simple questions.

14.3.3 Question 1: Does the Problem this Technique Addresses Exist in this Service Environment?

Here the answer is obviously yes, since the professors were brought in specifically to address the reverse supply chain of unrepaired inventory. However, one can imagine other scenarios where an idea like this would be proposed:

- The idea may be published in a book like the one you are reading now.
- The concept may be published in a leading journal.

- The supply chain manager may present the concept and accomplishments post-implementation at an industry conference.
- The ideas may be incorporated into a commercial off the shelf software company.
- The method may be offered as a service by a transportation company with repair facilities or by a contract repair provider.

Let's assume that the professors completed the project and published their work in a leading journal. An inventory specialist at a major airline engine repair facility might have read the article with interest and compared it to the methods his company used to perform engine overhauls for their own and other airlines. While the method may not have been 100% applicable to his own engine repair operation, there certainly would be parallels. Because of the high cost of engines, the airlines may already move engines directly from aircraft to the repair facility and the process would begin almost immediately. If however the engine repairs consisted both of planned and unplanned activity, the concept of a model like the one that the professors proposed to optimize the order of repairs would be applicable using the same or a similar approach.

14.3.4 Question 2: How Does this Organization Currently Deal with this Problem?

For our medical equipment manufacturer we have already described how they currently manage the problem. Other approaches in practice are similar to the situation we have described and others are radically different.

An information technology hardware provider has a strategy of outsourcing all of its manufacturing to contract manufacturers. They also require that these manufacturers supply service parts, thereby bypassing the need to have a significant investment in a service parts management with the exception of an investment in field parts inventory. In this type of environment, each replaced assembly might be returned to the manufacturer for return credit, then the challenge of managing the part repair process becomes the contract manufacturer's challenge.

Regardless of the contractual arrangement however, the fundamental process of return and repair for resale still exists, and one can envision the adoption of the technique by the contract manufacturer. One can imagine the manufacturer maintaining one or several repair facilities and a parts warehouse where it refurbishes returned parts from all its customers and returns them to stock for resale.

In the aircraft engine example, the maintenance facility may receive the engines from its customers, perform an initial inspection to determine the parts required, order any missing parts, and begin the refurbishment operation only once all component parts are received and assigned to the repair order, and a workstation is available to begin the work.

14.3.5 Question 3: Is the Idea Appropriate for this Environment?

The idea would generally be applicable to the medical equipment manufacturer. Of course problems could arise. The professors may have received some misinformation or faulty data. The existing systems may not be able to provide the required data in real-time. The optimization model may be overly computationally intensive to accommodate the volume of transactions experienced. The existing transactional systems may not have the technical capability for integration with the decision modules, requiring a total replacement at a cost not viable, and so on.

For the contract manufacturer, the idea may be generally applicable. Unless the manufacturer integrates its systems with its customers systems however, it may not be able to activate the scrap, retain or repair decision until items are received from its customers. It may use the scrap decision to discontinue granting of return credit to its customers since they would no longer be of value. The optimization of the order of repairs with the objective of minimizing customer delay times may be directly applicable since it is solving the same problem, albeit by another organization.

For the aircraft engine repair facility, the concept may apply but the optimization model may be overly simplistic because aircraft engines may have many more parts used in an overhaul procedure than the field replaceable assemblies from the medical devices. The aircraft engines are typically made up of subassembly components themselves that must be refurbished, necessitating a more complex multi-indenture optimization approach such as those described in Muckstadt (2005) and Sherbrooke (2004).

14.3.6 Question 4: Are there Other Methods that would Lead to an Improvement in this Area?

There may be alternatives to the approach recommended by the professors that the medical equipment manufacturer may consider. Rather than make the higher cost investment in the development and integration of the two real-time decision support models into their existing systems, it may instead consider a less costly simpler approach that would achieve a portion of the results. For example, it could establish some heuristic thresholds to automate the decision to scrap, retain, return or air ship the return from the field. An example might include the following:

- If the supply exceeds the forecasted remaining life cycle requirements adjusted for yield rates and some level of safety stock, scrap the item.
- If >X months supply on hand exist, retain the item for future repair.
- If on hand inventory < safety stock, air ship the return to the repair facility.
- If an outstanding backorder exists, expedite the repair upon receipt.

Of course, both ideas might make sense, the rules-based heuristics listed above for a "quick fix", followed by a more elegant solution to realize more cost reduction and service level improvement.

Another possibility might be consideration of more strategic alternatives. One alternative might be to discontinue the medical device manufacturer's internal repair operation and outsource it to a contract repair provider that has a higher scale operation, such as the example of the IT hardware provider's relationship with the contract manufacturer who maintains supplementary repair capabilities. A contract repair provider may have the scale to leverage information technology investments to achieve a higher return on investment and share a portion of the savings with their customers while increasing its own profitability.

An alternative method that the aircraft engine manufacturer might pursue could arise through a root cause analysis of its repair cycle times and causes of delays. For example, perhaps much of its cycle times are due to times awaiting parts to complete engine overhauls. A closer look may reveal that component part supplies are managed using an independent-demand forecasting and ordering model optimized to a fill rate objective. Perhaps time awaiting parts could be reduced through the incorporation of a dependent-demand forecast approach coupled with an inventory optimization model whose objective function is to minimize repair cycle times.

14.3.7 Step 3: Quantify and Rank Each Idea by Value

The four-question approach while simple, may be useful for reducing the number of investment alternatives under consideration, and may lead to investigation of other alternatives that would not have been considered previously. Examples include the case of the outsourcing alternative for the medical device provider or the dependent-demand approach for the aircraft engine repair facility.

It can be helpful to keep a list of improvement ideas that survive the evaluation process described in Step 2. First, consider service level improvements. I have proposed that if service levels fall below requirements, fix them first.

One way to prioritize competing approaches to improve service level is to identify what service metric will be used, and then rank ideas by cost divided by service improvement. For example, a retail parts supplier may use fill rate as the performance metric if customers take their purchases elsewhere if orders cannot be filled immediately. For each competing idea, you may evaluate the project cost per expected increases in orders filled.

For the medical equipment provider, backorder time may be a superior service metric to fill rate, since its customers may have no alternatives to placing a backorder when stock-outs occur. Here for every day the customer must wait for backorders to be filled, their medical device may be unavailable for use. Therefore, competing projects might be ranked by their investment cost per expected reductions in backorder days.

If cost reduction is the objective, then a similar method could be used in measuring investment required per dollar of annual spend reduction. Computation of net present value for each project using time-phased expenditures and time-phased return on investment will enable comparison of projects with the greatest value. If investment funds are limited, internal rate of returns should also be considered.

To project inventory savings, it will often be helpful to apply Little's Law where $L = \lambda * w$ where L = inventory, λ = demand rate and w = cycle times (see Muckstadt 2005; Fredendall and Hill 2001). Little's Law is helpful in translating cycle time improvements to inventory reductions. (It is also useful in translating the number of backorders to backorder delay times.)

Because service parts supply chains often perform very inefficiently due to the types of mistakes that have been outlined here, many improvement projects will often result in service level improvement *and* cost reduction. Here it may be useful to add the value of service improvement and cost reduction per dollar of investment. It is not straightforward to value service improvement due to the behavioral questions of what levels of service improvement will increase customer loyalty, improve the company's reputation to attract more business, and similar arguments. For simplicity's sake, it can be helpful to arbitrarily assign a value to the service metric, such as the value of a lost order or the value of a backorder day. Don't forget that losing an order may also involve loss of repeat business. While there may be no "right" value to place on these service metrics (don't ask the accountants to answer it for you), you can construct your spreadsheet with a parameter for service value, then increase and decrease it until it seems to make the appropriate trade-off between cost reduction and service level improvement. In other words, ask yourself how much you would be willing to invest in a project for service level improvement to derive a reasonable value. In this manner, it can be convenient to rank projects that improve service, reduce cost, or both.

14.4 Conclusion

Service parts supply chains can yield dramatic improvements for reasonable levels of investments because it is an area where current practice does not incorporate the processes and technologies that are currently available to run them more effectively and efficiently.

The initial and perhaps most significant challenge for the service parts manager, practitioner or consultant is to get the attention of senior management and show convince them of the magnitude of the opportunities that exist. Part of that selling process should include consideration for whether the senior managers are afflicted with the misperceptions I've identified here. It should also look at the motives of the manager and translation of those motives into the implications and opportunities for service parts management. The two most common are to maintain the uptime of the customer's equipment, and to provide the consistently profitable annuity streams that aftermarket support businesses represent.

The consultant must craft a convincing argument that *educates* senior management on the link between their objectives and the improvement opportunities you are requesting permission to pursue. The education must be quick and convincing. "Deep dive" discussions into arcane inventory theory can quickly lose the interest of many. Focus your proposal and presentation on the conclusions you want the executive to draw. Here are three desirable reactions that may be worthwhile aiming for:

- That's my problem!
- These ideas will go a long way to helping solve my problem.
- This proposal seems low-risk.

Finally, nothing breeds confidence more than success. Smaller projects tackled first that deliver results quickly will build confidence both by the project sponsors and by the teams implementing them.

Acknowledgments To Robert G. Brown, whose teaching, consulting and text *Advanced Service Parts Inventory Management* (Brown 1982) when I was a young analyst working with Eastman Kodak's service parts organization inspired me to apply analytical methods and would one day lead to my assignment there as Service Parts Supply Chain Director. To my friends Peter L. Jackson and John A. Muckstadt of Cornell University whose knowledge and understanding of the elements involved in optimizing supply chains under the uncertain conditions of service environments convinced me that "It's not rocket science, it's much harder than that!" Many of the issues I've highlighted in this chapter would have been overlooked by me were it not for the mentoring of Professor Muckstadt and working prototypes developed by Professor Jackson.

References

Brown R (1982) Advanced service parts inventory control. Materials Management Systems, Norwich

Fredendall L, Hill E (2001) Basics of supply chain management. St. Lucie Press, Boca Raton

Kotter J (1996) Leading change. Harvard Business School Press, Boston

Lee H, Padmanabhan V, Whang S (1997) The bullwhip effect in supply chains. Sloan Management Review 38:93–102

Muckstadt J (2005) Analysis and algorithms for service parts supply chains. Springer, New York

Sherbrooke C (2004) Optimal inventory modeling of systems. Springer, New York

United States General Accounting Office (2001) Air Force inventory: parts shortages are impacting operations and maintenance. Report: 01-587

Index